Environmental Catalysis

Edited by G. Ertl, H. Knözinger, J. Weitkamp

Further Reading from WILEY-VCH

Gerhard Ertl / Helmut Knözinger / Jens Weitkamp (Eds.)
Preparation of Solid Catalysts
1999, 630 pages with 325 figures. Wiley-VCH.
ISBN 3-527-29826-6

Gerhard Ertl / Helmut Knözinger / Jens Weitkamp (Eds.)
Handbook of Heterogeneous Catalysis
5 Volume Set
1997, 2469 pages with 1836 figures and 488 tables. Wiley-VCH.
ISBN 3-527-29212-8

Boy Cornils / Wolfgang A. Herrmann (Eds.)
Catalysis from A to Z
A Concise Encyclopedia
1999, approx. 900 pages with approx. 450 figures and approx. 8 tables. Wiley-VCH.
ISBN 3-527-29855-X

Jens Hagen
Industrial Catalysis
A Practical Approach
1999, 416 pages. Wiley-VCH.
ISBN 3-527-29528-3

John Meurig Thomas / W. John Thomas
Principles and Practice of Heterogeneous Catalysis
1996, 669 pages with 375 figures and 25 tables. Wiley-VCH.
ISBNs 3-527-29239-X (Softcover), 3-527-29288-8 (Hardcover)

R. I. Wijngaarden / K. R. Westerterp / A. Kronenberg
Industrial Catalysis
Optimizing Catalysts and Processes
1998, 268 pages. Wiley-VCH.
ISBN 3-527-28581-4

Environmental Catalysis

Edited by G. Ertl, H. Knözinger, J. Weitkamp

WILEY-VCH

Weinheim · New York · Chichester · Brisbane · Singapore · Toronto

Prof. Dr. G. Ertl
Fritz-Haber-Institute
of the Max Planck Society
Dept. Physical Chemistry
Faradayweg 4–6
D-14195 Berlin
Germany

Prof. Dr. H. Knözinger
Ludwig-Maximilian-University
Institute of Physical Chemistry
Sophienstr. 11
D-80333 München
Germany

Prof. Dr. J. Weitkamp
University of Stuttgart
Institute of Technical Chemistry I
Pfaffenwaldring 55
D-70569 Stuttgart
Germany

Title page: View through an exhaust gas catalyzer, photo courtesy Archiv Degussa-Hüls AG.

Library of Congress Card No. applied for.

A catalogue record for this book is available from the British Library.

Deutsche Bibliothek Cataloguing-in-Publication Data:
A catalogue record for this book is available from the Deutsche Bibliothek

Printed on acid-free and chlorine-free paper.

Composition: Asco Typesetters, Hong Kong
Printing: Strauss Offsetdruck, D-69509 Mörlenbach
Bookbindung: J. Schäffer, D-67269 Grünstadt
Printed in the Federal Republic of Germany

Foreword

The Handbook of Heterogeneous Catalysis was published in 1997. This book is part of the Handbook, now published as a monograph. Publisher and Editors felt that the Handbook of Heterogeneous Catalysis, which is only available as a full set of five volumes covering almost all aspects of heterogeneous catalysis, might not always be accessible to individuals interested in narrower areas of this field of chemistry. Therefore, the chapters dealing with aspects of environmental catalysis were selected and put together in this monograph. Catalysis is a rapidly growing field of both academic and technological interest; this Handbook aims to cover the concepts without an encyclopedic survey of the literature, so – although the chapters chosen could not be updated for the present volume – we believe that it will prove most useful to all readers interested in the chemical and physiochemical basis of environmental catalysis and its technological application.

Contents

Contributors

Z. Ainbinder
Dupont
Central Research und Development
Experimental Station
P.O. Box 80262
Wilmington, DE 19880-0262/USA
(*Chapter 4*)

R. A. Dalla Betta
Catalytica Combustion Systems
430 Ferguson Drive
Bldg 3
Mountain View, CA 94043-5272/USA
(*Chapter 3*)

B. H. Engler
International Catalyst Technology ICT, Inc.
65 Challenger Road
Ridgefield Park, NJ 07660/USA
(*Chapter 1*)

R. L. Garten
Catalytica Combustion Systems
430 Ferguson Drive
Bldg 3
Mountain View, CA 94043-5272/USA
(*Chapter 3*)

F. J. Janssen
Arnhem Institutions of the Dutch Electricity Utilities
N. V. KEMA, R & D Division
Department of Chemical Research
P.O. Box 9035
6800 ET Arnhem/The Netherlands
(*Chapter 2*)

E. S. J. Lox
Degussa
Inorganic Chemical Product Division
Postfach 1345
63403 Hanau (Wolfgang)/Germany
(*Chapter 1*)

L. E. Manzer
Corporate Catalysis Center
Central Science and Engineering
DuPont Company
Experimental Station
Wilmington, DE 19880-0262/USA
(*Chapter 4*)

M. Nappa
Corporate Catalysis Center
Central Science and Engineering
DuPont Company
Experimental Station
Wilmington, DE 19880-0262/USA
(*Chapter 4*)

V. N. Parmon
Boreskov Institute of Catalysis
Prospect Akademika Lavrentieva, 5
Novosibirsk 630090/Russia
(*Chapter 5*)

J. C. Schlatter
Catalytica Combustion Systems
430 Ferguson Drive
Bldg. 3
Mountain View, CA 94043-5272/USA
(*Chapter 3*)

K. I. Zamaraev
Boreskov Institute of Catalysis
Prospect Akademika Lavrentieva, 5
Novosibirsk 630090/Russia
(*Chapter 5*)

1 Environmental Catalysis – Mobile Sources

E. S. J. Lox, Degussa, Inorganic Chemical Product Division, Postfach 1345, 63403 Hanau (Wolfgang)/Germany

B. H. Engler, International Catalyst Technology ICT, Inc, 65 Challenger Road, Ridgefield Park, NJ 07660/USA

1.1 Introduction

1.1.1 Origin of Emissions

The industrialization of the Western world was accompanied by a drastic increase in the consumption of fossil fuels. The energy stored in fossil fuels was freed mostly by flame combustion, which is the reaction between the carbon containing constituents of the fossil fuel and the oxygen of the air, according to the reaction

$$C_mH_n + (m + n/4)O_2 \rightleftharpoons (m)CO_2 + (n/2)H_2O \tag{1}$$

Carbon dioxide and water are the main products of this reaction. However, incomplete combustion causes some emissions of unburned hydrocarbons, as well as intermediate oxidation products such as alcohols, aldehydes and carbon monoxide. As a result of thermal cracking reactions that take place in the flame, especially with incomplete combustion, hydrogen is formed and emitted, as well as hydrocarbons that are different from the ones present in the fuel.

Most fossil fuels have some amount of sulfur-containing and nitrogen-containing constituents as well, that will yield some emissions of sulfur oxides – mainly SO_2 – nitrogen oxides, commonly denoted as NO_x but consisting mainly of NO, and a small amount of N_2O.

During flame combustion, temperatures in excess of 1700 K occur. At those temperatures, the reaction between the air constituents nitrogen and oxygen is thermodynamically favored, resulting in the formation of nitrogen oxides, according to the overall reaction equation

$$N_2 + O_2 \rightleftharpoons 2NO \tag{2}$$

With fuels used in internal combustion engines, the reaction of eq 2 is the major cause of nitrogen oxides emissions. Of course, the amount of carbon monoxide, hydrocarbons and nitrogen oxides that are emitted is dependent on the detailed composition of the fuel as well as on the way the combustion is performed. But as an order of magnitude, the exhaust gas of a gasoline-powered spark-ignited internal combustion engine will have the composition shown in Fig. 1.

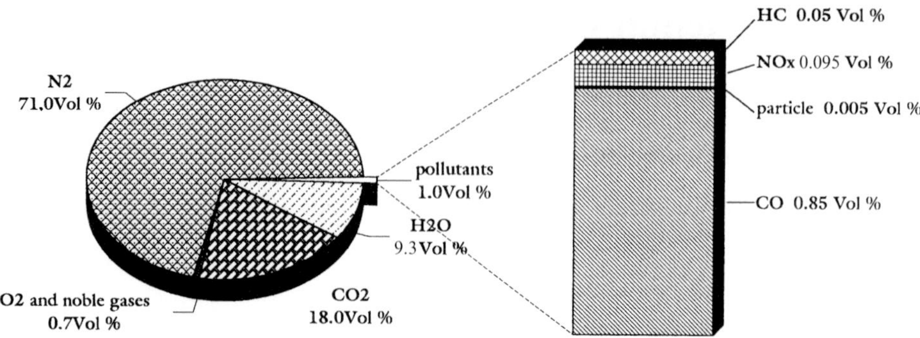

Figure 1. Typical composition of the exhaust gas of a gasoline-powered spark ignition internal combustion engine.

1.1.2 Importance of Traffic

To date the majority of on-road traffic is performed by vehicles equipped with internal combustion engines that use fuels derived from fossil fuels. The vehicles range from passenger cars to heavy duty trucks, and the engines used are either spark ignited or compression ignited reciprocating internal combustion devices.

An order of magnitude of the relative importance of traffic in the total emissions of carbon monoxide, hydrocarbons, nitrogen oxides, sulfur oxides and dust is shown in Fig. 2. From these data, it is apparent that road traffic is one of the major sources of carbon monoxide, hydrocarbons and nitrogen oxides emissions [1].

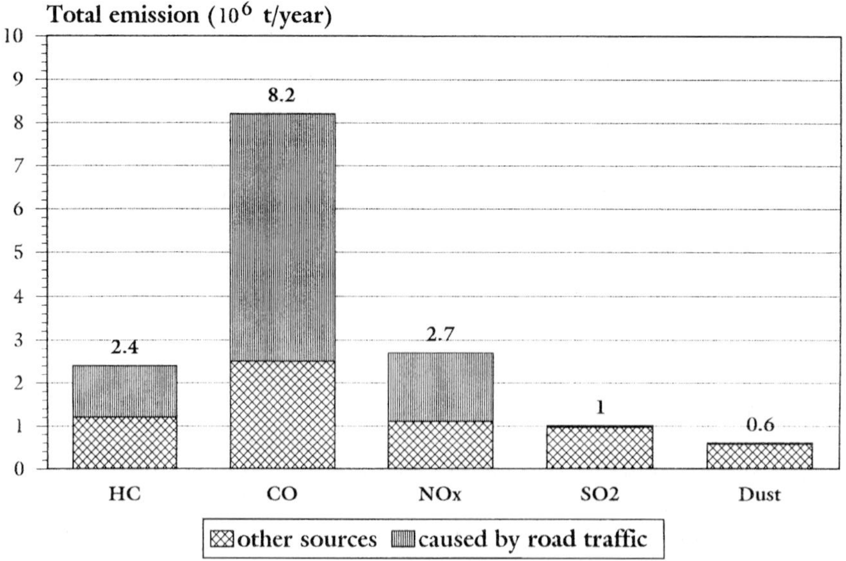

Figure 2. Total emissions of hydrocarbons (HC), carbon monoxide, nitrogen oxides (NO_x), sulfur dioxide and dust in the Federal Republic of Germany in 1989, and relative importance of traffic.

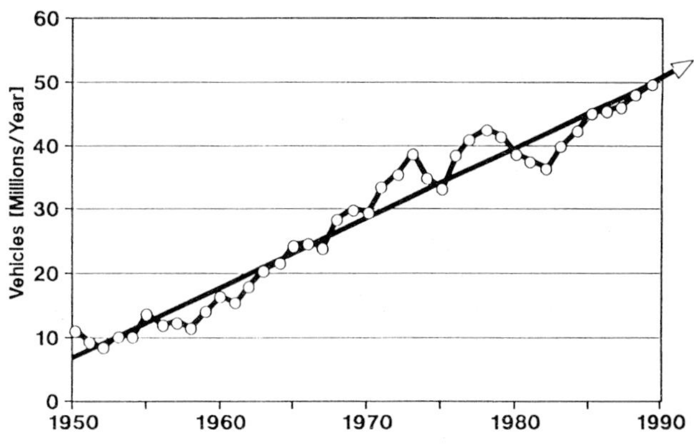

Figure 3. Annual worldwide production of vehicles. Adapted from [2].

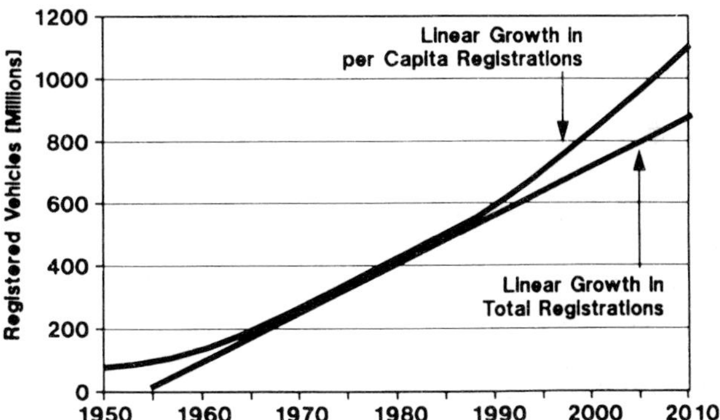

Figure 4. Trends and projections of worldwide motor vehicle registration. Adapted from [2].

In 1988, there were about 500 million vehicles on the road, of which about 400 million were in North and Central America and Europe alone. The annual worldwide production of vehicles amounted in 1990 about 50 million, the historical evolution of which is shown in Fig. 3. The extrapolation shown in Fig. 4 predicts a vehicle population of about 800 million by the year 2000 [2]. Also, their increasing use, expressed as average distance driven per vehicle and per year, should be noted].

1.2 Legislation

1.2.1 History

From the data presented above, it is apparent that traffic has an important impact on the air quality, which affects both the environment and human health. Therefore, legislation was introduced to limit the emission of carbon monoxide, hydrocarbons, nitrogen oxides and particulates caused by traffic.

The initiative to introduce legislation to limit exhaust gas emissions from vehicles was taken in the USA. In 1966, California introduced limits for the exhaust gas

Table 1. Historical overview of the limits for CO, HC and NO_x emissions from passenger cars in the USA.

Year	Area	Pass levels (g $mile^{-1}$ except where stated)			Test method
		CO	HC	NO_x	
1966–67	California	1.5%	275 ppm	–	7 Mode
1968–69	Federal & California	1.5%	275 ppm	–	
1970	Federal & California	23	2.2	–	
1972	California	39	3.2	3.2	FTP-72
	Federal	39	3.4	–	
1975	California	9	0.9	2	
	Federal	16	1.5	2	
1980	California	8	0.41	1	
	Federal	7	0.41	2	FTP-75
1981	Federal & California	3.4	0.41	1	
1993	California	3.4	0.25[a]	0.4	

[a] Non-Methane Hydrocarbons

emission of carbon monoxide and hydrocarbons from passenger cars equipped with spark ignition engines. Similar regulations were then extended to other states.

The US Federal Clean Air Act, introduced in 1970, prescribed a further drastic reduction in the allowable exhaust gas emissions from passenger cars. An historical overview is given in Table 1.

Similar measures were then taken in Japan, Australia and Switzerland. In 1985, the European Community passed respective strict legislation for passenger cars with spark ignition engines, to be followed by South Korea in 1987 and Brazil in 1988.

1.2.2 Present and Future

The definition of emission legislation is a continuous process. At present, a variety of new steps are being taken worldwide to further improve the air quality.

The first step is that an increasing number of countries beyond the USA, Japan and the European Union are introducing stringent emission standards for passenger cars. The emission standards and the procedures to measure them are in most cases based upon those practised in the USA. Some countries prescribe a gradual decrease of emissions, starting with emission limits that were valid about ten years ago in the USA. Other countries, however, base their legislation on the best available technology and consider the introduction of emission limits identical to those currently applicable in the USA. Generally, emission legislation applies to new cars only, but sometimes fiscal incentives are given to modify existing cars so as to reduce their emissions as well. These are the so-called retrofitting programs. The major effect of retrofitting programs is that a much quicker reduction of the country fleet emission is achieved, because passenger cars have a lifetime of typically at least ten years and, in the countries that now consider the introduction of emission legislation, sometimes even considerably longer.

Table 2. Projected legislation on exhaust emission limits (g mile^{-1}) for passenger cars and light duty trucks in California (test weights ⩽1730 kg).

Legislation[4]	Year	50 000 miles/5 years				100 000 miles/10 years			
		HC	CO	NO_x	PM[3]	HC	CO	NO_x	PM[3]
Present California/ US Tier 1	1993/94	0.25	3.4	1.0 diesel 0.4 gasoline	0.08	0.31	4.2	1.25 diesel 0.6 gasoline	0.1
California TLEV	1994[1]	0.125[2]	3.4	0.4	–	0.156[2]	4.2	0.6	0.08
California LEV	1997[1]	0.075[2]	3.4	0.2	–	0.09[2]	4.2	0.3	0.08
California ULEV	1997[1]	0.04[2]	1.7	0.2	–	0.055[2]	2.1	0.3	0.04
California ZEV	1998[1]	0[2]	0	0	0	0[2]	0	0	0

[1] Phase-in beginning in year shown.
[2] Non-methane organic gas (Non-Methane Hydrocarbons + aldehydes + alcohols).
[3] Diesel vehicles only.
[4] TLEV transitional low emission vehicle; LEV low emission vehicle; ULEV ultralow emission vehicle; ZEV zero emission vehicle.

The second step is that those countries that already have an emissions legislation for passenger cars in effect introduce more stringent limits. Table 2 shows as an example the new legislation on emission limits for passenger cars introduced in California by the Clean Air Act in 1990. Discussion is going on to extend this legislation to other states. The European Union has recently introduced new emission limits for gasoline – and diesel fueled passenger cars, valid from model year 1996 onwards. Table 3 summarizes these limits.

Thirdly, the procedures used to quantify the exhaust emissions are revised. For example, the European Union is studying the effect of the temperature at which the emission test is performed on the emission values obtained. Also, the vehicle driving procedure used in the emission test has been changed over the years.

As a fourth step, discussion continues for the introduction of emission limits for other exhaust gas components, and for particulate matter of diesel powered vehicles. For example, there has been discussion in the USA and some European countries on separate – additional – emission limits for carbon dioxide, benzene and/or aldehydes. In the USA there is a project to consider an additional ozone-formation factor to be allocated to the tailpipe emission of passenger cars. This is because each exhaust gas component has a different potential to contribute to atmospheric ozone formation. This potential is quantified according to the theory of Carter by the maximum incremental reactivity (MIR) factor, expressed as grams of

Table 3. Emission limits for passenger cars in the European Union, 1996.

Engine type	Carbon monoxide (g km^{-1})	Total hydrocarbons and nitrogen oxides (g km^{-1})	Particulate matter (g km^{-1})
Gasoline spark ignition engine	2.2	0.5	–
Direct injection diesel engine	1.0	0.9	0.1
Indirect injection diesel engine	1.0	0.7	0.08

Table 4. Maximum incremental reactivity factor for some selected exhaust gas components.

Exhaust gas component	Maximum incremental reactivity factor (g ozone per g component)
Methane	0.015
Ethane	0.290
n-Hexane	0.960
2-Methylpentane	1.530
Methylcyclopentane	2.820
Ethene	7.29
1-Hexene	4.42
2-Methyl-2-pentene	6.60
Cyclohexene	5.87
Benzene	0.48
Toluene	2.73
Formaldehyde	7.18
Methanol	0.80

ozone that might be formed per gram of a given exhaust gas component. Table 4 gives this MIR factor for some typical exhaust gas emission components [3].

The procedure is to make a detailed quantitative analysis of the exhaust gas and then to multiply the emission of each of those components with the corresponding MIR factor. The summation of these values leads to the amount of ozone that might be formed by the particular vehicle per unit distance driven.

The fifth step is to extend emission legislation to transportation other than passenger cars. On one side are motorbikes, equipped with small two-stroke or four-stroke single cylinder gasoline engines that have engine displacements typically of about 50 cm^3. On the other side are trucks and busses, equipped with large multi-cylinder gasoline or diesel engines with an engine displacement of 6 dm^3 or more, and trains and ships that are powered by even much larger diesel engines. Legislation for emission sources such as small utility equipment (e.g. chain saws and lawn mowers) or larger construction equipment (e.g. cranes and bulldozers) is also being considered.

Finally, widespread awareness for environmental concerns has sparked initiatives for new car propulsion systems, such as electrical engines in conjunction with batteries and/or fuel cells, and small turbines and hybrid engine systems that include both an electrical engine together with batteries as well as a small internal combustion engine used to reload the battery. Currently, it is believed that electrical engines powered by full cells are most likely to succeed.

1.2.3 Measurement of Emissions

A Passenger Cars and Light Duty Vehicles

The exhaust gas emissions from passenger cars and light duty vehicles are generally measured in a vehicle test, in which a vehicle is mounted on a chassis dynamometer

Figure 5. Vehicle dynamometer for the measurement of exhaust emissions.

(Fig. 5). The temperature of the vehicle is stabilized by keeping the vehicle in an air conditioned room for 12 h, in which the temperature is maintained at 298 K.

The vehicle is driven according a prescribed velocity pattern, during which time the exhaust gas is collected. To avoid condensation of water, the exhaust gas is diluted according to a prescribed procedure before it is stored in one or more expandable plastic bags. The number of bags used depends on the respective legislation. After finishing the test, the stored exhaust gas is analyzed. Table 5 summarizes the gas components that need to be analyzed and the corresponding types of analyzers used.

To measure the mass of particulates emitted by diesel engines, an additional dilution tunnel is used, in which the exhaust gas is diluted with air so as to keep the temperature of the mixed gas below 325 K. Figure 6 shows such a dilution tunnel. This mixed gas is then led through a device containing a filter paper of known weight on which the particulates are collected (Fig. 7). After collection, the filter paper is equilibrated to constant humidity, and weighed again. The weight difference gives the mass of particulates emitted.

The driving cycles used differ from country to country, as they are designed to simulate typical driving patterns, which differ from country to country. Figure 8 shows the current test cycles used in the USA, the European Union and in Japan. The corresponding key features are summarized in Table 6.

Table 5. Exhaust gas components measured and the corresponding analyzers used.

Gas component	Analytical method
Carbon monoxide	Nondispersive infrared analyzer, on-line
Total hydrocarbons	Flame ionization detector, on-line
Nitrogen oxides (nitrogen monoxide and nitrogen dioxide)	Chemiluminescence detector, on-line
Particulates	Gravimetric after collection on a filter paper, off-line
Carbon dioxide	Nondispersive infrared analyzer, on-line
Oxygen	Paramagnetic detector, on-line

Figure 6. Dilution tunnel to measure the particulate mass emission from diesel powered passenger cars.

Figure 7. Detail of a filter to collect particulates during a vehicle test with a diesel powered passenger car.

It is not easy to correlate the results obtained in each of the different cycles, because the technology mounted on a vehicle to achieve given emission targets might differ from country to country. Keeping this in mind, Fig. 9 shows the correlation between the emissions obtained in the US-FTP 75 cycle and in the European cycle, for some selected gasoline passenger cars equipped with closed-loop-controlled three-way catalysts [4].

As discussed above, the test cycles and the corresponding test procedures are also subject to changes, that might affect the emission result obtained. Figure 10 shows the different generations of the European test cycle. Table 7 gives the corresponding emission results, at fixed vehicle technology, for some selected gasoline passenger cars [5].

Also the temperature at which the vehicle is conditioned and/or tested has a major influence on the emission result obtained (Fig. 11) [6].

Finally, it should be mentioned that for development purposes, chassis dynamometers are also equipped with modal analytical facilities, that allow the on-line

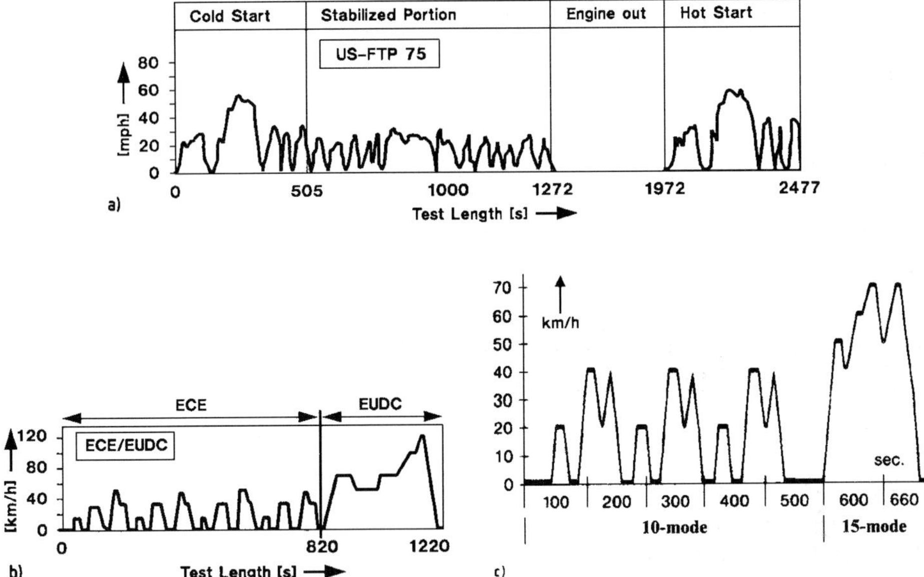

Figure 8. Vehicle test cycles in use in 1995 to measure the exhaust gas emissions from passenger cars: (a) USA; (b) the European Union; (c) Japan.

analysis of the exhaust gas up- and downstream from the catalyst with a resolution in the order of magnitude of seconds. This leads to knowledge on the catalyst performance at each of the driving conditions during the test cycle.

B Heavy Duty Trucks and Busses

Contrary to passenger cars and light duty vehicles, current legislation does not require the measurement of emissions from heavy duty trucks and busses using a vehicle dynamometer. The main reasons for this are the costs of the testing equip-

Table 6. Key features of the test cycles used in 1995 to measure exhaust emissions from passenger cars and light duty vehicles in the USA, the European Union and Japan.

Feature	Unit	USA	European Union	Japan
Name	–	US Federal Test Procedure	MVEG-A	10.15 Mode
Number of cycles per test	–	1	1	1
Number of different bags used to collect the exhaust gas	–	3	1	1
Total test length	km	17.9	11.0	4.16
Total test duration	s	2477	1220	660
Average speed	$km\,h^{-1}$	34.1	33.6	22.7
Maximum speed	$km\,h^{-1}$	91.2	120	70.0

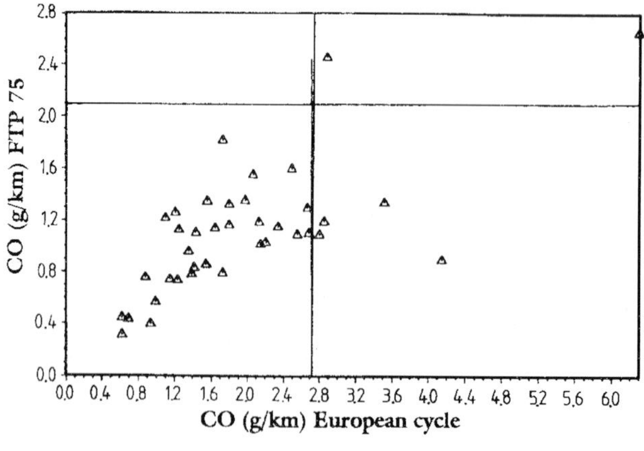

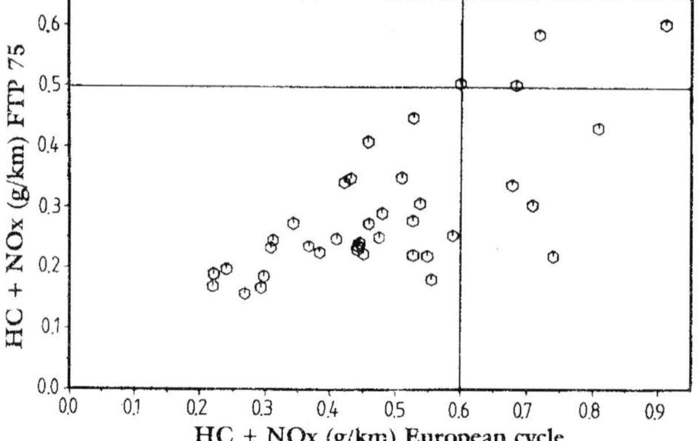

Figure 9. Correlation between the exhaust emission values obtained in the US FTP 75 and the European vehicle test cycle (MVEG-A), with gasoline-powered passenger cars. Adapted from [4].

ment and that the same type of engine is mounted on a much broader range of vehicle types, that are each produced in much smaller quantities.

Therefore, the legislation in the USA and the European Union is based upon tests with engines mounted on large engine dynamometers. Again, with the current status of the legislation, major differences exist between the test procedures applied in the various countries. The engine test cycle used in the USA is shown in Fig. 12. This so called transient test cycle has a highly dynamic nature, as not only a large number of engine rotational speed and engine load setpoints are used, but also the effect of transitions between those experimental set points is accounted for. The final result is one value for the emissions of the gaseous components carbon monoxide, hydrocarbons and nitrogen oxides and one filter result for the emission of particulates.

In contrast, the test procedure used in the European Union up to the year 2000 is more static in nature. It uses 13 prescribed combinations of engine rotational speed and engine load as set points (Fig. 13). The gaseous emission components are measured separately, after which a weighted average is calculated. The particulate

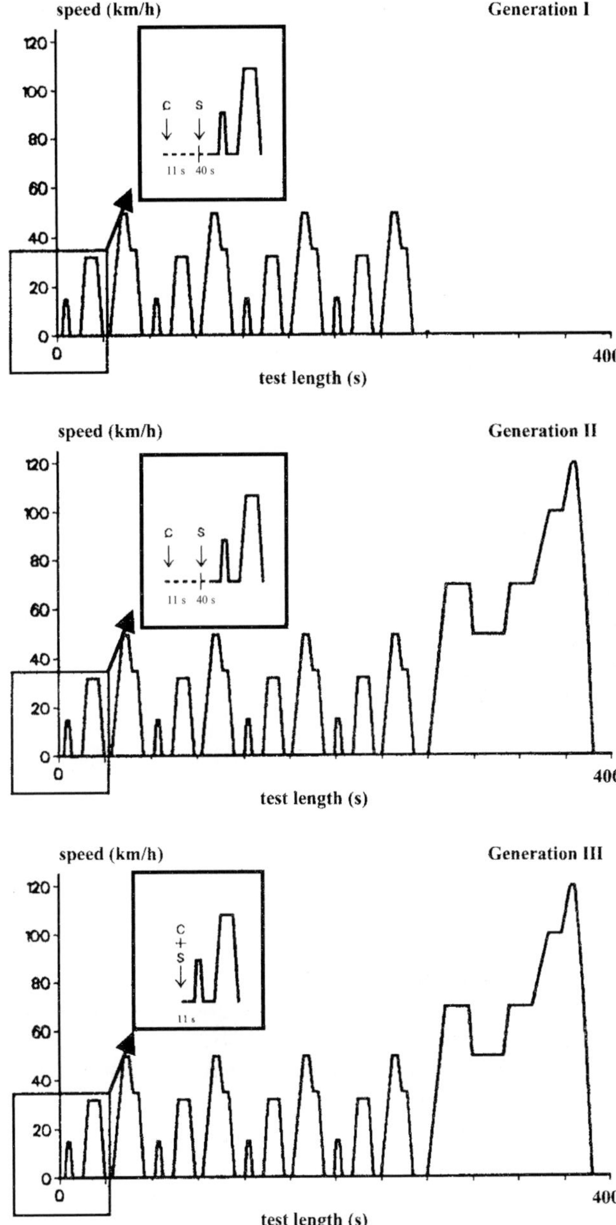

Figure 10. Different generations of the European test cycle to measure the exhaust emissions from passenger cars and light duty vehicles (c = cranking, s = sampling start).

matter is collected over the whole test. This test procedure does not take into account the effect of transitions between the thirteen engine operation points on the emission of gaseous components. The equipment needed to perform a transient test is much more expensive than that for the European 13 mode test.

Table 7. Exhaust emission results obtained with gasoline powered passenger cars in different versions of the European test cycle. Adapted from [5]

Test cycle description	Units	Gasoline-engine	
		CO	HC + NO_x
Urban driving cycle (ECE 15.04)	g $test^{-1}$	19.2	4.21
1995 European Union procedure (MVEG-A)	g km^{-1}	2.0	0.51
1995 European Union procedure (MVEG-A)	g km^{-1}	3.25	0.58
European Union procedure with sampling beginning at engine cracking (MVEG-B)	g km^{-1}	4.42	0.75

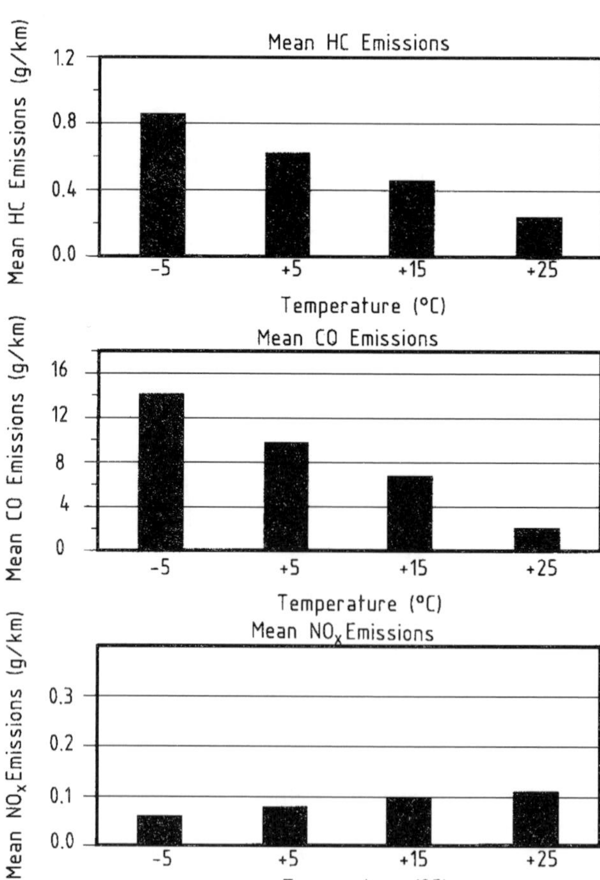

Figure 11. Effect of the preconditioning and the test temperature on the exhaust emission of gasoline fueled passenger cars in the European test cycle. Adapted from [6].

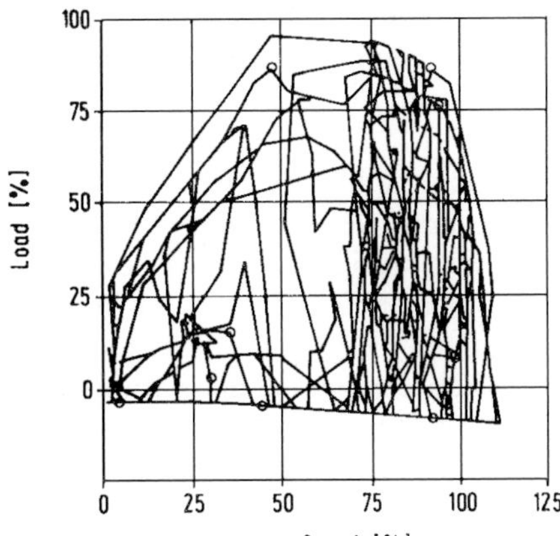

Figure 12. Engine load and revolution set points in the US transient cycle, to measure exhaust emissions from heavy duty engines. Reprinted from ref. [13] with kind permission of Elsevier

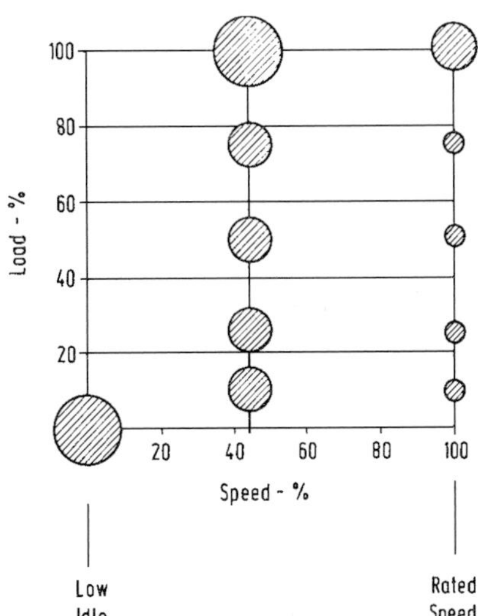

Figure 13. Engine load and revolution set points in the European 13-mode test cycle, to measure the exhaust emissions from heavy duty engines. Reprinted from ref. [13] with kind permission of Elsevier Science.

1.2.4 Technology to Meet Legislation

A Spark Ignition Engines

The technologies needed to achieve a given emission target depend on the emission level to be reached, and on the type and quality of the fuel used. These technologies can basically be subdivided into three categories:

(i) Modification of engine design – for example the fuel management system – to decrease the engine out, also called raw emission.
(ii) Aftertreatment of the engine exhaust by heterogeneous catalysts, to convert the engine raw emission.

(iii) A combination of engine exhaust aftertreatment by heterogeneous catalysts with engine design modification and/or controlled engine operation, to allow optimal functioning of the aftertreatment device.

Taking the history of the USA legislation as a reference, it appears that the first drastic improvements in exhaust emissions from passenger cars with gasoline engines were achieved by modification of the engine design.

Between 1960 and 1975, when legislation called for a further drastic reduction in the emission levels of carbon monoxide and hydrocarbons, there was some competition between engine design modification and exhaust aftertreatment by heterogeneous catalysis. As the majority of that generation catalysts relied on base metals, serious deactivation occurred, caused by the high catalyst operation temperature and sulfur present in the fuel. Therefore, the emission targets were in most cases again met by engine modification only.

The breakthrough for heterogeneous catalysis came in 1975, because of the further reduction in exhaust emissions required by new legislation. The advent of precious metals based aftertreatment catalysts, in conjunction with the elimination of lead in the gasoline fuel was the key. Catalysts of that time targeted the conversion of carbon monoxide and hydrocarbons through oxidation reactions. The further steps in legislation, in 1977, called for a reduction in nitrogen oxides. This was the breakthrough for the combination of exhaust gas aftertreatment together with engine modification/control. Present, and most probably future legislation, is best met by this approach [7–12].

B Compression Ignition Engines

Since the mid-1980s, aftertreatment for compression ignition engines has concentrated on the reduction of particulate matter. The emissions of CO, gaseous hydrocarbons and NO_x were so low that they easily satisfied the legislation.

A number of different technologies has been proposed to eliminate the particulate matter, although few have reached the marketplace. The first group of technologies uses filtering devices. One such device is the ceramic wall flow filter (Fig. 14). The filter is a monolithic structure made from porous cordierite, traversed by a multitude of straight channels. A channel which is plugged at its inlet side is adjacent to a channel which is plugged at its outlet side. This forces the exhaust gas to pass through the monolithic structure; the particulate matter remains on the walls of the filter.

Another device is the wire mesh filter (Figs. 15 and 16). Both devices enable a high removal efficiency, typically about 90% of the particulate matter. Their disadvantage is discontinuous operation. First, the particulate matter is collected, after which the loaded filter needs to be regenerated. Regeneration is usually done by burning off the carboneous fraction of the particulate matter, which requires a temperature of about 720 K. Such temperatures are rarely met during the typical operation of the diesel engines.

The absence of regeneration at controlled intervals causes excess particulate matter to be accumulated, which leads to an increased back-pressure. As a result, the exhaust gas temperature is raised, and a sudden burn off of the heavely loaded

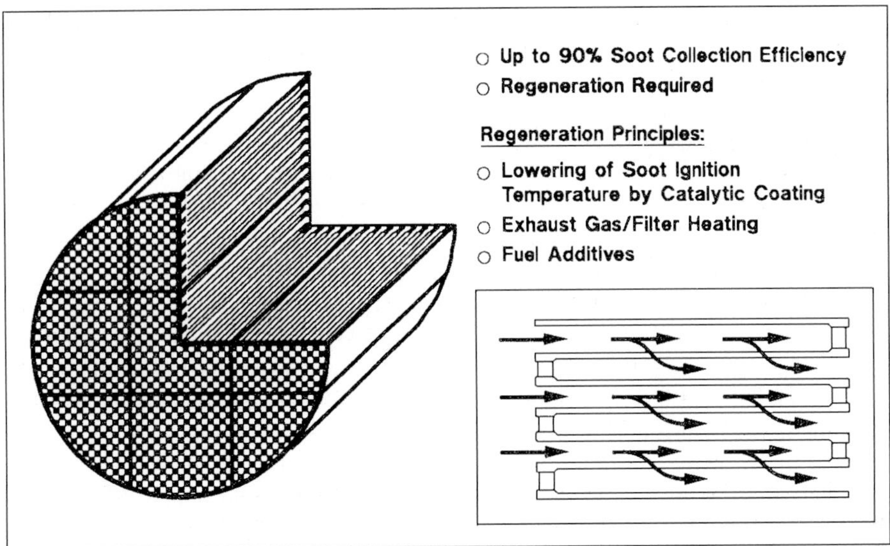

Figure 14. Operation principle of the ceramic wall flow diesel particulate filter. Reprinted from ref. [13] with kind permission of Elsevier Science.

filter can occur. The heat generated by this combustion can raise the temperature of the filter to above 1600 K, with consequent melting of the filter. Several approaches, to guarantee controlled regeneration at regular intervals, are summarized in Fig. 17 [13].

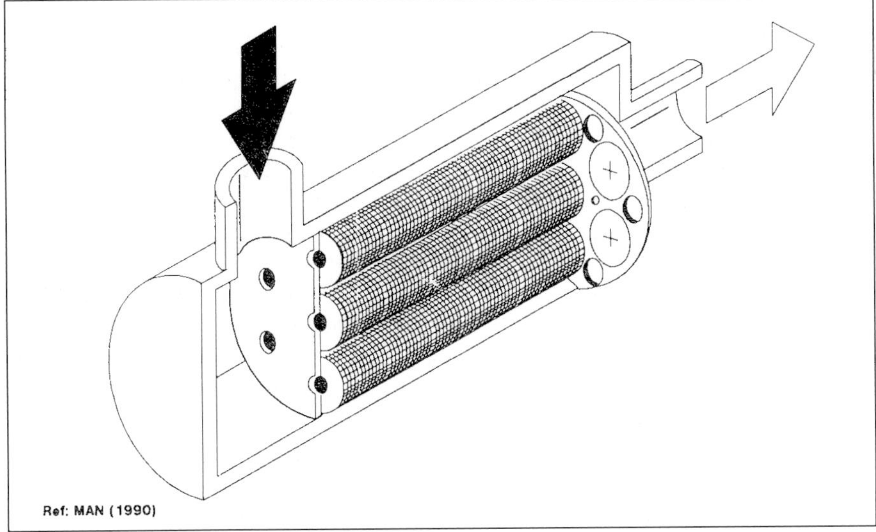

Figure 15. Operation principle of the wire mesh diesel particulate filter.

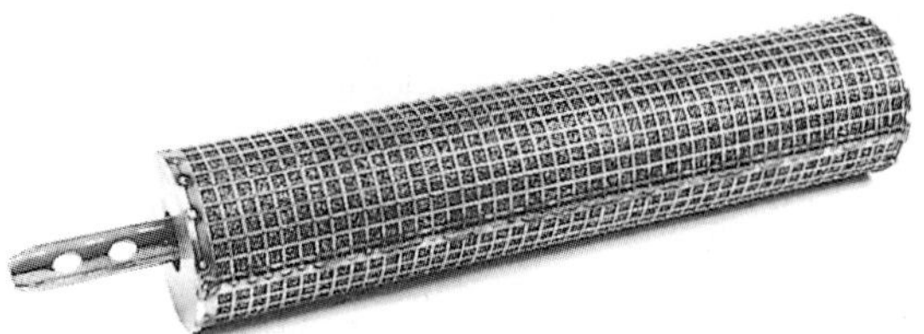

Figure 16. Example of a wire mesh diesel particulate filter.

In these approaches, either the temperature of the filter and/or of the exhaust gas is increased at regular intervals above the minimum temperature to start the burn off, or the minimum temperature required for the burn off is lowered by a catalyst. These different approaches can be combined. The temperature of the exhaust gas can be raised by controlled throttling of the engine, or by applying external heating.

The filter temperature can be raised by direct heating with a burner, placed in front of the filter, or by equipping the filter with resistive heating elements.

Several procedures exist to lower the minimum temperature at which the burn off starts. One alternative is to add a catalyst precursor to the diesel fuel. Examples of such homogeneous catalysts are organic derivatives of cerium or iron. They are burned together with the fuel in the engine, so that the resulting iron or cerium oxide is built into the particulate, thus ensuring intimate contact with the carbonaceous matter. Another alternative is to apply a soot combustion catalyst as a coating on the filter. Several oxides, especially vanadium oxide, are effective catalysts for the combustion of soot. Their disadvantage is that they need to be mobile, to some extent, at typical diesel exhaust temperatures, to ensure good contact with the

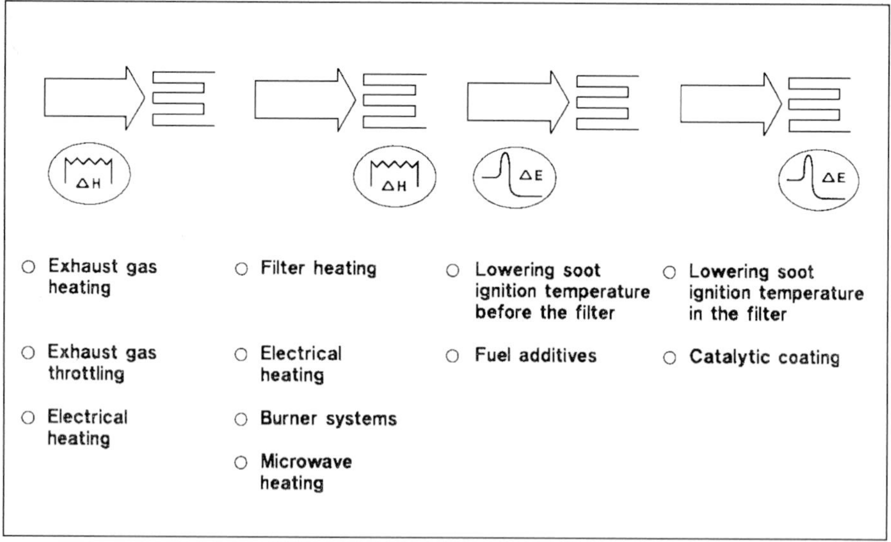

Figure 17. Various principles for the regeneration of diesel particulate filters. Reprinted from ref. [13] with kind permission of Elsevier Science.

particulate matter. This mobility can cause a migration of the catalytic material during use, resulting in an inhomogeneous distribution over the filter and, therefore, in a lowering of the regeneration efficiency.

Neither the fuel additive nor the catalytic soot combustion coating have the ability to substantially reduce the emission of carbon monoxide or aldehydes. To achieve this, a different, precious metal based, catalytic coating can be applied to the filter, or a precious metal based catalyst can be added in line with the filter system. To date, filtering systems have not reached widespread application, mainly because of the high costs associated with their complexity.

A simpler way to achieve a considerable reduction in the mass of particulates emitted, is the use of special diesel oxidation catalysts (Section 1.4). They offer advantages over filtering devices as they are simple, cost effective and can also reduce the emissions of CO, gaseous hydrocarbons, aldehydes and polynuclear aromatic hydrocarbons. Their main disadvantage is that their particulate mass reduction efficiency is only about half that achieved by filtering devices.

Finally, it is worthy of mention that the patent literature discloses many unconventional ways to reduce tailpipe emission of particulate matter; for example, the use of electrostatic precipitation, cyclones or gaswashing devices. However, none of these has reached practical application [13].

1.3 Catalysts for Gasoline Fueled Spark Ignition Engines

1.3.1 Introduction

Since the mid-1970s, the concepts and the design of catalysts for gasoline engines have changed, in parallel with changes in engine design and legislation.

1.3.2 Engine-out Emissions

The engine-out emission of carbon monoxide (CO), hydrocarbons (HC) and nitrogen oxides (NO_x) depends on the engine air-to-fuel ratio (A/F), defined as

$$\mathrm{A/F} = \frac{\text{mass of air consumed by the engine}}{\text{mass of fuel consumed by the engine}} \tag{3}$$

For gasoline engines it can be calculated that the air-to-fuel ratio at stoichiometry, i.e. where there is just enough air for complete combustion of all hydrocarbons in the fuel, is 14.7. If the air-to-fuel ratio is below this value, then the engine operates under excess fuel conditions, giving raise to incomplete fuel combustion. The exhaust gas will then contain more reducing reactants (CO, HC) than oxidizing reactants (O_2, NO_x), and is called rich. If the air-to-fuel ratio exceeds 14.7, then the engine operates under excess air conditions, giving rise to an exhaust gas that contains more oxidizing reactants than reducing reactants, and the exhaust gas composition is called lean.

Table 8. Composition of exhaust gas of a gasoline fueled spark ignition engine, as a function of the engine air-to-fuel ratio.

air-to-fuel ratio	CO (vol. %)	CO_2 (vol. %)	HC (vol. ppm)	O_2 (vol. %)	NO_x (vol. ppm)
13.684	2.536	13.24	442	0.17	2015
14.185	2.455	12.70	411	0.95	1786
14.334	2.135	12.90	391	0.96	1909
14.498	1.793	13.16	370	0.97	2066
14.623	1.534	13.30	360	0.98	2171
14.688	1.411	13.36	350	0.99	2196
14.741	1.309	13.42	341	0.99	2245
14.820	1.171	13.47	333	1.01	2334

Table 8 gives an example of the engine-out gas composition as a function of the air-to-fuel ratio. Note that the exhaust gas also contains hydrogen, in the molar ratio $y_{CO} : y_{H_2} = 3:1$ (mol : mol), almost independent of the engine air-to-fuel ratio. The engine-out exhaust gas also contains SO_2 in an amount which is practically independent of the air-to-fuel ratio within the typical operation ranges of a modern spark ignition engine, but which depends on the amount of sulfur in the gasoline, according to

$$y_{SO_2} = 1.1837 \times x_s/(A/F) \tag{4}$$

where y_{SO_2} is the SO_2 content in the exhaust gas in vol ppm, and x_s is the sulfur content of the fuel in milligrams of sulfur per liter of fuel.

A common way to classify the engine-out exhaust gas composition is the lambda (λ) value:

$$\lambda = \frac{\text{actual engine A/F}}{\text{stoichiometric engine A/F}} \tag{5}$$

In this way, the exhaust gas composition can be described independently of the exact fuel composition (Fig. 18).

It has been a common practice in the design of emission control technologies to calculate the exhaust gas lambda value from the exhaust gas composition. On the basis of some simplifying assumptions, the following can be derived:

(i) For stoichiometric and lean exhaust gas composition,

$$\lambda = 1 + \frac{D \times (4.79m + 0.9475n)}{(100 - 4.79 \times D)(m + 0.25n)} \tag{6}$$

$$D = y_{CO_2} + 0.5y_{NO} - 0.5y_{CO} - 0.5y_{H_2} - (m + 0.5n) \times y_{C_mH_n} \tag{7}$$

where y is the concentration in the exhaust gas in vol %, m is the number of

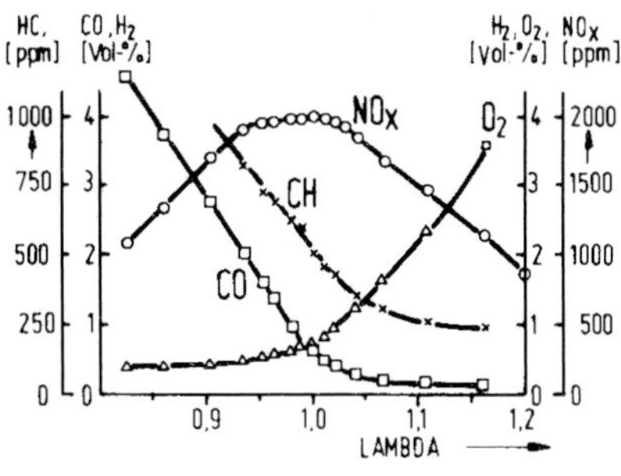

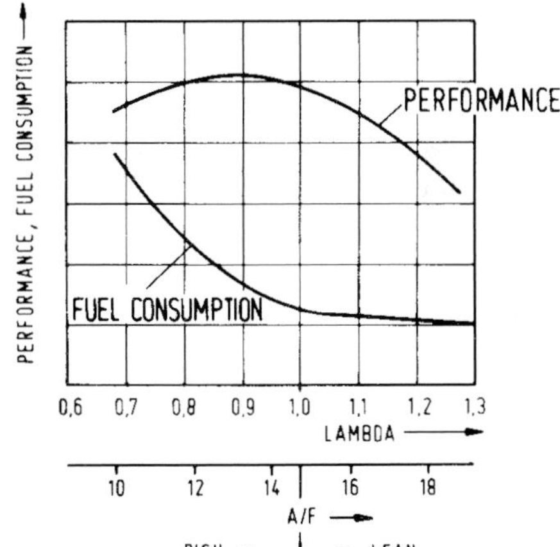

Figure 18. Engine out emission of CO, HC and NO_x: fuel consumption and power output of a gasoline fueled spark ignition engine as a function of the lambda value.

carbon atoms in the exhaust gas hydrocarbon molecule C_mH_n, and n is the number of hydrogen atoms in the exhaust gas hydrocarbon molecule C_mH_n.

(ii) For rich exhaust gas composition,

$$\lambda = 1 - \frac{A \times (4.79m + 0.9475n)}{(200 - 3.79 \times A)(m + 0.25n)} \tag{8}$$

$$A = y_{CO} + (2m + 0.5n)y_{C_mH_n} - y_{NO} - 2y_{O_2} + y_{H_2} \tag{9}$$

with y, m and n defined as before.

Equations 6–9 can be easily applied in the case of catalyst activity evaluation using model gas reactors, because then the composition of the hydrocarbon C_mH_n is exactly known. In the case of engine exhaust gas, a weighted average of the exhaust gas hydrocarbon composition can be used. More detailed procedures, that also take the fuel composition into consideration, can be found in the literature [14].

1.3.3 Reactions

A multitude of reactions can occur between the engine-out exhaust gas constituents (Fig. 19). The main reactions that lead to the removal of CO, hydrocarbons and nitrogen oxide are:

(i) Reactions with oxygen (oxidation)

$$C_mH_n + (m + 0.25n)O_2 \rightleftharpoons mCO_2 + 0.5nH_2O \tag{10}$$

$$CO + 0.5O_2 \rightleftharpoons CO_2 \tag{11}$$

$$H_2 + 0.5O_2 \rightleftharpoons H_2O \tag{12}$$

These reactions occur especially with stoichiometric and with lean exhaust gas compositions.

(ii) Reactions with nitrogen oxide (oxidation/reduction)

$$CO + NO \rightleftharpoons 0.5N_2 + CO_2 \tag{13}$$

$$C_mH_n + 2(m + 0.25n)NO$$

$$1(m + 0.25n)N_2 + 0.5nH_2O + mCO_2 \tag{14}$$

$$H_2 + NO \rightleftharpoons 0.5N_2 + H_2O \tag{15}$$

These reactions occur with stoichiometric and with rich exhaust gas compositions.

Especially with rich exhaust gas compositions, the water-gas shift reaction

$$CO + H_2O \rightleftharpoons CO_2 + H_2 \tag{16}$$

and the steam reforming reaction

$$C_mH_n + 2mH_2O \rightleftharpoons mCO_2 + (2m + 0.5n)H_2 \tag{17}$$

may contribute to the removal of carbon monoxide and hydrocarbons.

Depending upon the catalyst operating conditions, a number of reactions may

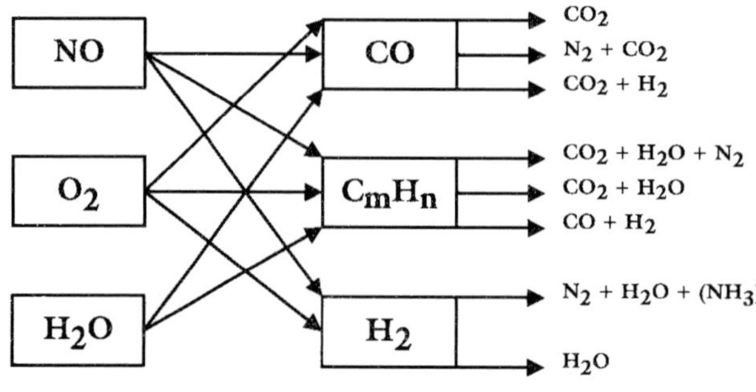

Figure 19. Overview of the possible reactions between some components in the exhaust gas of an internal combustion engine.

occur giving rise to what are called the secondary emissions. The most common of these are:

(i) Reactions, with SO_2

$$SO_2 + 0.5O_2 \rightleftharpoons SO_3 \tag{18}$$

$$SO_2 + 3H_2 \rightleftharpoons H_2S + 2H_2O \tag{19}$$

(ii) Reactions with NO

$$NO + 0.5O_2 \rightleftharpoons NO_2 \tag{20}$$

$$NO + 2.5H_2 \rightleftharpoons NH_3 + H_2O \tag{21}$$

$$2NO + CO \rightleftharpoons N_2O + CO_2 \tag{22}$$

It is the skill of the emission control system designers to selectively promote the desirable reactions 10–17 and to prevent the secondary reactions 18–22.

The extent to which each of the main reactions contributes to the removal of carbon monoxide, hydrocarbons and nitrogen oxides depends on the catalyst formulation and the catalyst operating conditions. Detailed kinetic data for these reactions are rarely found in the literature. Some fundamental data do exist for the oxidation of carbon monoxide, reaction 11, and some overall kinetic data exist for the other reactions [15–19].

However, it can be stated that with a stoichiometric exhaust gas composition, the overall rate of carbon monoxide removal is one to two orders of magnitude higher than the overall rate of either hydrocarbon and nitrogen oxide removal.

1.3.4 Emission Control Concepts

About five basic catalytic concepts have been used in the history of catalytic emission control (Fig. 20(a)).

The first concept is the closed-loop-controlled three-way catalyst. In this, one type of catalyst, which is placed in the exhaust gas stream, is able to promote all the main reactions that lead to the simultaneous removal of carbon monoxide, hydrocarbons and nitrogen oxides. To balance the extent of the oxidation and the reduction reactions, the composition of the engine-out exhaust gas is maintained at or around stoichiometry. This is achieved by a closed-loop engine operation control, in which the oxygen content of the engine-out exhaust gas is measured up-stream of the catalyst with an electrochemical oxygen sensor, also called lambda sensor.

This component is used by the engine management system to regulate the amount of fuel fed into the engine, and so to regulate the engine operation around the stoichiometric A/F ratio. As shown in Fig. 20(b) the optimum in the simultaneous removal of carbon monoxide, hydrocarbons and nitrogen oxides is reached under those conditions. Also, the extent of the secondary reactions is minimal under these conditions. The feedback control of the engine causes a small cyclic variation of the engine exhaust gas composition. This variation occurs in less than one second,

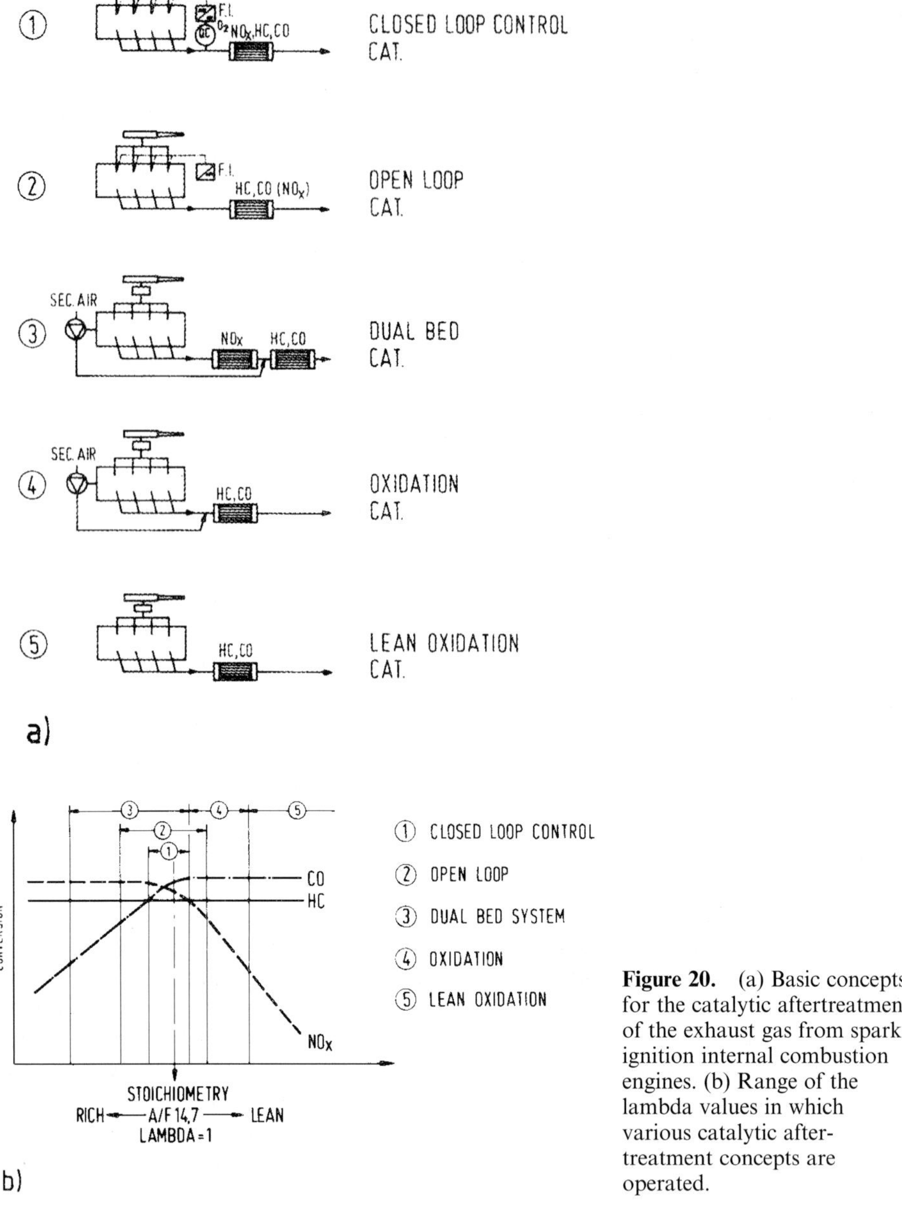

Figure 20. (a) Basic concepts for the catalytic aftertreatment of the exhaust gas from spark ignition internal combustion engines. (b) Range of the lambda values in which various catalytic after-treatment concepts are operated.

which means a frequency of one to three Hertz, and with an amplitude of 2–5% of the A/F-set point. This transient operation of the catalyst, however, has a significant effect upon its performance, as will be described below.

There exist a multitude of engine management systems with various degrees of complexity and refinement, affecting the speed and the amplitude range of control of the engine A/F ratio at each of the engine load and speed operation conditions.

The refinement of the engine management system affects both the performance and the durability of the emission control catalyst.

Because of its performance in promoting the main reactions to reach completion and at the same time minimizing the extent of the secondary reactions, the closed-loop-controlled three-way catalyst has become the most widely applied technique for catalytic emission control.

The second concept is the open-loop three-way catalyst. This is a simplification of the first, as again a multifunctional catalyst is used, that is able to promote all the reactions that lead to the removal of carbon monoxide, hydrocarbons and nitrogen oxides. However, the composition of the exhaust gas is not controlled and therefore varies over a wide range. This wider operation range results in an overall lower simultaneous conversion of the three exhaust gas constituents (Fig. 20(b)). This concept is used if the legislative limits can be reached with a conversion of about 50%, or for the retrofitting of engines that were not designed to be equipped with catalytic emission control devices.

The third concept is the dual-bed emission control catalyst. In this, the catalytic converter is made of two different types of catalyst. The first is either a multifunctional catalyst or at least one capable of promoting NO_x reduction reactions. The engine is calibrated so as to guarantee a net reducing exhaust gas composition. Under these conditions, the first catalyst will lead to an elimination of the nitrogen oxides. The second catalyst is an oxidation catalyst. Extra air is injected in front of the second catalyst to assist the removal of carbon monoxide and hydrocarbons. The secondary air can be added either by mechanically or by electrically driven air pumps.

The dual-bed concept allows for a wider engine A/F range, while still maintaining a high conversion efficiency for the three exhaust gas constituents under consideration. Therefore, a less sophisticated engine management system can be used.

The fourth emission control concept is the oxidation catalyst. Secondary air is added to the exhaust gas to assure a lean composition, independent on the engine operation condition. The catalyst is designed to promote reactions between oxygen and both carbon monoxide and hydrocarbons, which can be removed to a high extent. However, nitrogen oxides cannot be removed in this manner.

Finally, the fifth concept is also an oxidation catalyst, but is applied to engines that operate under lean conditions, i.e. lean burn engines. The A/F of these engines reaches values up to 26, corresponding to a lambda value of about 1.8. As shown in Fig. 18 under these conditions the engine-out emission of nitrogen oxides decreases drastically. The function of the catalyst could be limited to converting carbon monoxide and hydrocarbons. Because of the dilution effect in lean combustion, the exhaust gas is colder than for closed-loop-controlled engines. Therefore, special catalysts with good low temperature activity for the oxidation reactions are needed. To date, this concept has not yet achieved widespread application. However, it is a topic of intensive development activity because it appears that catalytic NO_x conversion will be necessary to reach the current and future emission limits. This requirement is the origin of some new additional emission control concepts for lean burn spark ignition engines, that are discussed later.

1.3.5 Emission Control Catalyst Design

A General Aspects

In the early days of emission control catalyst development, various catalysts and catalytic reactor designs were proposed. One of the first consisted of bead shaped catalysts, in analogy with the shape of many catalysts used in the chemical and petrochemical industries. However, in parallel, completely different designs of catalytic reactors were developed, including monolithic structures in various shapes made out of ceramics or metals. The driving forces behind these developments were, in addition to the main task of achieving legislative requirements, the boundary conditions. Examples are minimal additional pressure drop, minimal additional weight, minimal space requirement (or at least a space requirement that does not interfere negatively with the vehicle platform design), limited complexity (to allow for minimal additional costs to the vehicle) and a production technology to allow for production of several million units per year.

Examples of early designs are given elsewhere [20]. Ultimately, three different designs emerged for widespread application: the bead catalyst reactor, the ceramic monolith reactor and the metal monolith reactor (Fig. 21).

B Bead Catalysts

The bead catalytic reactor has been used in USA until about 1992 and is well described in the literature (Fig. 22). It consists of two sieve plates mounted in a more

Figure 21. Basic designs of emission control catalysts (beads, ceramic monoliths, metal monoliths).

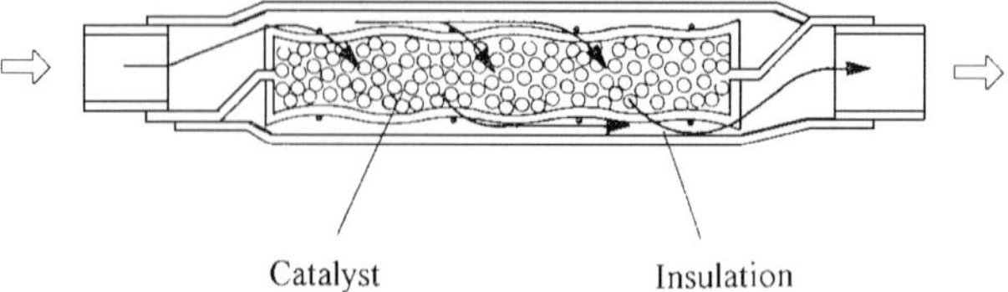

Figure 22. Design principle of a bead catalytic reactor.

or less diagonal fashion inside a flat, broad reactor. The space between the sieve plates is filled with bead shaped catalysts. The exhaust gas traverses the sieve plates from top to bottom. The reactor itself has a double shell, with insulation material between the two shells. On one side of the reactor, a fill plug connects the outer shell with the sieve plate, which also allows for easy exchange of the catalyst (Fig. 23).

The beads consist of transitional alumina, such as gamma, delta or theta alumina, with a Brunauer–Emmett–Teller (BET) surface area of about $100\,m^2\,g^{-1}$. The bead diameter range applied is the result of a compromise between mechanical requirements, such as high crushing strength and low pressure drop, and the requirements of a quick mass transfer from the gas phase to the catalyst surface. Most of these requirements can be expressed mathematically [21]. The crushing strength is proportional to the square of the bead diameter. The pressure drop increases with decreasing bead diameter according to the Ergun equation, and the exhaust gas flow through the reactor is turbulent. The mass transfer rate from the gas phase to the catalyst is proportional to the geometric surface area of the catalyst, which depends on the bead diameter as shown in Table 9. These boundary conditions result in the choice of beads with diameters between about 2.4 mm and about 4.0 mm.

Bead porosity is the result of a compromise between mechanical requirements, the intraparticle mass transfer requirements (both with respect to the reactants which are to be converted, and also with respect to the poisoning elements), and physical requirements such as a low thermal mass of the converter system. It has been shown that the crushing strength decreases with increasing porosity, but also that the thermal mass is significantly reduced by increasing the porosity.

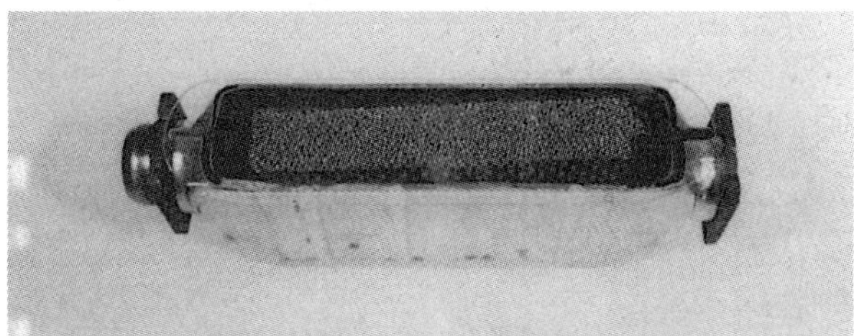

Figure 23. Cutaway view of a bead catalytic reactor.

Table 9. Geometric surface area of bead shaped catalysts, as a function of the radius of the beads and of the type of bead.

Bead type	Sieve fraction (mm)	Average radius (mm)	Weight per bead (mg)	Volume per bead (μl)	Geometrical surface area per bead (mm^2)	Bulk density of packed beads ($kg\,l^{-1}$)
A	1.7–2.0	0.925	4.3	3.3	10.7	0.54
B	2.36–2.80	1.29	10.3	8.9	20.9	0.50
B	2.80–3.35	1.54	15.2	15.2	29.7	0.50
C	4.75–5.60	2.58	68.0	72.5	84.1	0.52

The porosity of the beads used is the result of a lot of optimization, and is formed both by macropores with pore diameters exceeding 0.1 μm and by micropores with a pore diameter less than 20 nm. The micropores give the high BET surface area, whereas the macropores assure a high intraparticle mass transfer rate as well as a resistance against deactivation by poisoning.

The result of this compromise are beads with a pore volume between $0.5\,cm^3\,g^{-1}$ and $1.0\,cm^3\,g^{-1}$ and a bulk density between $670\,kg\,m^{-3}$ and $430\,kg\,m^{-3}$.

It has been shown that, due to vehicular vibrations, it is necessary to ensure a high resistance against attrition. Best resistance is obtained with spherical beads, that can be made by pan pelletizing or the oil drop process.

The catalyst is made by impregnating the beads with aqueous solutions of salts of some rare earth metals and of salts of the desired precious metals such as Pt, Pd and Rh; these impregnated beads are then dried and calcined. The distribution of precious metals over the bead radius must be achieved with care, to balance the mass transfer requirements with the poison resistance requirements (Figs. 24–26). The distribution of the active component over the pellet radius can be measured by an Energy Dispersive X-ray (EDX) scan on an individual pellet. However, since in the application a relatively broad distribution in diameters occurs, special procedures have been developed to determine some kind of average distribution of the active components over the pellet radius. The most common procedure is the attrition test, in which a known mass of pellets of known diameter distribution is immersed in a liquid that neither dissolves the active components nor the carrier. The pellets are stirred for a defined time, and are separated from the attrited powder. The powder mass is determined, and its chemical composition analyzed by sensitive methods,

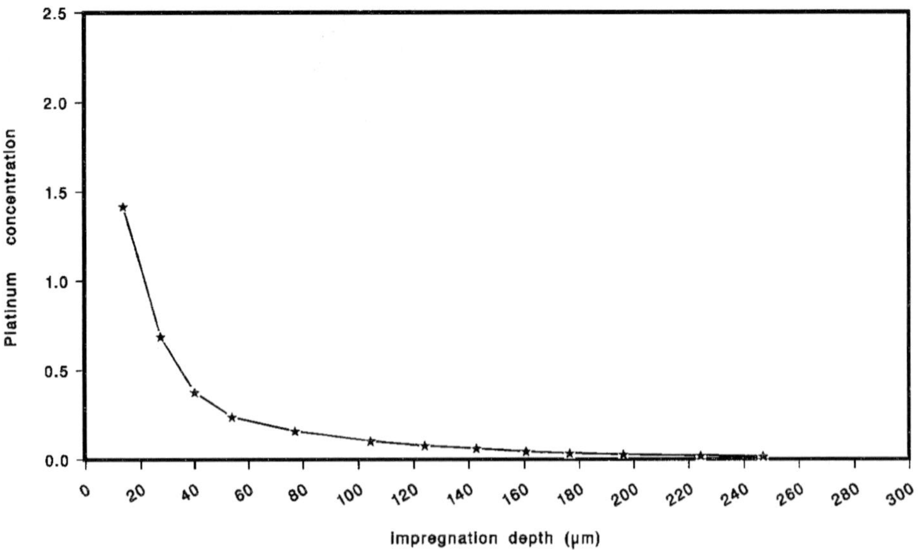

Figure 24. Concentration of platinum at various depths in a bead shaped three-way catalyst.

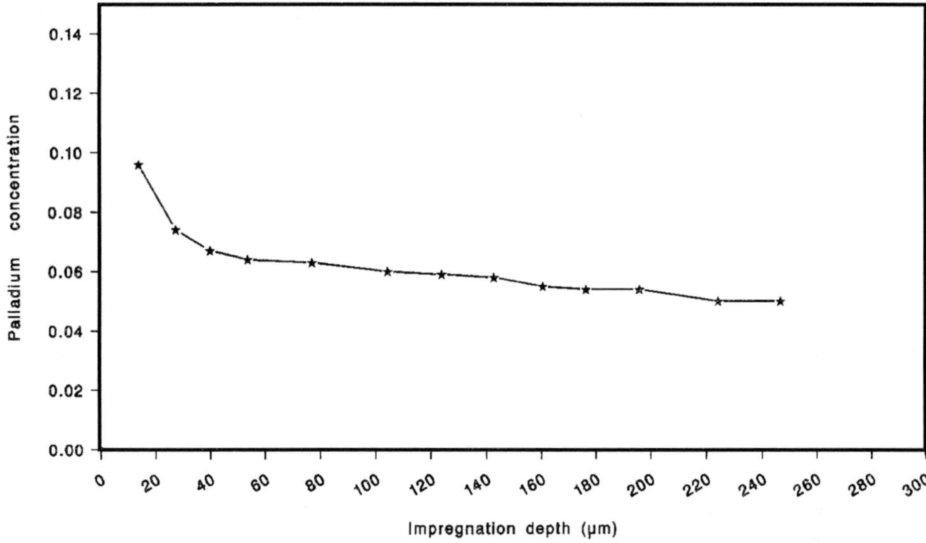

Figure 25. Concentration of palladium at various depths in a bead shaped three-way catalyst.

such as X-ray fluorescence (XRF) or proton-induced X-ray emission (PIXE). This procedure is repeated until at least 50% of the original pellet radius is removed.

The impregnation depth of the active components is calculated after each attrition step using models that take into account the shape of the original pellet. If the

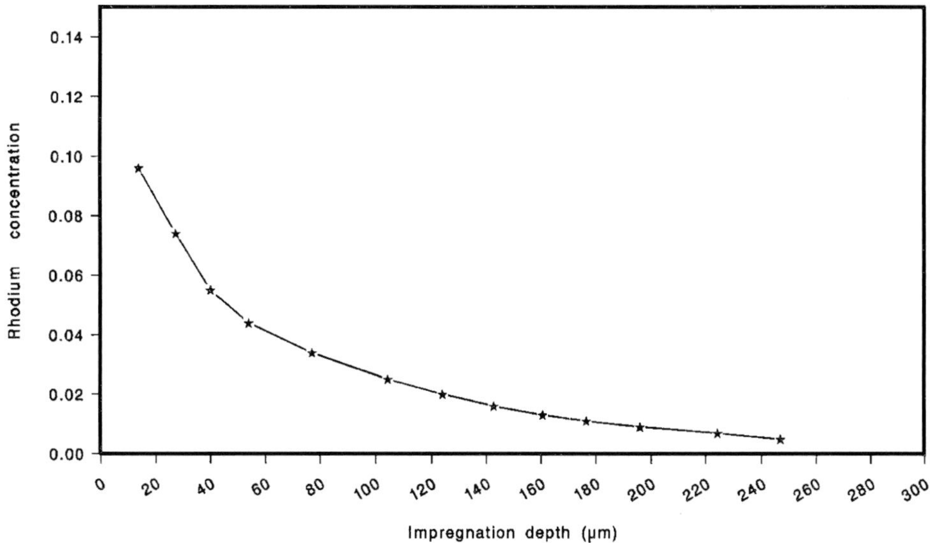

Figure 26. Concentration of rhodium at various depths in a bead shaped three-way catalyst.

pellet is assumed to be a perfect sphere, then the impregnation depth follows from [22]:

$$r_n = r_0\left(1 - \left[\frac{W_n}{W_0}\right]^{1/3}\right) \tag{23}$$

where r_n is the impregnation depth of precious metals in the bead, calculated from the nth attrition run (m), r_0 is the original radius of the bead (m), W_0 is the original mass of the bead (kg), and W_n is the mass of the bead after the nth attrition run (kg).

Some authors used a refined model, in which the original pellet shape is assumed to be an ellipsoid [23]. In this case, the impregnation depth follows from

$$1 - \left(\frac{W_n}{W_0}\right) = (0.5a - r_{n-1})^2(0.5b - r_{n-1}) - (0.5a - r_n)^2(0.5b - r_n) \tag{24}$$

where r_n is the impregnation depth of precious metals in the bead, calculated from the nth attrition run (m), a is the major axis of the ellipsoid (m), b is the minor axis of the ellipsoid (m), W_0 is the original mass of the bead (kg), and W_n is the mass of the bead after the nth attrition run (kg).

The main advantages of the pellet converter in the automotive emission control application over the monolithic converter are the following:

(i) The exhaust gas flow in the pellet converter is turbulent, thus giving an enhanced gas to solid mass transfer rate.
(ii) A more uniform deposition of poisoning elements occurs over the converter, especially as it has been shown that the pellets move slowly around in the converter during use.
(iii) The risk of destroying the support by so-called thermal shocks, which are sudden changes in catalyst temperature, is very low.

However, the pellet converter also has some disadvantages. The most important are a higher pressure drop and a more complex construction of the converter housing as compared to the monolithic reactor. Also, the risk of losing some catalyst by attrition is to be considered.

As a result, the pellet converter is nowadays used only on a limited number of vehicles. The majority of passenger cars is now equipped with monolithic converters.

C Monolithic Catalysts

a Ceramic Monolithic Converters

An example of an automotive converter based upon a ceramic monolith is shown in Fig. 27. It consists essentially of three parts, see Fig. 28:

(i) A monolithic ceramic support that carries the catalyst.
(ii) A mat, that surrounds the monolithic support, made either out of ceramic

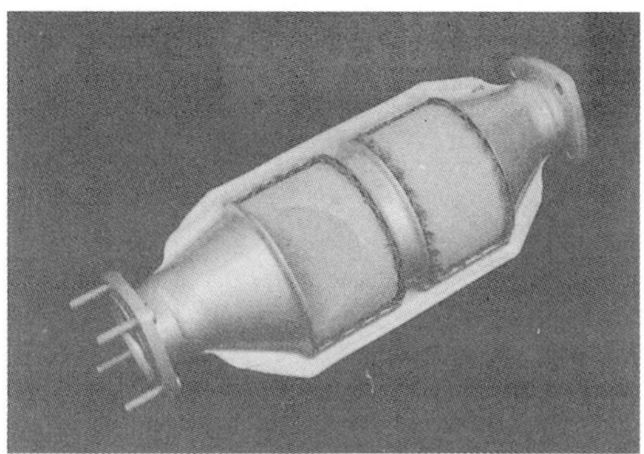

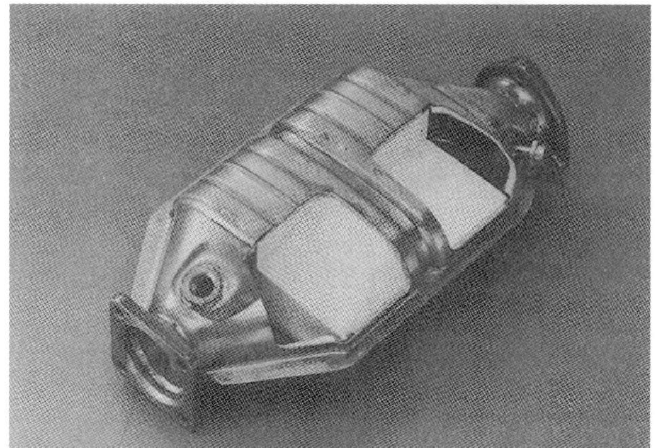

Figure 27. Cutaway view of ceramic monolith based converters for the catalytic after-treatment of exhaust gases.

material, or out of a metallic wire mesh. The mat ensures that the ceramic monolith is tightly packed inside the converter at all operating temperatures. It protects the ceramic monolith against mechanical impact and also serves as a thermal insulation.

(iii) A converter housing made out of high quality, corrosion-resistant steel. The converter is the piece that is inserted in the exhaust pipe. The converter holds the ceramic monolith and ensures that the exhaust gas flows through the ceramic monolith.

There exist numerous designs for the converter. The boundary conditions for the converter design are the space which is available in the vehicle platform to mount the converter and the overall volume of catalyst needed for the catalytic function. A single converter can be packed with one single or with more, typically two or sometimes three, single pieces of ceramic monolith. When multiple pieces are used, they are mounted with a well defined distance between them, to effect turbulent flow conditions at the inlet to each piece.

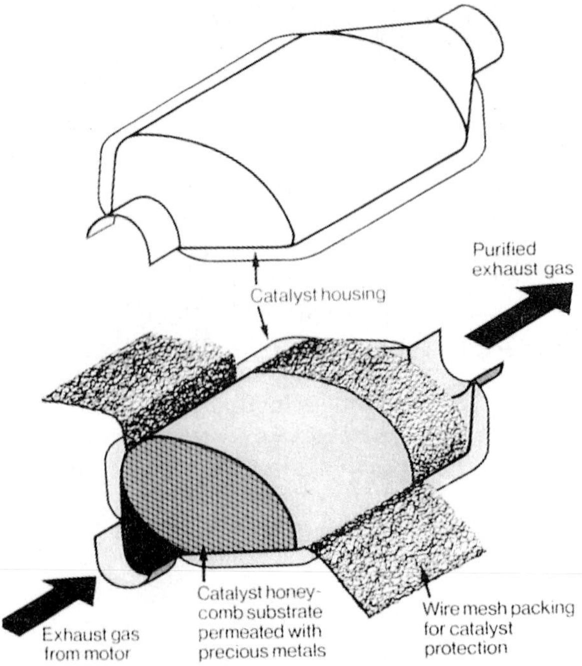

Figure 28. Design principle of a ceramic monolith based converter for the catalytic aftertreatment of exhaust gases.

The monolithic support is a ceramic cylindrical body traversed by a multitude of straight channels, typically 5000 or more (Fig. 29). The catalyst volume is defined as the geometrical volume of this cylindrical body. Again, a multitude of body shapes are used, the most common being either circular or oval-like (Fig. 30).

The ceramic material most commonly used is porous cordierite, ($2MgO \cdot 2Al_2O_3 \cdot 5SiO_2$). Its overall chemical composition is about 14 wt % MgO, 36 wt % Al_2O_3 and

Figure 29. Example of a ceramic monolithic support for exhaust gas aftertreatment catalysts.

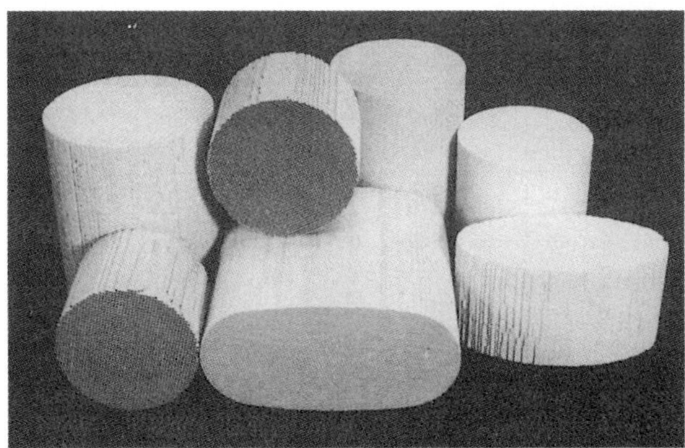

Figure 30. Various designs of ceramic monolithic supports.

50 wt% SiO_2, with minor amounts of substances such as Na_2O, Fe_2O_3 and CaO. The porosity of the monolith walls is between 20 vol.% and 40 vol.%, mainly formed by macropores with an average diameter of a few micrometers [24].

The monolith channels are the empty spaces between the walls. During the development of monolithic ceramic supports, several channel shapes have been considered. Examples are circular, hexagonal, square and triangular. Only the last two have reached widespread commercial use, with the majority being square channels (Fig. 31).

A variety of monoliths are offered to the market which differ in the size of the channels and in the thickness of the walls. Together, these parameters fix what is called the cell density, the number of cells per unit frontal surface area. With ceramic monoliths, these two parameters can be varied to some extent independently. As of today, the most commonly used supports have about 62 channels per square centimeter, which corresponds to a channel width of about 1 mm and a wall thickness of about 0.15 mm. The bulk density of such a support is about 420 kg m^{-3}. The supports currently used have, within the production variance, homogeneous channel

Figure 31. Close-up view of a ceramic monolithic support with 62 cells cm^{-2}.

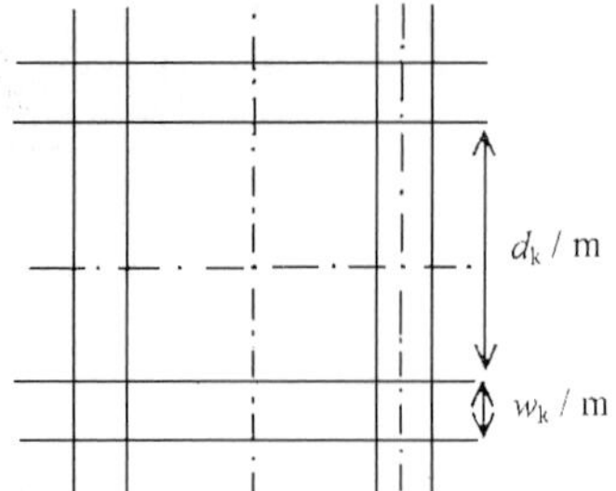

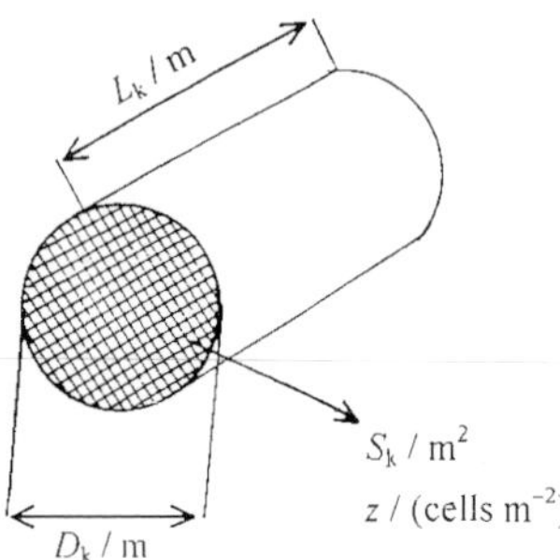

Figure 32. Design parameters for cylindrical ceramic monolithic support with square channels.

properties over the whole structure, although other alternatives, such as structures with a cell density varying over the monolith radius, also have been described in the literature.

Considering a round cylindrical monolith structure with homogeneous square channels, a number of relationships can be derived for the design parameters (Fig. 32).

The cell density z is given by

$$z = \frac{1}{(d_k + w_k)^2} \tag{25}$$

where z is the cell density (m^{-2}), d_k is the monolith channel width (m), and w_k is the monolith wall thickness (m). The total frontal area S_k follows from

$$S_k = \frac{\pi \times D_k^2}{4} \tag{26}$$

or

$$S_k = \frac{\pi \times D_k^2}{4} \times (z) \times (d_k + w_k)^2 \tag{27}$$

where S_k is the total frontal area (m^2), and D_k is the diameter of the monolith (m).

The open frontal area S_k^0 is calculated as

$$S_k^0 = \frac{\pi \times D_k^2}{4} \times \frac{d_k^2}{(d_k + w_k)^2} \tag{28}$$

where S_k^0 is the open frontal area (m^2).

The geometrical monolith volume V_k is given by

$$V_k = L_k \times S_k \tag{29}$$

where V_k is the geometrical monolith volume (m^3), and L_k is the length of the monolith (m).

The geometrical surface area per unit monolith volume S_g can be calculated from eqs 30–32:

$$S_g = [S_g^0]/[V_k] \tag{30}$$

$$S_g = 4 \times z \times d_k \tag{31}$$

$$S_g = 4 \times \frac{d_k}{[d_k + w_k]^2} \tag{32}$$

where S_g is the geometrical surface area per unit catalyst volume ($m^2\,m^{-3}$), and S_g^0 is the geometrical surface area per monolith (m^2).

Finally, the number of channels per support (N_k) follows from

$$N_k = S_k \times z \tag{33}$$

or

$$N_k = S_k/(d_k + w_k)^2 \tag{34}$$

Values for these key design parameters are given in Table 10 for some typical commercial products.

The design parameters influence the key catalyst operating parameters [25, 26].

The gas flow per channel (G_k) is calculated from

$$G_k = \frac{G}{S_k \times z} \tag{35}$$

Table 10. Typical values for some design parameters of ceramic monolithic supports.

Geometrical surface area ($cm^2\,cm^{-3}$)	Bulk density ($kg\,l^{-1}$)	Open frontal area (%)	Wall thickness (mm)	Cell size (mm)	Cell density (cells cm^{-2})
18.5	0.52	69	0.31	1.48	31
19.1	0.44	74	0.25	1.54	31
22.0	0.57	64	0.29	1.18	46
23.5	0.43	74	0.20	1.27	46
27.4	0.42	75	0.17	1.10	62

where G_k is the gas flow per channel ($Nm^3\,h^{-1}$), and G is the total exhaust gas flow ($Nm^3\,h^{-1}$).

The linear gas velocity per channel v_k ($m\,h^{-1}$) is given by

$$v_k = \frac{G}{S_k} \times \frac{(d_k + w_k)^2}{d_k{}^2} \tag{36}$$

The residence time of the exhaust gas in the channel t_k (h) follows from

$$t_k = \frac{L_k}{v_k} \tag{37}$$

or

$$t_k = \frac{V_k}{G} \times \frac{d_k{}^2}{(d_k + w_k)^2} \tag{38}$$

The Reynolds number Re inside the channels is given by

$$Re = \frac{d_k \times V_k}{v} \tag{39}$$

or

$$Re = \frac{G \times (d_k + w_k)^2}{v \times S_k \times d_k} \tag{40}$$

where v is the kinematic viscosity ($m^2\,h^{-1}$).

Table 11 summarizes the value of some of these parameters for some of the commercially available supports.

The production procedure for ceramic monoliths has been described in the literature. Basically, they are made by extrusion of a paste that contains the cordierite precursors together with processing aids, followed by drying and reactive calcination [24].

Table 11. Typical value of the key catalyst reaction conditions as a function of the design parameters of the monolithic support for operating conditions relative to a monolith with 62 cells per square centimeter and wall thickness 0.17 mm.

Abundance of entry effect (%)	Length of entry effect (%)	Nusselt (%)	Reynolds (%)	Pressure drop in a channel (%)	Residence time in a channel (%)	Linear velocity in a channel (%)	Wall Thickness (mm)	Cell Density (cells cm^{-2})
50	198	198	147	60	128	110	0.31	31
50	198	198	142	52	139	101	0.25	31
75	132	132	125	101	99	116	0.29	46
75	132	132	116	75	115	100	0.20	46
100	100	100	100	100	100	100	0.17	62
151	66	66	84	173	76	107	0.17	93
151	66	66	79	132	87	94	0.11	93

Figure 33. Examples of metallic monolithic supports.

b Metallic Monolithic Converters

Monoliths made out of metal were introduced to the market to some extent as an alternative to ceramic supports. As shown in Fig. 33, these supports consist of a metallic outer shell, in which a honeycomb-like metallic structure is fixed. The honeycomb is formed by alternate flat and corrugated thin metal foils (Fig. 34). These foils are made out of corrosion- and high-temperature-resistant steel, and are about 0.05 mm thick.

Honeycomb structures offered to the market have a similar cell density as the ceramic ones. After welding inlet and outlet cones to the outer shell, the metallic monolithic converter can be inserted directly into the exhaust gas pipe, which means that the canning procedure used for the ceramic monoliths is not needed anymore.

There has been a long debate about the relative advantages of metallic monoliths over ceramic monoliths. From the standpoint of automotive catalyst manufacturing, durability and performance, the following aspects are of relevance [27]:

Figure 34. Close-up view of a metallic monolithic support.

(i) The metallic support is nonporous. To assure an optimal adherence of the catalyst coating to the metal foil, thermal and/or chemical pretreatments could be necessary.
(ii) Metal foils have a considerably higher thermal conductivity than ceramic structures. This might affect the light-off properties of the catalyst, and also the durability, for example in the case of hot spots during use.
(iii) At a similar cell density, the open frontal area of a metallic monolith is higher than that of the current ceramic monoliths, because the wall thickness of the metallic foil is considerably lower than that of the current ceramic structures. At same thickness of the catalytic coating, this guarantees a lower pressure drop over the metallic monolith structure.

Finally, it should be noted that metallic monoliths allow for some additional design freedom over ceramic monoliths. Numerous examples of this are found in the literature (Fig. 35) [28]. One example shows a metallic structure consisting of a macroscopic corrugated foil and a microscopic corrugated foil. The microscopic corrugation considerably increases the geometrical surface area of the structure. A further example shows a metallic monolith in which the channels interconnect to

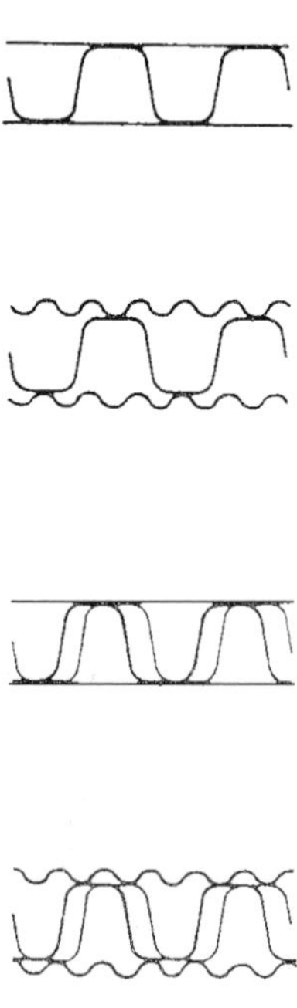

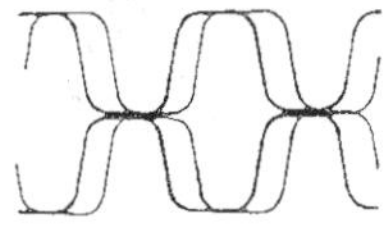

Figure 35. Examples of the matrix design of metallic monoliths. Adapted from [28].

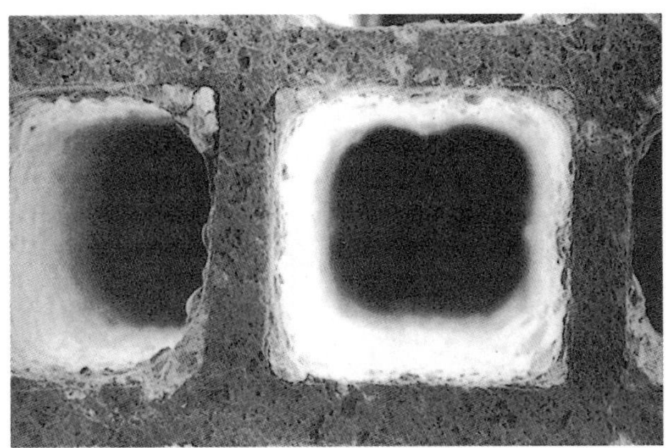

Figure 36. Close-up view of a channel of a washcoated ceramic monolith.

each other. This allows for a homogenization of the gas flow inside the monolith, even at suboptimal design of the inlet cones of the converter.

c Washcoat

Ceramic and metallic monolith structures have a geometrical surface area in the range 2.0–4.0 $m^2 l^{-1}$ support volume. This is much too low to adequately perform the catalytic conversion of the exhaust gas components. Therefore, these structures are coated with a thin layer of a mixture of inorganic oxides, some of which have a very high internal surface area. This mixture is called the washcoat (Fig. 36).

The washcoat increases the surface of the catalytic structure to 10 000–40 000 $m^2 l^{-1}$ support volume. This high internal surface area is needed to adequately disperse the precious metals and to act as an adsorbent for the catalyst poisoning elements present in the exhaust gas.

The washcoat components support the catalytic function of the precious metals and even take part in the catalytic reaction. Furthermore, they guarantee the resistance of the precious metal components against deactivation processes occurring at high temperature catalyst operation conditions. As shown in Fig. 37, the washcoat layer typically has a thickness of about 10–30 μm on the walls of the support and of about 100–150 μm in the corners of the support. These values are valid for ceramic supports with square shaped channels.

The washcoat is made up of so-called secondary washcoat particles, with a diameter of about 2 μm up to about 30 μm. The voids between the washcoat particles cause the washcoat macroporosity. These voids have a diameter of a few micrometers (Fig. 38). The secondary washcoat particles are responsible for the washcoat meso- and microporosity, and consist of either pure inorganic oxides or a microscopic mixture of several inorganic oxides. These inorganic oxides constitute the primary washcoat particles which have, in the fresh state, diameters of 10–20 nm. The precious metals are then deposited onto these primary washcoat particles (Fig. 39).

The amount of washcoat used depends on the properties of the support, the type of catalyst, and the catalyst performance and durability requirements. On a

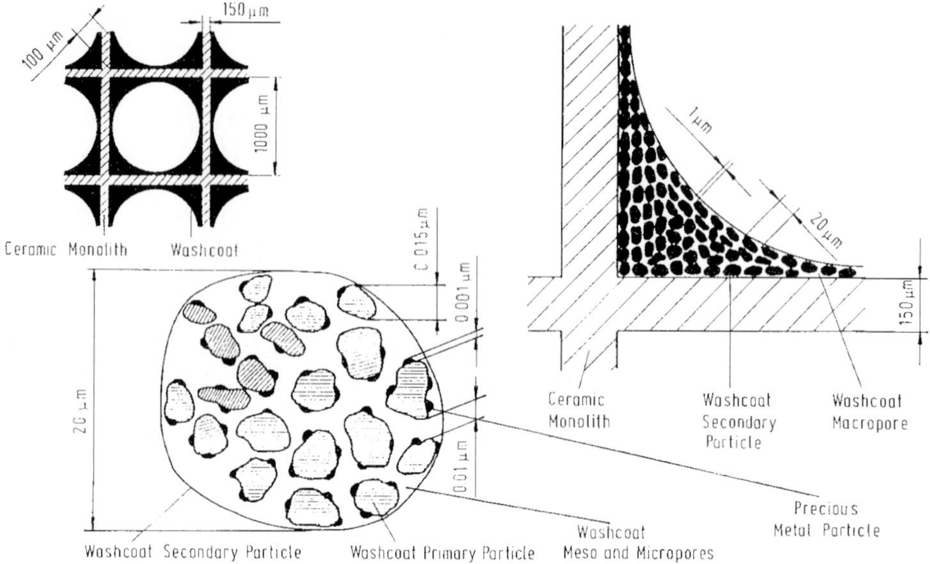

Figure 37. Dimensional relationships of a washcoat on a ceramic monolith.

62 cell cm^{-2} ceramic monolith, washcoat loadings of 100–400 g l^{-1} monolith volume are presently used. The washcoat can be present as a single layer, or as multiple layers with different chemical compositions. The amount of washcoat and its chemical composition are typically uniform in the radial and the axial direction of the monolith structure, although different designs have been described in the patent literature.

The chemical composition of the washcoat belongs to the core know-how of the catalyst manufacturers. The most common washcoats contain aluminum oxides, cerium oxides and zirconium oxides as major constituents. The minor constituents

Figure 38. Scanning electron microscope view of a washcoat layer.

Figure 39. Transition electron microscope picture of precious metals on a washcoat particle.

are oxides that stabilize the internal surface area of some or all of the major washcoat constituents, that support the catalytic function of some of major washcoat constituents and/or that influence their interaction with the precious metals. Examples of minor washcoat constituents are CaO and MgO, as well as the oxides of some rare earth elements other than cerium, such as lanthanum. Some washcoats also contain oxides of transition elements, such as nickel and iron. They have specific tasks, such as preventing the emission of hydrogen sulfide, which can be formed under specific operation conditions. Until recently, the high internal surface area of the washcoat was reached by the properties of its aluminum oxide component. Typically, γ-, δ- and θ-Al_2O_3 are used with an internal surface area of 50–250 $m^2\,g^{-1}$. Their morphanlogy and their chemical composition, especially the type and amount of impurities, are designed so as to conserve the high internal surface area, even during prolonged operation ($\approx$2000 h), at catalyst temperatures of about 1000 K up to about 1300 K. Also, their pore structure is designed to balance the requirement for a high internal surface area with the requirement for a limited intraparticle diffusion resistance.

The aluminum oxide is typically formed in the desired modification and with the desired chemical composition before it is added to the washcoat.

Nowadays, some of the washcoat internal surface area is provided by cerium and zirconium oxides. Those oxides are available in a modification that exhibits a moderately high internal surface area, typically 20–100 $m^2\,g^{-1}$, together with an appreciable stability of this internal surface area at typical catalyst operating temperatures.

The main task of the cerium oxide washcoat component is oxygen storage, because the cerium ion is easily reduced and oxidized under typical catalyst operating conditions, formally according to the reaction

$$2CeO_2 \rightleftharpoons Ce_2O_3 + 0.5O_2 \tag{41}$$

The oxygen storage ability widens the range of exhaust gas compositions at transient operation under which the closed-loop-controlled three-way catalyst has optimal conversion for CO, HC and NO_x simultaneously. Indeed, considering a catalyst that was first operated under a slightly lean exhaust gas composition, then the cerium ion will be present in the highest valence state (+4), thus storing oxygen. If the exhaust gas composition becomes slightly rich, the stored oxygen is released and becomes available for a few seconds to support the conversion of carbon monoxide and hydrocarbons under this oxygen-deficient exhaust gas composition. The cerium ion, now in the lower (+3) valence state, will store oxygen again for a few seconds, as soon as the exhaust gas composition changes back to slightly oxidizing. In doing so, the conversion of nitrogen oxides is enhanced at the slightly oxidizing exhaust gas composition during this transition period. The cerium washcoat component also influences the stability of the dispersion of some of the precious metals and thus the catalytic function of the precious metal component.

Zirconium oxides are the preferred supports for the precious metal component rhodium. The cerium oxide and/or the zirconium oxide are added to the washcoat either as preformed oxides or as oxide precursors, such as their respective carbonates or nitrates – the oxides are then formed in situ during washcoat drying and calcination.

Various methods exist for the application of the washcoat to the support. With ceramic monolithic supports, the washcoat is typically applied as an aqueous slurry by a dipping process, followed by drying and calcination [29].

The same procedure can be used for preshaped metallic supports, the only difference being that some metal foil surface treatments may be used prior to the washcoat application. For some metallic supported catalysts, however, the washcoat is applied to the metal foil by a continuous dipping or spraying process, prior to the formation of the monolithic structure. This has the advantage that there is no handling of the many individual support units throughout the washcoating step and, eventually, the precious metal impregnation step. The major disadvantage is that it is much more difficult to form a very rigid monolith structure out of an already coated metal foil, as brazing or welding processes cannot be applied easily.

d Precious Metals

The early days of automotive catalytic converter research and development, targeted the use of base metal catalysts. Numerous publications describe the results obtained with catalysts that contain, for example, the oxides of Cu, Cr, Fe, Co and Ni [7, 8].

Despite this effort, no real breakthrough was achieved. One reason for this is that the precious metals have a much higher intrinsic activity, i.e., the activity per gram component, for the simultaneous conversion of CO, HC and NO_x than base metal catalysts. Furthermore, it is easier to obtain and to keep precious metals in a very finely divided state. Precious metal based catalysts are also much more resistant to sulfur poisoning at temperatures below 750 K than base metal catalysts.

Figure 40 shows the conversion of CO, HC and NO_x at typical automotive catalyst operation conditions for a precious metal based catalyst on a ceramic monolith with an extremely low precious metal loading, and for a precious metal free catalyst in which the same ceramic monolith support was used but with a washcoat consisting of a typical base metal catalyst formulation. The extremely poor con-

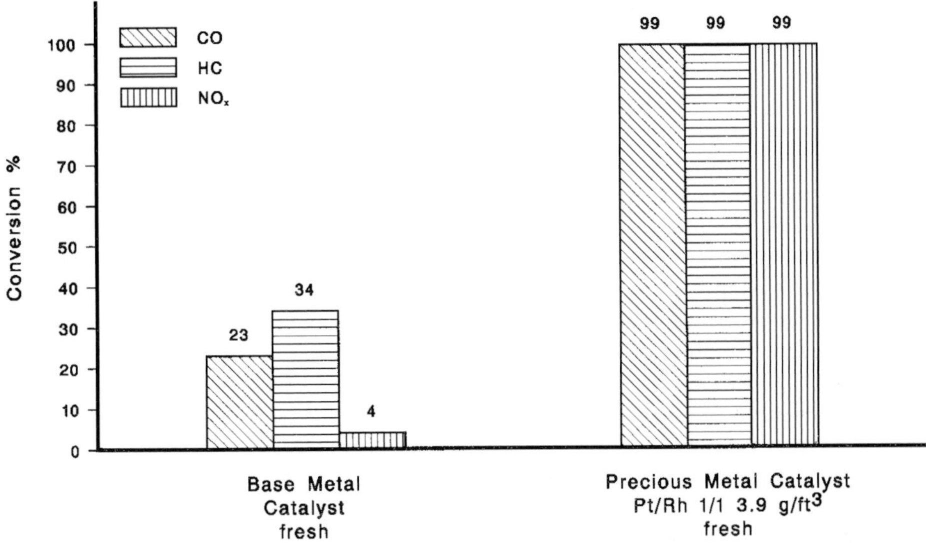

Figure 40. Conversion of CO, HC and NO_x on a monolithic base metal catalyst and on a low loaded monolithic precious metal catalyst, both in the fresh state (space velocity 60 000 $Nl\,l^{-1}h^{-1}$; gas temperature 723 K; test procedure A/F scan-values shown at lambda-0.999; frequency 1 Hz; amplitude 1 A/F unit). Reprinted with permission from ref. [34], © 1991 Society of Automotive Engineers, Inc..

version of nitrogen oxides demonstrates that base metal catalysts as of today are not able to meet the catalyst performance requirements within the present boundary conditions of catalyst volume, operating conditions and durability requirements.

The precious metals currently used in three-way catalysts are platinum, palladium and rhodium. In the past, also ruthenium and iridium have been tested, but because of the volatility and/or the toxicity of these metals or their oxides, neither has found practical application.

Until 1995, most of the monolithic three-way catalysts contained platinum and rhodium, in mass ratios of about 5–20 : 1 Pt : Rh. The total precious metal loading is typically 0.9–2.2 $g\,l^{-1}$ catalyst volume. These are only typical values, as the amount of precious metals and the mass ratio of platinum to rhodium depends on the specific application of the catalyst and is governed by factors such as the composition of engine-out emissions, the emissions targets to be reached, the catalyst operating conditions and the properties of the fuels used.

Palladium was a common element in oxidation catalysts where it was used together with platinum, in a mass ratio of about 5 : 2 Pt : Pd at a total precious metal loading of about 1.5 $g\,l^{-1}$ catalyst volume. Some three-way catalysts used palladium together with platinum and rhodium, in which the palladium was a partial replacement of platinum. The loadings used were 0.9–3.1 g Pt, 0.0–3.1 g Pd and 0.15–0.5 g Rh per converter [9]. Nowadays, three-way catalysts are available with all or the majority of the platinum replaced by palladium. These catalysts have a higher total precious metal loading, typically in the range 2–5.5 $g\,l^{-1}$ catalyst volume, and mass ratio of Pt : Pd : Rh of about 0–1 : 8–16 : 1 [30].

In addition, catalysts have been developed that use only palladium as the precious metal component, with Pd loadings of about 1.8–10.6 $g\,l^{-1}$ catalyst volume. These catalysts are mainly used as light-off catalysts, which are typically small catalysts located close to the engine outlet, and used in combination with a main catalyst that contains rhodium together with platinum which is located in the vehicle underbody downstream of the light-off catalyst. However, these palladium-only catalysts are sometimes also used as the main catalyst, in conjunction with particular engine management systems.

The precious metal composition is typically uniform in the radial and axial directions of the monolith structure, although different designs have been described in the patent literature and have even been used in some selected applications. However, much more common is a nonuniform distribution of the precious metals within the washcoat layer. One – macroscopic – example of nonuniform distribution is that the amount of one precious metal component decreases from the part of the washcoat that is in contact with the gas phase towards the part of the washcoat that is in contact with the monolith wall and eventually vice-versa for the second precious metal component. Another – microscopic – example of nonuniform distribution within the washcoat is that each precious metal component is selectively deposited on different washcoat components. These nonuniformities are intentional and are desirable for kinetic reasons or because of specific beneficial interactions between the precious metals and the washcoat oxides. The type of nonuniformity that can be achieved depends strongly on the production procedure of the catalyst.

Within a single secondary washcoat particle, the distribution of the precious metals can be assumed to be relatively homogeneous. The precious metals are typically present in a highly dispersed state. Dispersions measured by CO chemisorption methods are typically in the range 10–50% or even higher, for fresh catalysts. This means that the precious metals are present as single atoms or as small clusters of about ten atoms. For a catalyst with about 1.8 g precious metal per liter of catalyst volume, this corresponds to a precious metal surface area in the range of about 3–30 $m^2\,l^{-1}$ catalyst volume.

Compared to a washcoat surface area of about 20 000 $m^2\,l^{-1}$ catalyst volume, it is apparent that the precious metal surface area is several orders of magnitude lower than monolayer coverage of the washcoat surface.

These values are only rough estimates of the order of magnitude and are valid only for fresh catalysts and are different for each of the individual precious metals components.

The precious metals are generally introduced in the catalyst by wet chemical methods such as incipient wetness impregnation, typically using aqueous solutions of the precious metal salts, followed by a drying step to remove the water, and then by a calcination step to decompose the precious metal salts. Sometimes, a reduction step is applied to convert the precious metal oxides into the metallic state. Other production procedures are used as well and are described in the patent literature.

Because of the large amount of automotive exhaust catalysts produced annually, they constitute an important fraction of the worldwide precious metal consumption

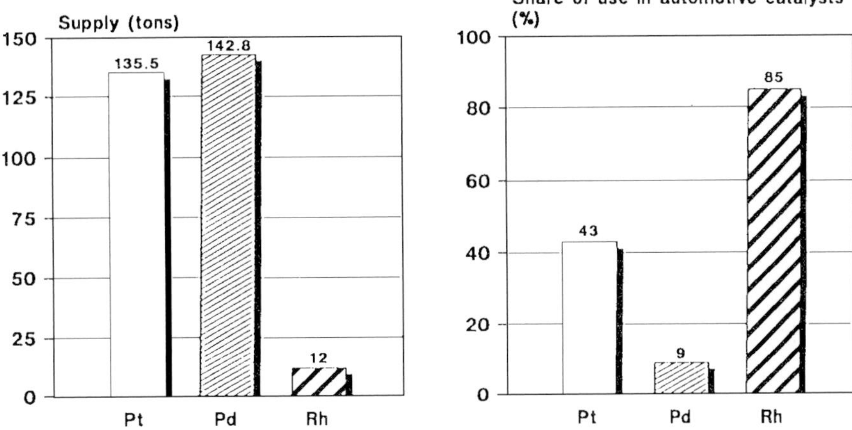

Figure 41. World supply of platinum, palladium and rhodium, and share of their use in automotive catalysts, in 1992. Reprinted with permission from ref. [30], © 1994 Society of Automotive Engineers, Inc.

(Fig. 41). Changes in emission legislation and in automotive emission control catalyst technology therefore might affect both the supply and the price of platinum, palladium and especially rhodium (Table 12).

Platinum, palladium and rhodium are mined in a limited number of countries, the most important being South Africa, the former USSR, the USA and Canada

Table 12. Historical overview of the price and the availability of platinum, palladium and rhodium.

Year	Metal	Total supply (tons)	Automotive catalyst use (%)	Average price (US$ per troy ounce)
1983	Pt	77	31	424
	Pd	78	11	135
	Rh	6[a]	11[a]	300
1988	Pt	107	34	537
	Pd	104	7	124
	Rh	10	85	1235
1990	Pt	130	39	472
	Pd	135	9	115
	Rh	9.8	75	3726
1993	Pt	144	39	380
	Pd	131	10	125
	Rh	12	85	1200
1997	Pt	158	28	400
	Pd	231	39	178
	Rh	14	82	300

[a] 1980 value.

Table 13. Relative importance of some countries in the supply of platinum, palladium and rhodium in 1990.

Country	Platinum (%)	Palladium (%)	Rhodium (%)
South Africa	74.6	34.6	54.4
USSR	18.7	52.7	41.0
USA and Canada	5.0	10.5	4.6
Other	1.7	2.2	0

Table 14. Relative importance of some countries in the supply of platinum, palladium and rhodium in 1997.

Country	Platinum (%)	Palladium (%)	Rhodium (%)
South Africa	75.0	29.9	83.3
Russia	15.2	55.8	12.6
USA and Canada	5.8	10.4	4.1
Others	4.0	3.9	0

Table 15. Reported total reserve and ratio of platinum to palladium to rhodium for some selected precious metal mines.

Place	Total reserve (10^6 troy ounce)	Ratio Pt : Pd : Rh (g : g : g)
South Africa (Bushveld Complex)		
Merensky Reef	491	19.6:8.3:1
UG2	885	5.2:4.4:1
Platreef	347	13.3:14.6:1
Canada (Sudbury)	8	3:4:1
USSR (Noril"sk)	198	–
USA (Stillwater)	33	2.3:7.7:1
Total	1966	8.25:7.1:1

(Tables 13 and 14) [31]. These countries report huge reserves, but with different mine ratios of Pt to Pd to Rh (Table 15). The concentration of Pt, Pd and Rh in the raw ore is reported to be in the order of magnitude of a few parts per million, whereas their concentration in automotive catalysts is in the order of magnitude of about 3 000 ppm for Pt and about 600 ppm for Rh. These aspects contribute to the increasing interest in the recycling of the precious metals out of used automotive emission control catalysts. By 1996 about 16% of the platinum demand for automotive catalysts, about 5% of the palladium demand and about 10% of the rhodium demand for automotive catalysts was supplied from the recycling of these catalysts [32, 33].

1.3.6 Three-way Catalyst Performance

A Introduction

The performance of three-way catalysts depends upon numerous factors (Fig. 42). These are basically the chemistry of the catalyst (such as the washcoat, precious metals, age and preparation), the physics of the catalyst (such as the support and converter design), and the chemical engineering aspects of the catalyst (such as reaction temperature, space velocity, gas composition and dynamic conditions). These factors are not independent and vary for each particular application of the three-way catalyst. It is therefore not possible to give a scientifically sound and complete description of all the factors that influence the catalyst performance or to derive general rules on the extent of their influence. Only some selected topics will be described, all dealing with ceramic monolithic three-way catalysts used in closed-loop-controlled gasoline spark ignition engines.

B Measurement of Catalyst Performance

The ultimate test of catalyst performance is the vehicle test described in Section 1.2. In such a vehicle test all the reaction conditions that influence the conversion reached over the catalyst vary simultaneously in a interdependent fashion, as they are fixed by the speed and the load of the vehicle at each moment of the test. For research and development purposes, however, it is useful to evaluate the catalyst performance using fewer parameters or parameters that can be varied independently. Several simplified test procedures have therefore been developed.

Engine tests are performed with an engine mounted on a computer-controlled brake, with an exhaust gas cooler or heater installed in between the engine outlet and the catalytic converter inlet. Several catalytic activity tests can be performed using this setup.

Light-off tests aim to measure catalyst performance at various settings of the exhaust gas temperature. One way of doing this is to fix the engine speed, load and A/F value, and then to adjust the exhaust gas temperature in front of the catalyst by

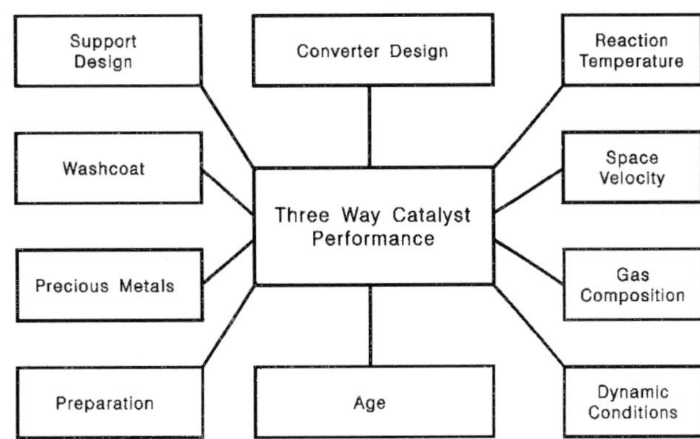

Figure 42. Factors that affect the performance of three-way catalysts. Reprinted with permission from ref. [42], © 1991 Society of Automotive Engineers, Inc.

Table 16. Major differences between vehicle and the engine test procedures used to measure the activity of emission control catalysts.

Operating conditions	Vehicle test	Engine tests	
		Light-off	A/F scan
Gas temperature	variable	variable	fixed
Space velocity	variable	fixed	fixed
Feedstock partial pressure	variable	fixed	fixed/variable
Redox ratio[a]	variable	fixed	variable
A/F modulation frequency[a]	variable	fixed	fixed
A/F modulation amplitude[a]	variable	fixed	fixed

[a] For three-way catalysts.

the exhaust gas cooler or heater. The catalyst performance is determined by measuring the conversion of CO, HC and NO_x at each exhaust gas temperature setting. The test can be repeated at various engine air-to-fuel settings, and allows measurement of the catalyst activity at fixed exhaust gas composition and fixed overall space velocity used.

In the A/F scan the engine is again operated at a fixed speed and load. Then, the composition of the exhaust gas at the catalyst inlet is varied. This test can be performed by at first successively increasing and then at successively decreasing set points of the A/F ratio to detect eventual hysteresis phenomena. Sometimes, the traverse of the A/F ratio range of interest is done in a few minutes, in which case the dynamics of the transition of the conversion between the various A/F settings is accounted for. Another way is to wait at each A/F set point till a stable conversion value is obtained. The A/F scan allows comparison of the catalyst performance at fixed space velocity and exhaust gas temperature. Table 16 summarizes the main differences between these two types of engine tests and the vehicle test.

Engine tests offer the additional advantage over vehicle tests that several catalysts can be evaluated at about the same time under real identical engine operation conditions. This is achieved with a multichamber converter, in which small cores drilled out of a full size converter are used (Fig. 43). The sizes of the cores is such that each is operated at the same space velocity as the full size catalyst would experience.

A further simplification in the procedure to measure the catalytic activity is the use of a model gas reactor (Fig. 44). A small piece of catalyst is mounted in a reactor tube, the temperature of which can be externally adjusted. Most commonly, a monolithic core of diameter 2.54 cm and length 7.5 cm is used. The desired exhaust gas composition is simulated by mixing either pure gases or mixtures of the desired exhaust gas component with nitrogen. Such a model gas reactor test gives the highest possible flexibility, as each of the characteristic parameters of the exhaust gas, such as composition, temperature and space velocity, can be varied in a truly independent fashion.

Figure 43. Engine dynamometer equipped with a multichamber catalytic converter, for the simultaneous aging of eight catalyst samples.

C Influence of Catalyst Operating Conditions

In the chemical and the petrochemical industries, heterogeneous catalysts are typically operated in a narrow range of reaction conditions. These reaction conditions are chosen so as to achieve an optimal feedstock conversion at minimal catalyst deactivation and are typically either constant over time or are slightly modified to compensate for feedstock conversion loss due to deactivation of the catalyst.

Quite the opposite is true for automotive catalytic converters. None of the operating conditions can be chosen to guarantee optimal conversion at minimal deactivation, as the operating conditions are fixed by engine speed and load. Further, the engine operation conditions depend upon the driving conditions. Table 17 gives a

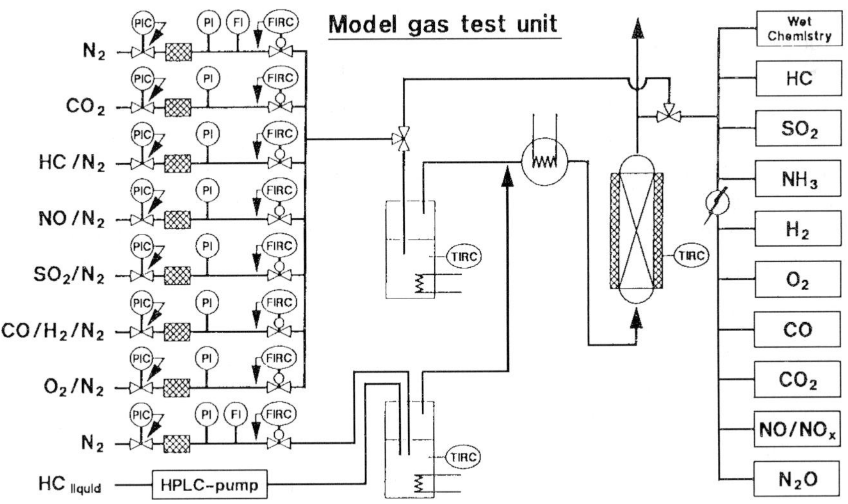

Figure 44. Model gas reactor to study the activity of automotive emission control catalysts.

Table 17. Influence of engine operation conditions upon the reaction conditions over a ceramic monolithic three-way catalyst (gasoline fueled spark ignition engine with four cylinders and a total displacement of 1.8 l; ceramic monolith catalyst with 62 cells cm^{-2}; total volume 1.24 l).

Operation Conditions	Idling	Partial Load			Full Load
Torque [Nm]	0	18	26	41	114
Rotational Speed [rpm]	900	2000	3000	4000	5000
Catalyst Inlet Temperature [K]	553	698	778	918	1183
Space Velocity [$Nl\,l^{-1}\,h^{-1}$]	5700	16100	27100	47100	123900
Linear Velocity [$m\,s^{-1}$]	0.54	1.92	3.62	7.54	25.2
Residence Time [s^{-1}]	0.3	0.075	0.045	0.021	0.006
Mass Flow [$kg\,h^{-1}$]	9.14	25.81	43.45	75.52	198.65
Gas Composition					
CO [vol. %]	0.76	0.55	0.68	0.78	1.05
HC [vol. ppm]	689	527	514	521	380
NO_x [vol. ppm]	162	980	1820	2820	2670
O_2 [vol. %]	1.17	0.69	0.70	0.67	0.43
λ-Value	1.0147	1.0047	1.0035	1.0015	0.9855
Dimensionless quantities					
Reynolds	13	36	61	108	282
Nusselt	0.11	0.21	0.31	0.46	0.896

typical example of the influence of engine operation upon the reaction conditions of a three-way catalyst on a ceramic monolith with 62 cells cm^{-2}. Figure 45 graphically represents in the reaction temperature versus reactant space time diagram the range of operation conditions of a three-way catalyst between an engine at idle and an engine at full load operation. For comparison purposes, the catalyst operation range for some major chemical and petrochemical processes is given as well [34].

Table 17 also shows that the exhaust gas flow is laminar inside the channels of a ceramic monolith with 62 cells cm^{-2} at all engine operation conditions. With a Reynolds number in the range of about 10 up to about 300, it is no surprise that both the limiting Sherwood and Nusselt numbers assume low values as well, which means that there is only a limited contribution of convection to the transfer of heat and mass from the gas phase to the catalyst surface.

However, as the exhaust gas flow is turbulent in front of the catalyst, there is a region of flow pattern transition at the inlet of the monolith. The length of this transition zone can be estimated from

$$l_e = 0.05 \times d_k \times Re \tag{42}$$

where l_e is the length of the flow transition zone (m), d_k is the width of the monolith channel (m), and *Re* is the Reynolds number in the monolith channel. It can be calculated with this equation that the length of the transition zone is typically less than 20% of the monolith length [35].

Because of the wide range of catalyst operation conditions, it is to be expected that a variety of kinetic regimes will occur during the use of the catalyst. This is exemplified in the Arrhenius diagram for the eqs 13 and 11 measured in a Berty reactor with a monolithic catalyst (Figs. 46 and 47) [36].

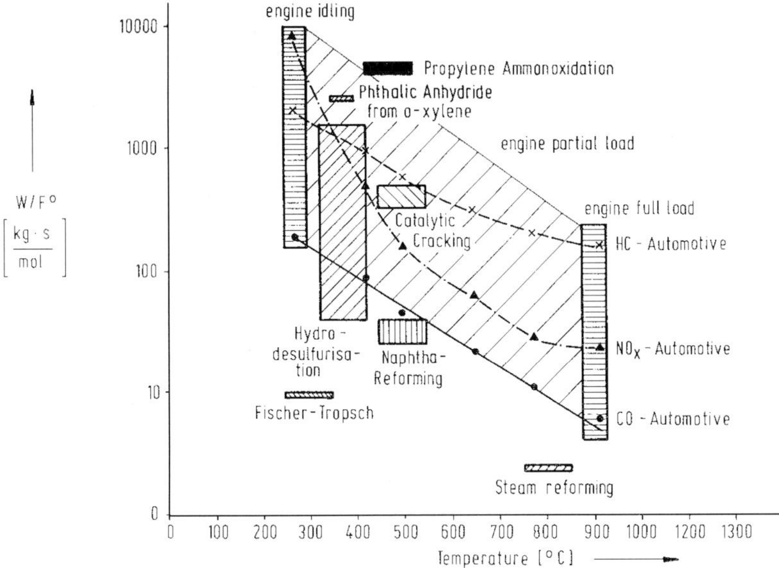

Figure 45. Comparison between the operating range of catalyst temperature and reactant space time for a three-way catalyst at two engine operation conditions (idling and full load operation), and for various heterogeneous catalytic processes used in the chemical and petrochemical industries. Reprinted with permission from ref. [34], © 1994 Society of Automotive Engineers, Inc.

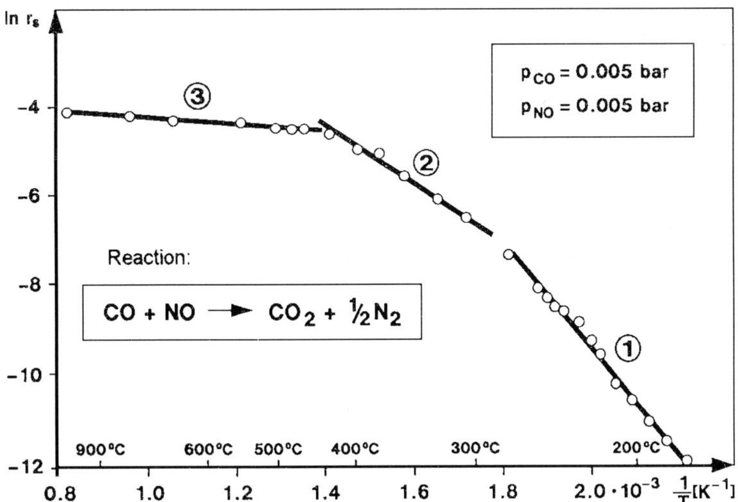

Figure 46. Arrhenius diagram for eq 13, recorded in a Berty reactor experiment with a fresh three-way catalyst (monolith catalyst with 62 cells cm^{-2}; partial pressure CO 0.005 bar, partial pressure NO 0.005 bar, balance N_2; Pt: 1.1 g l^{-1}, Rh: 0.2 g l^{-1}). Reprinted from ref. [36] with kind permission of Elsevier Science.

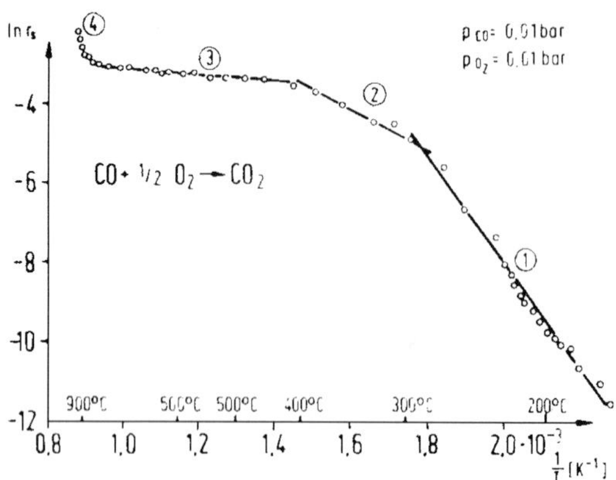

Figure 47. Arrhenius diagram for eq 11 recorded in a Berty reactor experiment with a fresh three-way catalyst (monolith catalyst with 62 cells cm^{-2}; partial pressure CO 0.01 bar, partial pressure O_2 0.01 bar, balance N_2; Pt: 1.1 g l^{-1}, Rh: 0.2 g l^{-1}). Reprinted from ref. [36] with kind permission of Elsevier Science.

Below a gas temperature of about 470 K, the reaction rate is so small that almost no conversion is reached over the catalyst. Above this temperature, and up to about 570 K, the extent of the conversion is governed by the rate of the chemical reaction, with an apparent activation energy of about 100 kJ $mole^{-1}$. The catalyst light-off occurs in this temperature range. In the temperature range 570 K to 770 K, the rate of the conversion is controlled by the rate of the intraparticle diffusion, which is the diffusion within the pores of the washcoat. The corresponding apparent activation energy is about 25 kJ $mole^{-1}$. Between 770 K and 1200 K, the catalyst is operated under interphase diffusion control, which is the rate of mass transfer between the gas phase and the washcoat boundary surface. The apparent activation energy is now about 6 kJ $mole^{-1}$. Finally, above 1200 K, for some reactions a noncatalytic, homogeneous gas phase reaction occurs [36].

As described above, the catalyst light-off, which is typically defined as the temperature at which under the chosen reaction conditions the reactant conversion reaches the value of 50%, occurs between about 470 K and 570 K. For a catalyst operated at a space velocity of 60 000 Nl l^{-1} h^{-1} the phenomenon of catalyst light-off is complex and still not well understood. Numerous simulations have shown that, depending on the difference between the temperature of the gas phase and the temperature of the catalyst, light-off can occur either at the inlet of the monolith and then progress to the downstream part of the catalyst, or that the light-off can occur in the middle of the catalyst, and then progresses both upstream and downstream through the catalyst [37, 38].

With the present generation of three-way catalysts, the light-off phenomenon is in many applications the factor that governs the overall conversion in the vehicle test procedures. This is demonstrated in Table 18. The raw emissions of CO, HC and NO_x are of the same order of magnitude in each of the three phases, but their conversion is below about 90% only in the first phase. The two main reasons for this are first, that the catalyst needs to be heated up from ambient temperature up till the light-off temperature during this phase, and second that the exhaust gas composition is rich during the first seconds following the cranking of the engine; the oxygen sensor also needs to exceed a minimum temperature before it starts to function. As

Table 18. Engine-out and tailpipe emissions of a passenger car equipped with a closed-loop three-way emission control catalyst in the three phases of the US-FTP 75 vehicle test cycle (engine aged catalyst).

Emission	Unit	Phase 1			Phase 2			Phase 3			Total		
		CO	HC	NO_x	CO	HC	NO_x	CO	HC	NO_x	CO	HC	NO_x
Engine-out emission	g $mile^{-1}$	18.6	2.0	3.8	11.6	1.9	2.8	10.0	1.5	4.2	12.6	1.8	3.4
Tailpipe emission	g $mile^{-1}$	9.6	0.62	0.55	0.14	0.02	0.10	0.72	0.07	0.14	2.25	0.16	0.21
Conversion	%	48.5	68.8	85.3	98.8	98.6	96.3	92.8	95.5	96.6	82.2	91.1	93.9

shown in Fig. 48, once the catalyst light-off happened, the exhaust gas temperature downstream of the catalyst will exceed the exhaust gas temperature in front of the catalyst, because of the heat released by the exothermic combustion of carbon monoxide and of the hydrocarbons. This exotherm is between about 50 K and 100 K. The catalyst temperature is rather constant during the second phase of the test cycle, and is maintained to some extent after the engine is stopped.

The time which the catalysts need to reach 50% conversion of the reactants strongly depends on the design of both the engine and the exhaust aftertreatment system, in addition to the catalyst formulation. The effect of the engine design is exemplified in Fig. 49. At a similar distance between the engine outlet collector and the catalyst inlet, the ramp of the exhaust gas temperature in front of the catalyst is about 180 K min^{-1} for vehicle I, and about 300 K min^{-1} for vehicle II, in which the engine was designed for quick heating of the exhaust gas following engine cranking. Furthermore, vehicle II was equipped with a so-called secondary air-pump, which

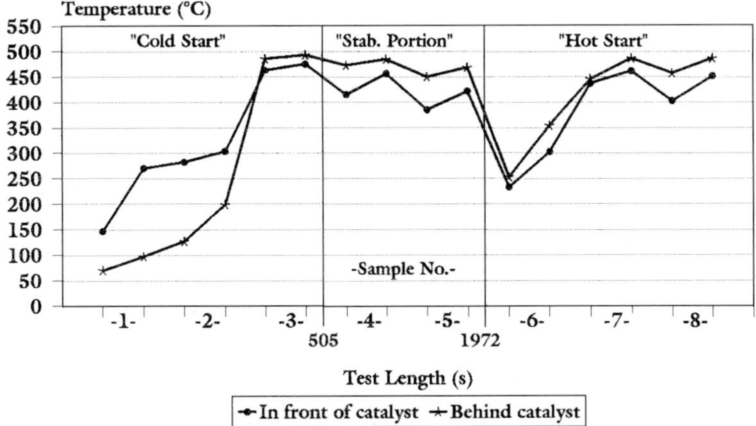

Figure 48. Evolution of the exhaust gas temperature in front of the catalytic converter, and behind the catalytic converter, as a function of the time in the US-FTP 75 vehicle test cycle.

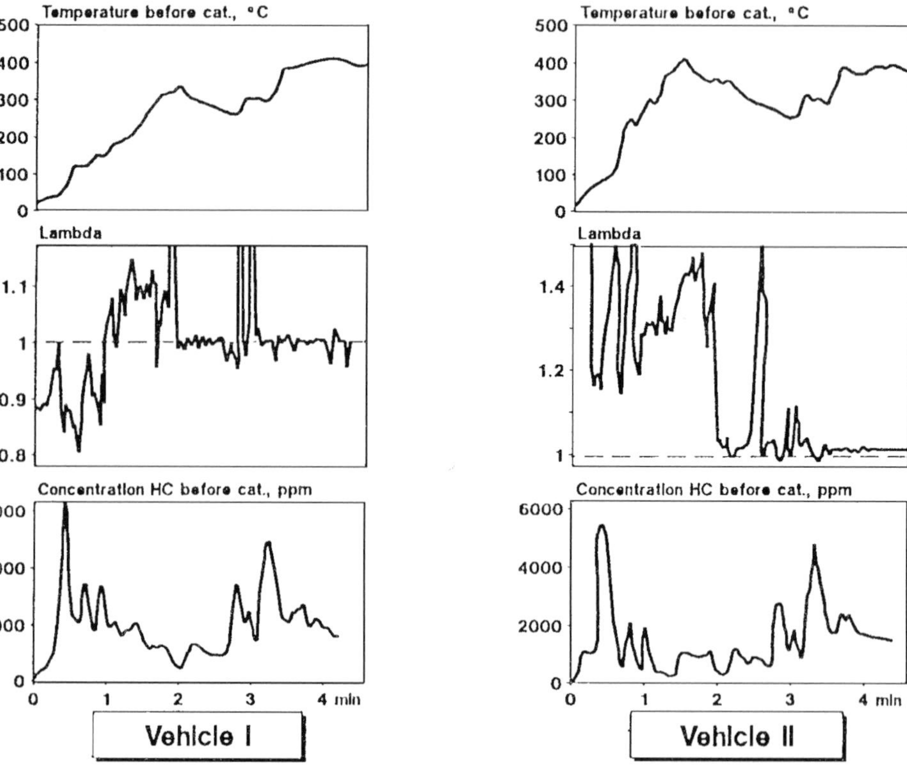

Figure 49. Evolution of the exhaust gas temperature in front of the catalytic converter, the exhaust gas lambda value and the hydrocarbon concentration in the exhaust gas, for two different engine designs, as a function of the time during the first four minutes of the US-FTP 75 test cycle. Reprinted with permission from ref. [39], © 1993 Society of Automotive Engineers, Inc.

adds air to the exhaust gas in front of the catalyst for some time after the cranking of the engine. Doing so, the composition of the exhaust gas in front of the catalyst is adjusted to lean, which favors the rate of the oxidation reactions, as will be shown below [39].

Another major difference between the operation of heterogeneous catalysts in chemical and petrochemical conversion processes, and in automobile catalytic converters, is that the exhaust gas composition in front of the three-way catalysts varies periodically. This is because of the feed-back control of the engine. This dynamic nature of the exhaust gas composition also affects the conversion reached over the catalyst, as is exemplified in Fig. 50. These phenomena are related to the dynamics of the coverage of the catalytic sites by the reactants and/or reaction products.

The overall composition of the exhaust gas has a drastic effect upon the conversion which can be reached over the catalyst. As stated above, the optimal simultaneous conversion of CO, HC and NO_x is reached with a stoichiometric exhaust gas composition. A typical example of the influence of the exhaust gas composition on

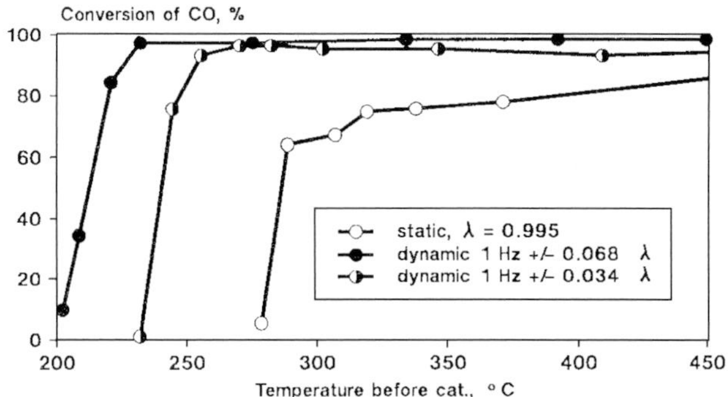

Figure 50. Influence of the dynamics in the exhaust gas composition on the conversion of carbon monoxide over a three-way catalyst at various settings of the exhaust gas temperature, (monolith catalyst 62 cells cm^{-2}; engine bench light-off test, space velocity 60 000 $Nl\,l^{-1}\,h^{-1}$; Pt: 1.42 $g\,l^{-1}$, Rh: 0.28 $g\,l^{-1}$).

the conversion is shown in Fig. 51. With a stoichiometric exhaust gas composition, the formation of eventual secondary emissions is minimal.

The operation of the engine outside of the stoichiometric point affects the amount of CO, HC, NO_x and O_2 emitted simultaneously, as well as the composition of the hydrocarbon fraction. The oxygen content of the exhaust gas is of particular interest as this is measured by the lambda sensor and is used to control the engine. Figure 52 shows as an example the influence of the exhaust gas oxygen content on the conversion of CO, HC and NO_x, at fixed CO, HC and NO_x content. This experiment was performed in a model gas reactor in which a stoichiometric composition is reached at about 1.0 vol% O_2. With an oxygen content above the stoichiometric

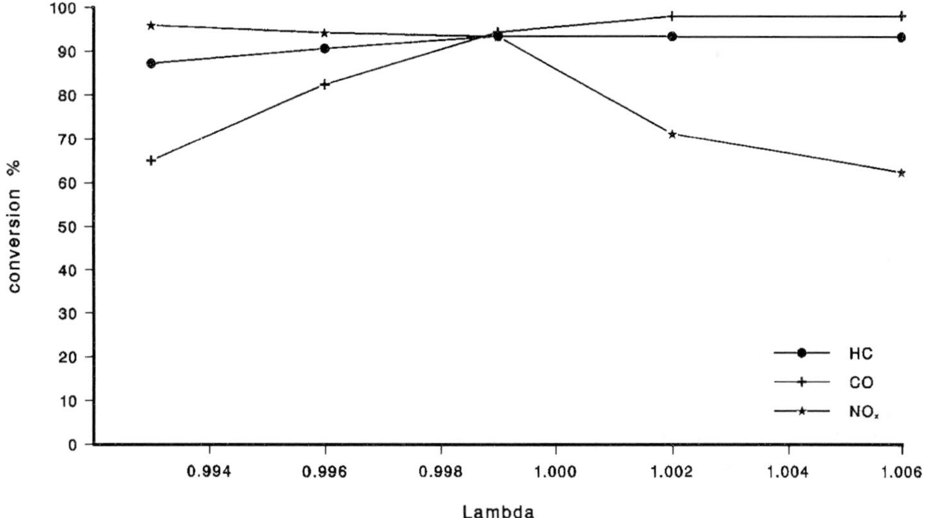

Figure 51. Conversion of CO, HC and NO_x over a three way catalyst as a function of the exhaust gas lambda value (monolith catalyst with 62 cells cm^{-2}, three-way formulation with Pt: 1.42 $g\,l^{-1}$, Rh: 0.28 $g\,l^{-1}$), fresh; engine bench test A/F scan at 623 K exhaust gas temperature; space velocity 60 000 $Nl\,l^{-1}\,h^{-1}$; dynamic conditions frequency 1 Hz; amplitude 0.5 A/F).

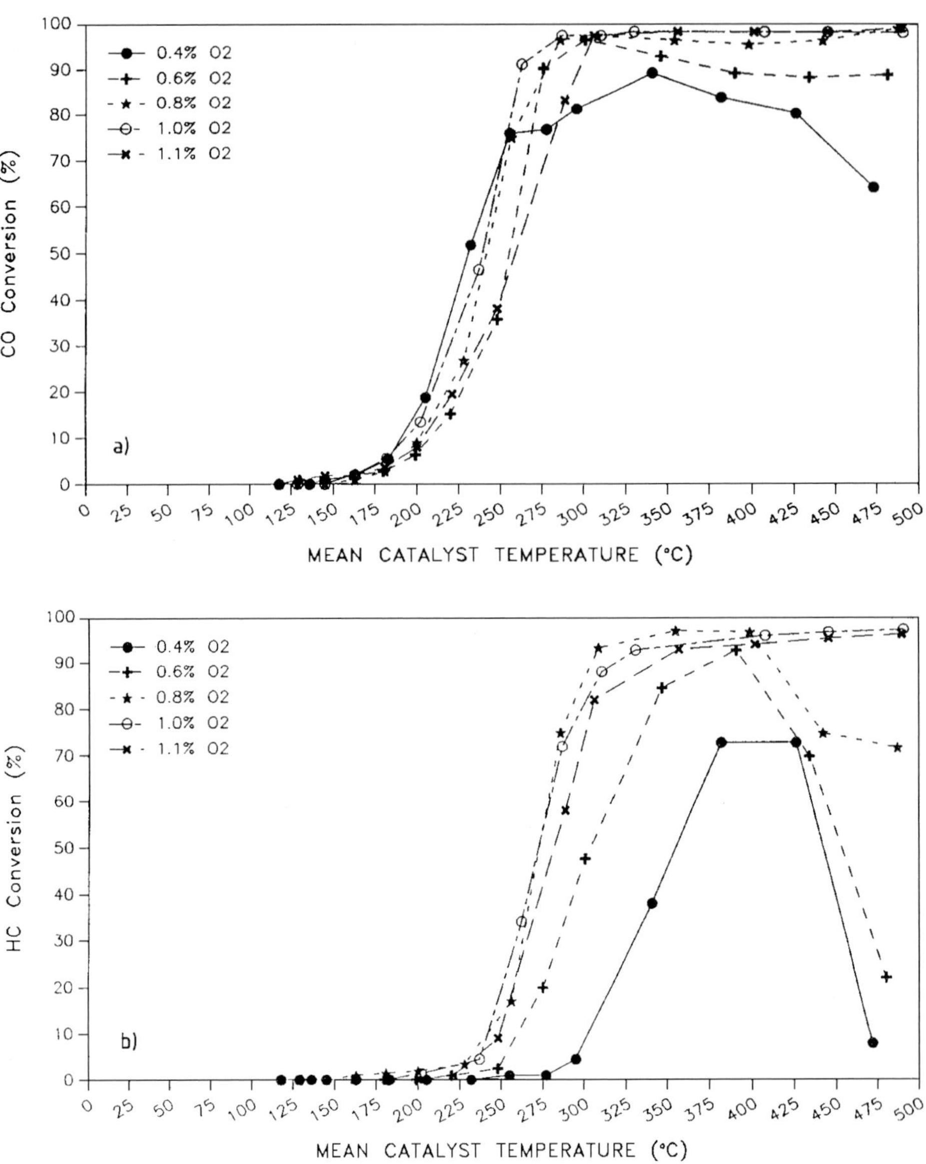

Figure 52. Influence of the exhaust gas oxygen content on the conversion of (a) CO (b) HC and (c) NO_x reached over a three-way catalyst at various settings of the catalyst temperature (monolith catalyst with 62 cells cm^{-2}, three-way formulation with Pt: 0.38 $g\,l^{-1}$, Rh: 0.16 $g\,l^{-1}$), fresh; model gas reactor; space velocity 60 000 Nl l^{-1} h^{-1}; static conditions).

value, the conversion of CO and HC at a given temperature is increased, whereas the conversion of NO_x is decreased with respect to the values obtained at stoichiometry. With an oxygen content below the stoichiometric value, the conversion

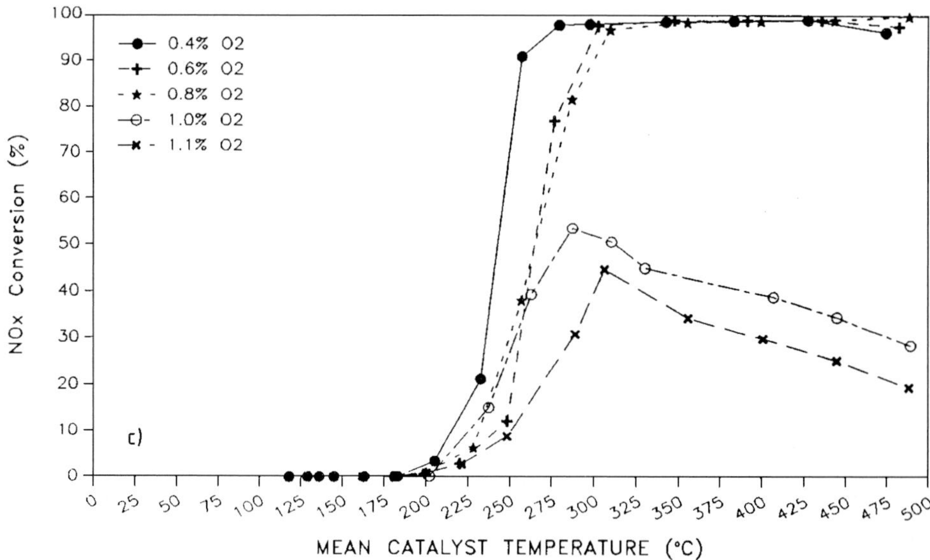

Figure 52 (continued)

of hydrocarbons shows a maximum as a function of the exhaust gas temperature. This maximum can be caused by the occurrence of hydrocarbon reforming reactions by which hydrocarbons are formed that have a drastically different response factor in the flame ionization detector (FID).

The extent to which this maximum occurs depends on the type of hydrocarbon and on the oxygen content (Fig. 53). It should be emphasized that the hydrocarbon fraction in the exhaust gas is composed out of typically more than 100 different individual hydrocarbons (Fig. 54). The composition of the hydrocarbon fraction depends on numerous parameters, such as the composition of the fuel, the engine operating conditions and the design of the engine (Fig. 55).

It was shown in Fig. 53 that the hydrocarbon conversion which is reached over the catalyst at fixed reaction conditions depends on the type of the hydrocarbon. Alkenic and aromatic hydrocarbons are more reactive than alkanic hydrocarbons. The reactivity of alkanic hydrocarbons increases with the number of carbon atoms in the molecule.

The nature of the hydrocarbon also effects the conversion of CO and NO_x (Fig. 56). With a lean exhaust gas composition, the level of 50% conversion of an alkanic hydrocarbon is reached at a higher temperature than the corresponding temperature for CO and NO_x. Alkenic hydrocarbons reach the 50% conversion level at the same temperature as CO. These phenomena strongly depend on the detailed kinetic mechanism of the conversion reactions and are therefore influenced by the composition and the design of the catalyst.

From the description given above, it is apparent that the tailpipe emission of CO, HC and NO_x as recorded in a vehicle test depend upon numerous factors, whose

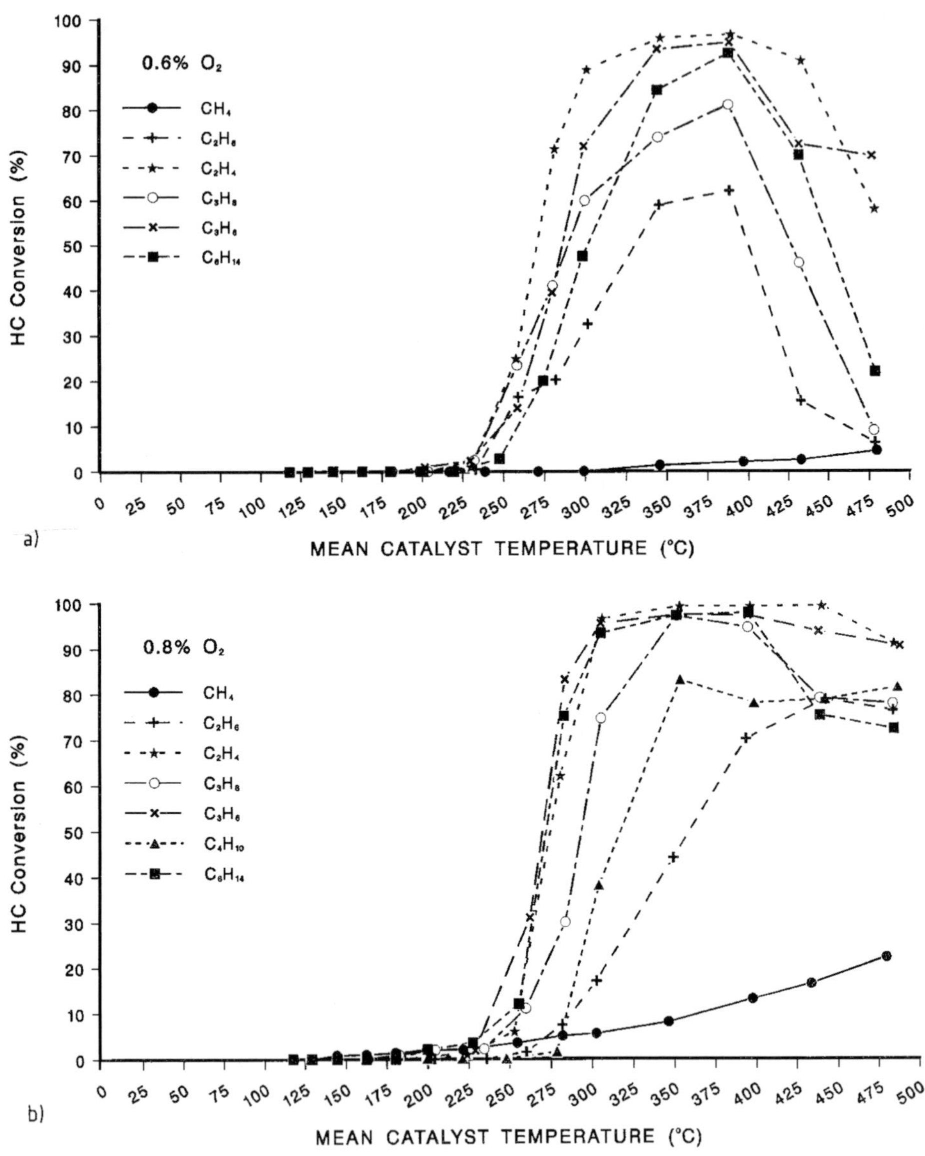

Figure 53. Influence of the exhaust gas oxygen content on the conversion of hydrocarbons, as a function of the type of hydrocarbon, at various settings of the catalyst temperature: (a) 0.6 vol.% O_2; (b) 0.8 vol.% O_2; (c) 1.0 vol.% O_2; (d) 1.1 vol.% O_2. (experimental details as for Fig. 52).

variation is caused by the varying operation conditions of the vehicle. Nevertheless, with the present generation of catalytic emission control systems, very low tailpipe emission values for CO, HC and NO_x are reached, at all the typical vehicle operation conditions (Fig. 57) [40].

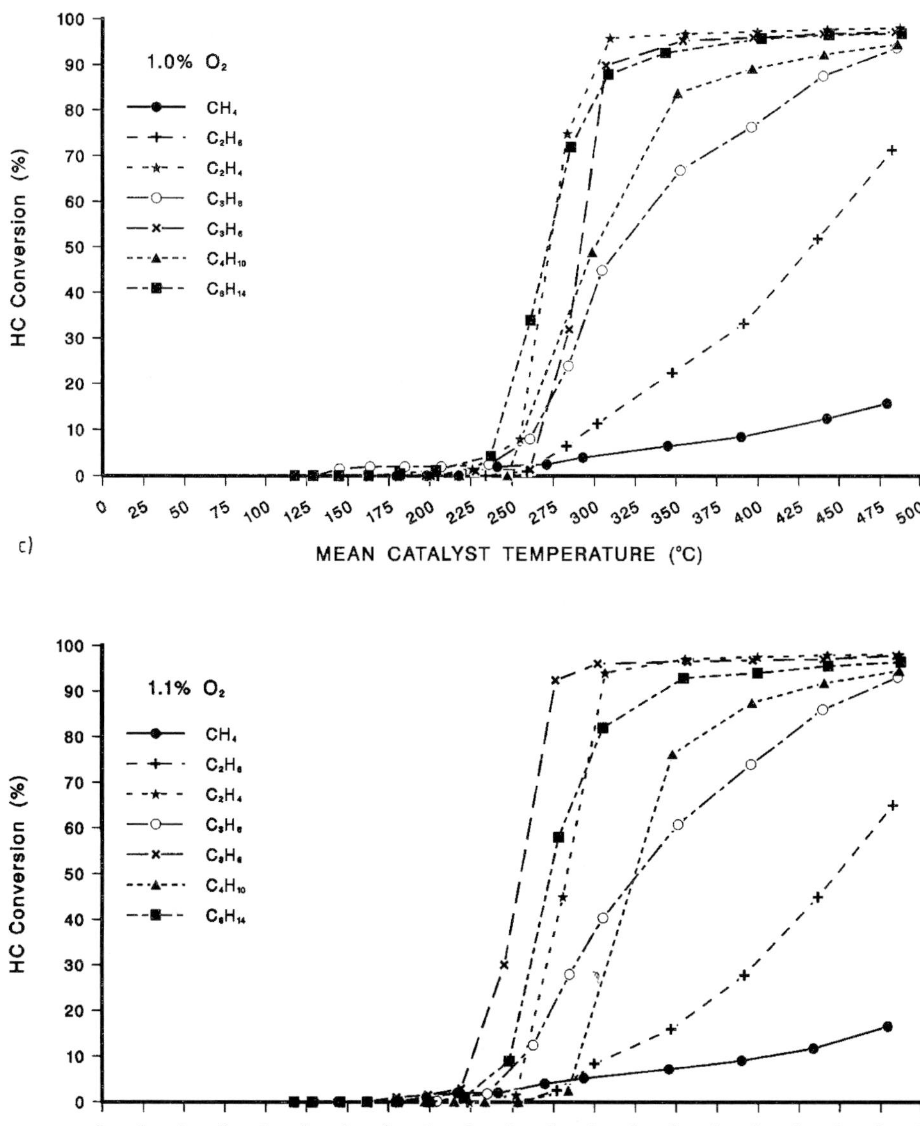

Figure 53 (continued)

From this description it is also to be expected that the performance of one and the same catalyst will depend on the particular application. This is exemplified in Fig. 58, which shows the tailpipe emission reached with two identical catalysts mounted on two different vehicles that have a similar engine displacement, a similar

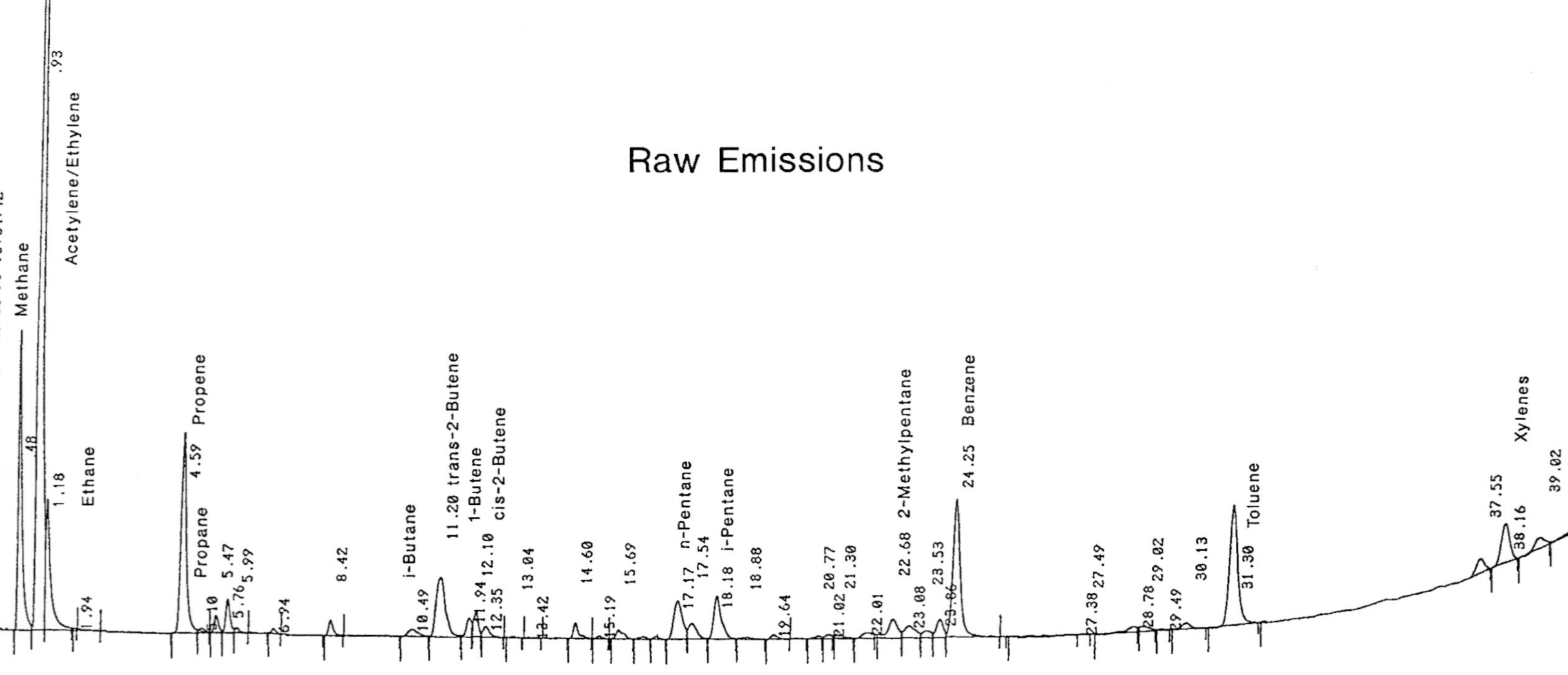

Figure 54. Gas chromatogram of the hydrocarbon fraction in the exhaust gas at the outlet of a gasoline fueled spark ignition engine. Reprinted with permission from ref. [34], © 1991 Society of Automotive Engineers, Inc.

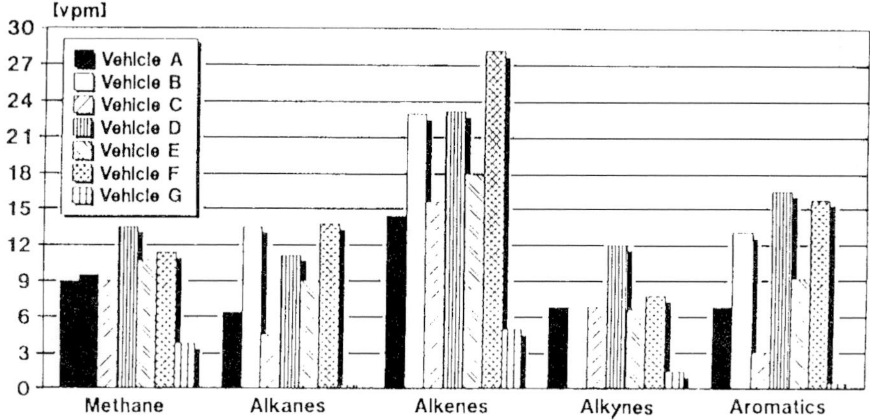

Figure 55. Concentration of various types of hydrocarbons in the exhaust gas of different gasoline powered spark ignition engines without catalysts (vehicles A–F) and one diesel engine without catalyst (vehicle G) during the first bag of the US-FTP 75 vehicle test cycle. Reprinted with permission from ref. [39], © 1993 Society of Automotive Engineers, Inc.

vehicle weight and a similar design of the exhaust gas system. This observation explains why each vehicle might need a different catalyst to meet the emission legislation in an optimal way.

D Influence of Converter and Support Design

In the application of catalytic converters to vehicles, an extremely broad range of different converter designs are used. The reasons for this are that each vehicle has different raw emissions, different catalyst operation conditions and different – in most cases limited – space available in the vehicle underbody to accommodate the catalyst.

As a rule of thumb, the volume of the converter is such that the geometric volume of the ceramic monolith corresponds approximately to the engine displacement. At average engine speed and load, this corresponds to a space velocity of about 60 000 Nl gas l^{-1} catalyst h^{-1}. The space velocity is generally defined as the ratio between the total exhaust gas volume stream at norm conditions (1 bar, 273 K) and the geometric volume of the monolithic support. It is therefore only a rough basis for comparison as it does not account for differences in the substrate characteristics and in the catalyst composition or for differences in the exhaust gas composition.

As expected, the space velocity and thus the catalyst volume has a major impact on the conversion reached over the catalyst. Especially with aged catalysts this is apparent both from the light-off temperature and the dynamic conversion at a fixed temperature (Figs. 59 and 60).

Even at a fixed space velocity, a different conversion is obtained depending on how the total catalyst volume is reached.

As is shown in Fig. 61, a lower tailpipe emission of CO is recorded when the total catalyst volume is reached by placing two smaller pieces at well defined distances

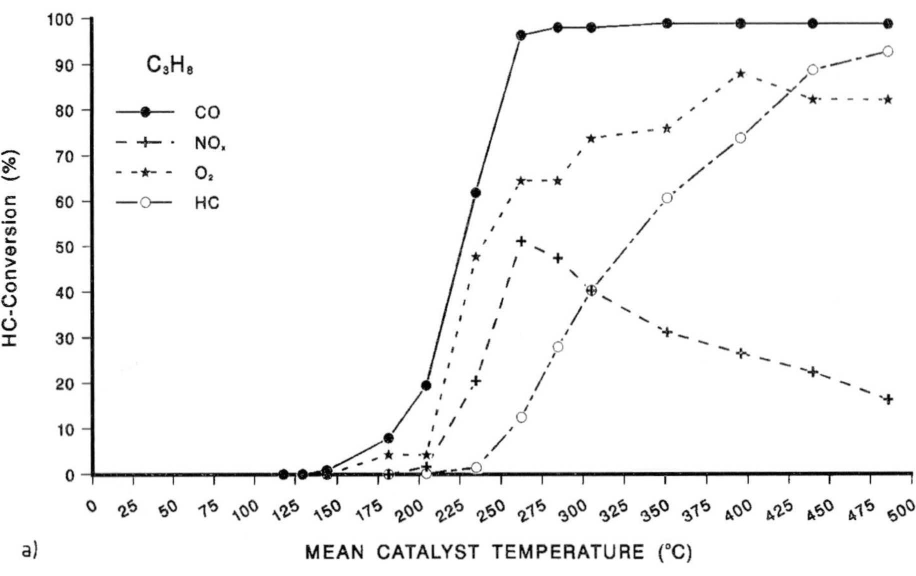

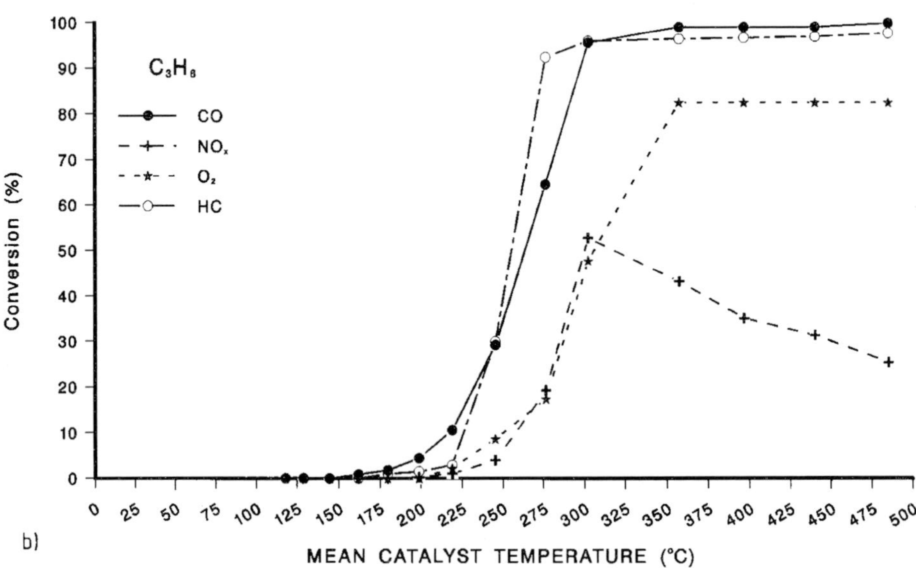

Figure 56. Conversion of CO, O_2, NO_x and the hydrocarbon component reached over a three-way catalyst at various settings of the catalyst temperature, as a function of the type of hydrocarbon: (a) C_3H_6; (b) C_3H_8 (experimental details as for Fig. 52). Reprinted with permission from ref. [34], © 1991 Society of Automotive Engineers, Inc.

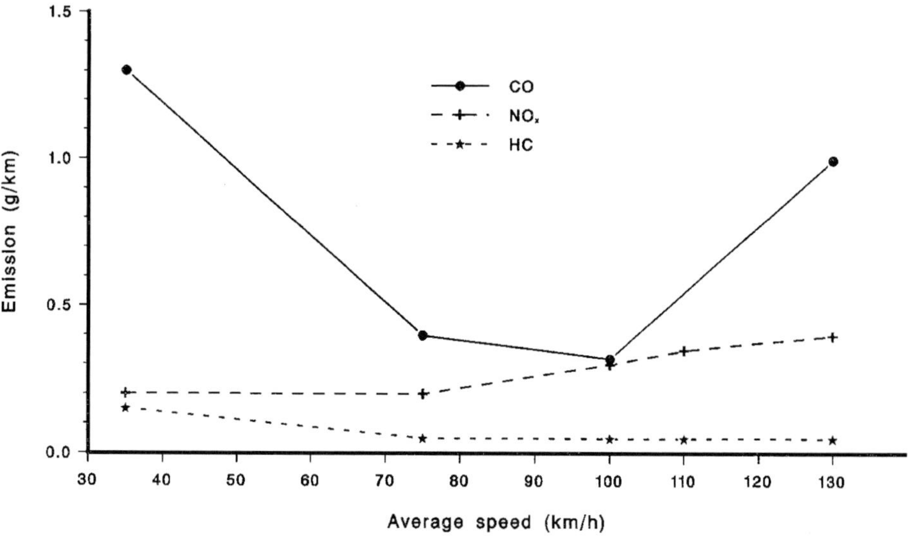

Figure 57. Tailpipe emission of CO, HC and NO_x for a vehicle with gasoline fueled spark ignition engine, equipped with a three-way catalyst, as a function of vehicle speed. Adapted from [40].

from each other, as compared to a single, longer piece of the same volume. The reason for this is that the enhanced heat and mass transfer during the transition from turbulent to laminar exhaust gas flow is used twice with the dual catalyst converter, provided that the distance between the two catalysts is sufficiently long to assure again turbulent flow conditions at the inlet of the second catalyst.

The same space velocity can be reached by simultaneously varying the diameter and the length of the monolith, keeping the total catalyst volume constant. As shown in Figs. 62 and 63, in accordance with numerical simulations, an increase in the catalyst diameter from 3.66 inch to 5.66 inch brings a slight improvement in the dynamic conversion, especially for NO_x with aged catalysts, but has a slight negative influence on the catalyst light-off temperature.

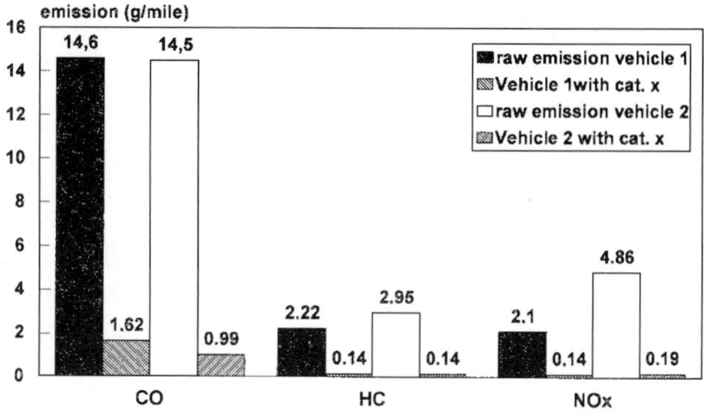

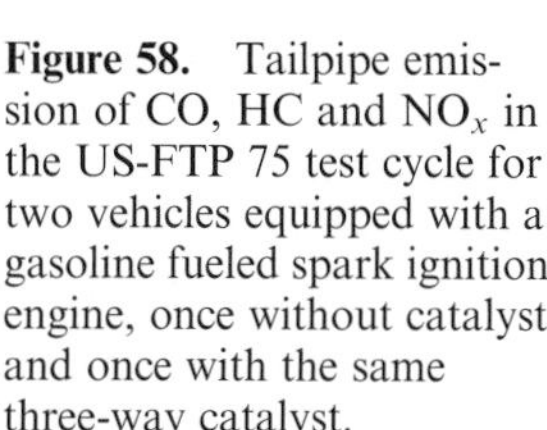
Figure 58. Tailpipe emission of CO, HC and NO_x in the US-FTP 75 test cycle for two vehicles equipped with a gasoline fueled spark ignition engine, once without catalyst and once with the same three-way catalyst.

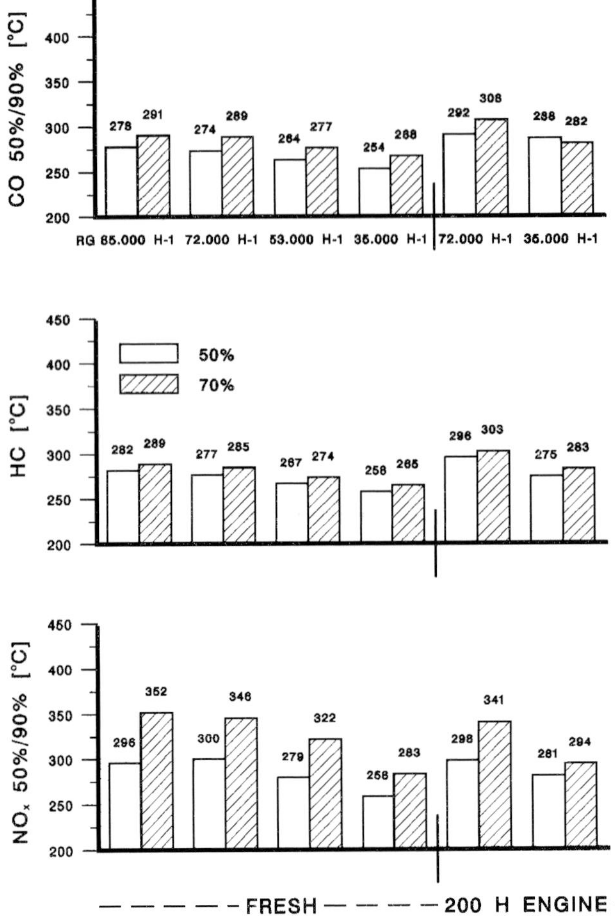

Figure 59. Influence of space velocity on gas temperature needed to reach 50% and 70% conversion of CO, HC and NO_x over a fresh and an engine aged three-way catalyst (monolith catalyst with 62 cells cm^{-2}, three-way formulation with Pt: 1.42 g l^{-1}, Rh: 0.28 g l^{-1}, engine bench light-off test at lambda 1.02 for CO and HC, and at lambda 0.986 for NO_x; engine bench aging during 200 h).

At identical catalyst volume, the pressure drop over the monolithic substrate is lower for the catalyst with the biggest diameter and therefore with the shortest length. Indeed, under laminar flow conditions, the pressure drop over the monolith follows from

$$\Delta p = 32 \times \mu \times V_k \times G \times \frac{(d_k + w_k)^2}{d_k^4} \times \frac{1}{S_k^2} \tag{43}$$

where Δp is the pressure drop over the monolith (Pa), μ is the dynamic viscosity (kg m^{-1}), V_k is the monolith geometrical volume (m^3), G is the gas flow (m^3/h), S_k is the monolith frontal area (m^2), d_k is the diameter of the monolith channel (m), and w_k is the thickness of the monolith wall (m). A given space velocity can also be reached with monoliths that differ in cell density, wall thickness and wall porosity.

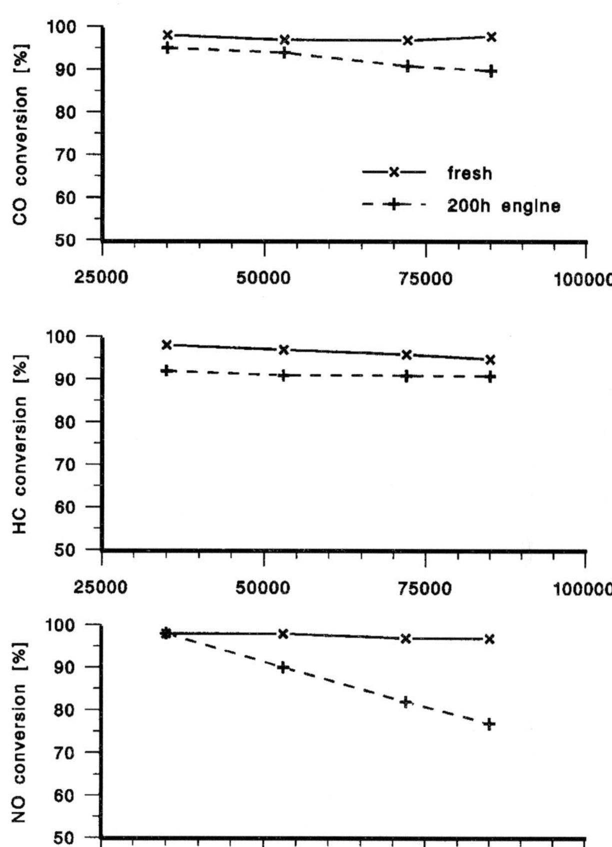

Figure 60. Influence of space velocity on the conversion of CO, HC and NO_x, reached over a three-way catalyst in the fresh state and after engine aging, at fixed exhaust gas temperature and exhaust gas composition (monolith catalyst with 62 cells cm^{-2}, three-way formulation with Pt: 1.42 g l^{-1}, Rh: 0.28 g l^{-1}; engine bench test at 723 K exhaust gas temperature; exhaust gas composition lambda 0.995; dynamic frequency 1 Hz; amplitude 1 A/F; engine bench aging during 200 h). Reprinted with permission from ref. [76], © 1990 Society of Automotive Engineers, Inc.

The cell density and the wall thickness of the monolith mainly affect the pressure drop over the monolith (eq 43).

These parameters, together with the porosity of the monolith walls, also affect the bulk density of the monolith, and therefore both the weight and the thermal mass of the converter. However, the data reported in the literature indicate that the influence of these parameters on the conversion reached over the catalyst is rather small, especially when considering the performance of aged catalysts.

Finally, the role of the design of the converter cones needs to be mentioned. Because of limitations in the space available at the vehicle underbody, catalytic converters are often fitted with short inlet and outlet cones. Short inlet cones tend to cause a nonuniform distribution of the exhaust gas flow over the inlet surface of the monolith (Fig. 64).

In accord with numerical simulations, a nonuniform distribution of the exhaust gas flow might have a positive influence on the catalyst light-off, as an increased fraction of the energy contained in the exhaust gas is concentrated on a smaller

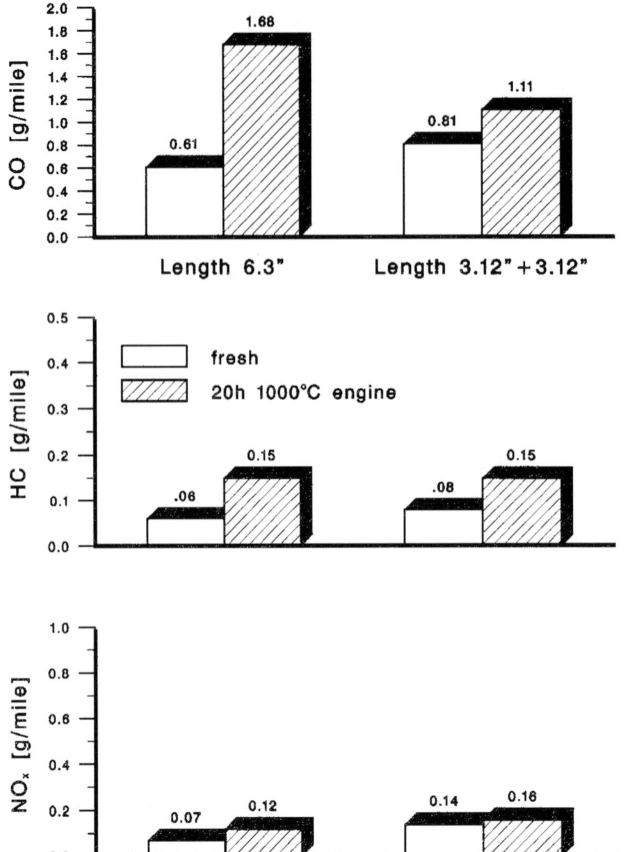

Figure 61. Tailpipe emission of CO, HC and NO_x from a gasoline fueled passenger car equipped with a three-way catalytic converter, in the US-FTP 75 vehicle test, as a function of the number of catalysts in the converter at fixed total catalyst volume (monolith catalyst with 62 cells cm^{-2}; three-way formulation with Pt: 0.83 $g\,l^{-1}$, Rh: 0.16 $g\,l^{-1}$; fresh condition and after high temperature aging for 20 hours on an engine bench).

portion of the catalyst. However, flow maldistributions have a negative impact on the conversion at a temperature above the catalyst light-off, as well as on the deactivation of the catalyst by deposition of poisoning elements. Finally, flow maldistributions increase the pressure drop over the converter. For example, it has been shown that the design of the inlet and outlet cones may account for up to 50% of the total pressure drop over the converter [41, 42].

E Influence of Catalyst Formulation

The catalyst formulation is fixed by the composition and the properties of its washcoat components, by the amount of the various precious metals used, and by the catalyst preparation procedure. The composition and the properties of the washcoat are the key factors that govern the performance and the durability of the catalyst.

The washcoat is the part of the catalyst, which is in direct contact with the gas

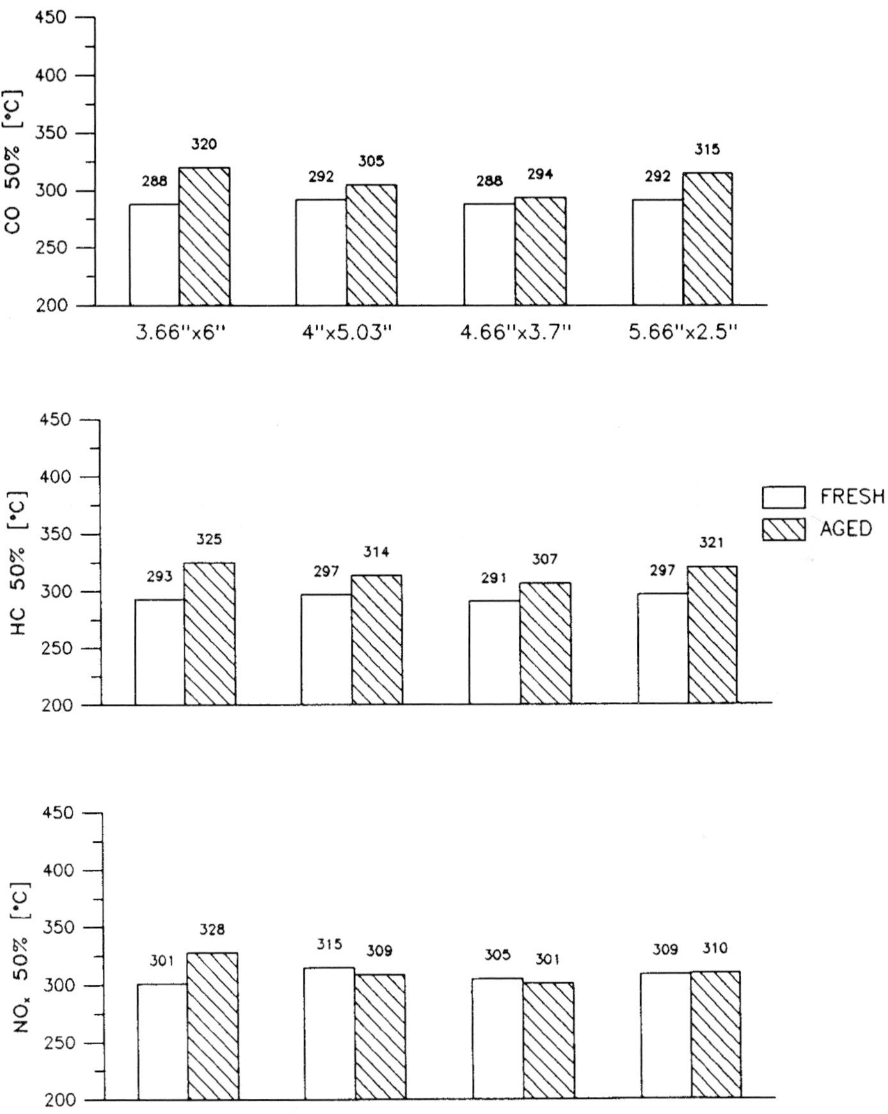

Figure 62. Influence of monolith diameter and length at fixed catalyst volume on the temperature needed to reach 50% conversion of CO, HC and NO_x for fresh and engine aged three-way catalysts (monolith catalyst with 62 cells cm^{-2}; three-way formulation with Pt: 1.42 g l^{-1}, Rh: 0.28 g l^{-1}; fresh and after engine bench aging during 200 h; engine bench test; light-off at a space velocity of 60 000 Nl l^{-1} h^{-1}; exhaust gas composition lambda 1.02 for CO and HC, 0.984 for NO_x).

phase. Its properties therefore control, for example, the rate at which the heat of the exhaust gas is transferred to the active sites of the catalyst (Fig. 65).

As the washcoat actively takes part in the catalytic function, it is to be expected

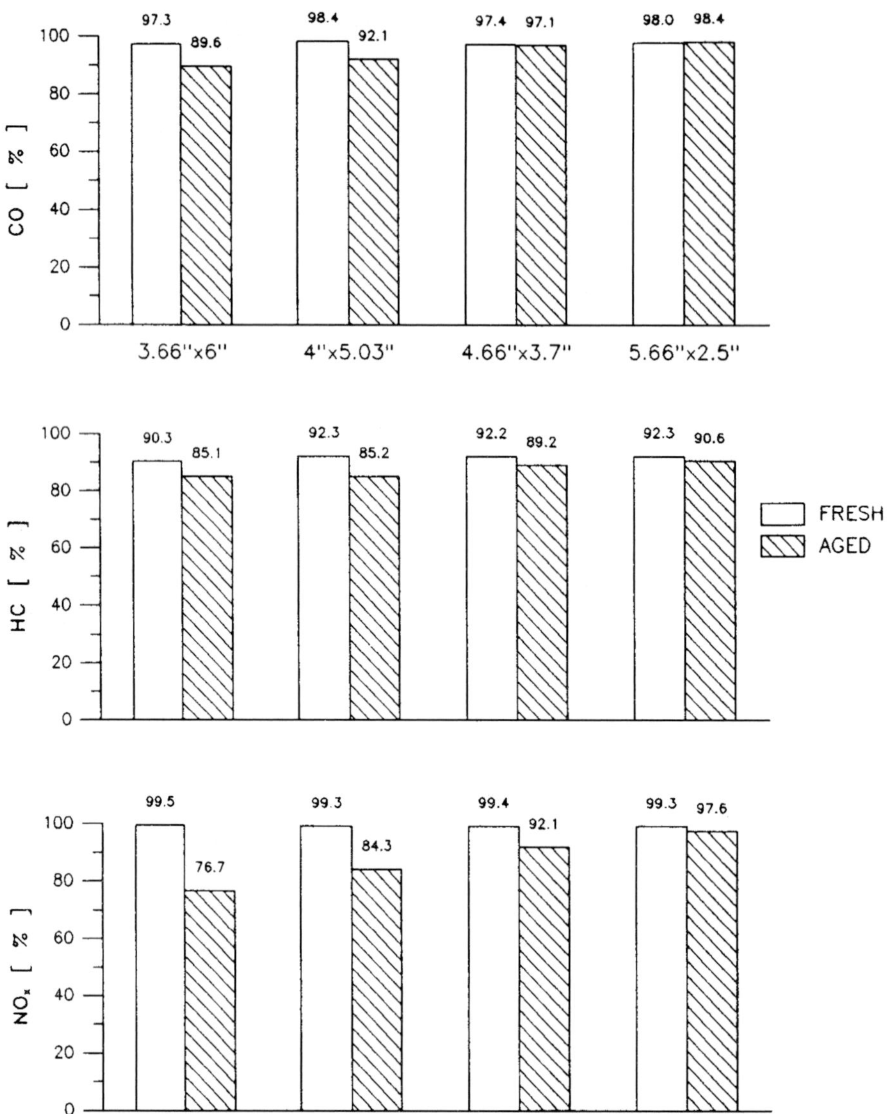

Figure 63. Influence of monolith radius at fixed catalyst volume on the conversion of CO, HC and NO_x reached over fresh and engine aged three-way catalysts at fixed exhaust gas temperature, space velocity and exhaust gas composition (monolith catalyst with 62 cells cm^{-2}; engine bench test; space velocity 60 000 Nl l^{-1} h^{-1} exhaust gas composition lambda 0.995; dynamic frequency 1 Hz; amplitude 1 A/F).

that the amount of washcoat used also has a significant influence on the performance of the catalyst. Figure 66 shows how NO_x conversion is affected by the amount of washcoat present on the monolith. The influence is particularly apparent

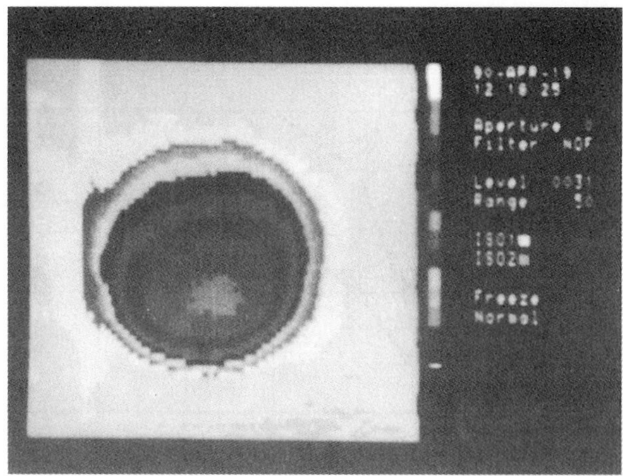

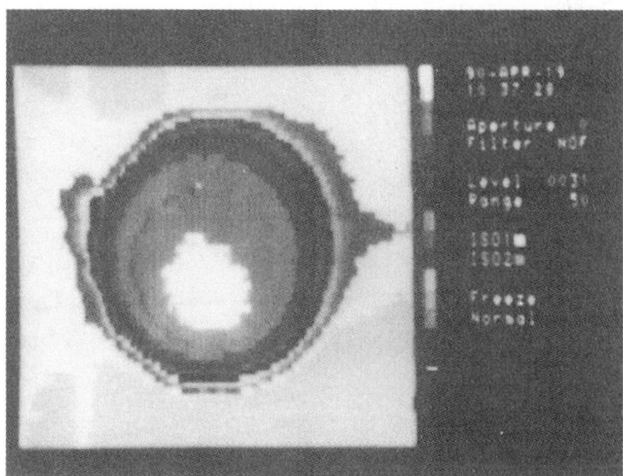

Figure 64. Influence of the design of the converter inlet cones on the distribution of the exhaust gas flow velocity over the radius of a monolith at the outlet frontal area; (above) converter with long inlet cones; (below) converter with short inlet cones.

for aged catalysts. For the same reasons the detailed composition of the washcoat will affect the catalyst performance in a significant way (Fig. 67).

It is difficult to separate the influence of the precious metals from the influence of the washcoat composition. Each of the precious metals platinum, palladium and rhodium, which have a specific task, interact with each other during the use and the aging of the catalyst, so that their influence on the catalyst performance is in most cases not additive. Figure 68 compares the performance of a fully formulated Pt/Rh catalyst to the performance of catalysts having either Pt or Rh alone in the same loadings and on the same washcoat as the fully formulated catalyst.

Comparing the performance of each of the precious metals Pt, Pd and Rh at an equimolar loading on the same washcoat, it is obvious from Fig. 69, that the hydrocarbon conversion reached changes with the reaction conditions in a different way.

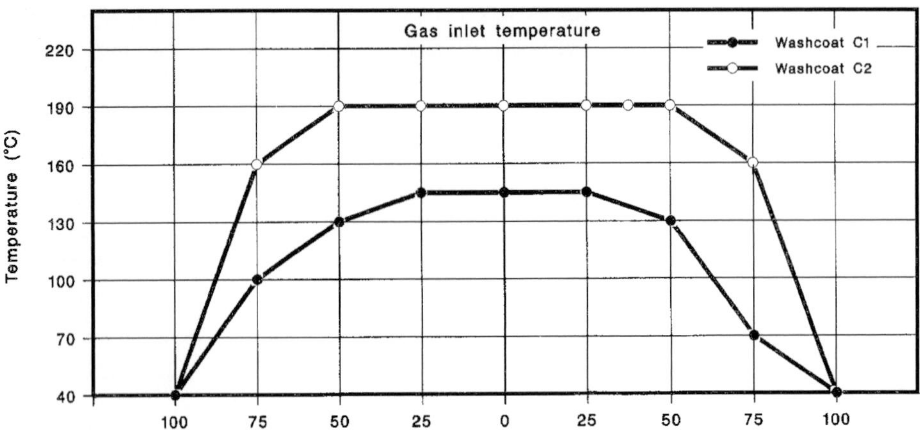

Figure 65. Effect of the gas to washcoat heat transfer rate on the distribution of solids temperature over the radius of a ceramic monolith at its outlet frontal area, for a washcoat with a high heat transfer rate (C2) and for a washcoat with a low heat transfer rate (C1). Reprinted with permission from ref. [34], © 1991 Society of Automotive Engineers, Inc.

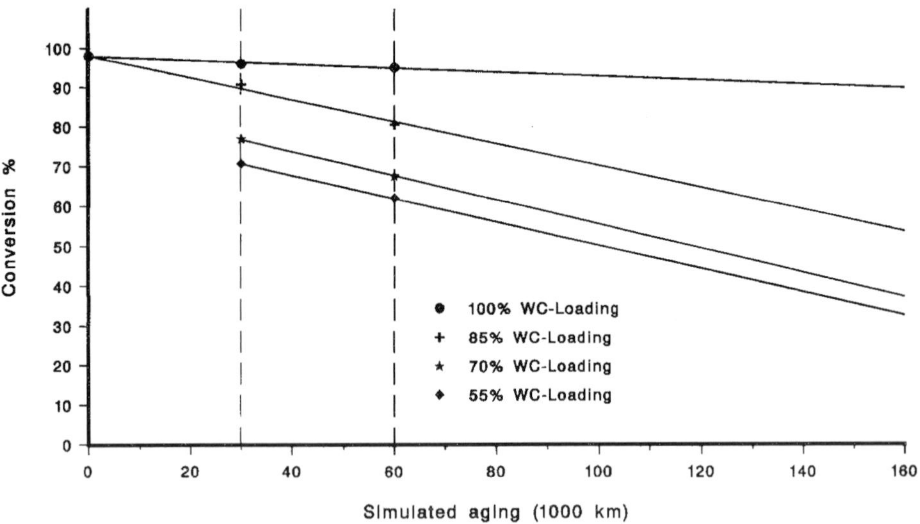

Figure 66. Influence of the washcoat loading of a ceramic monolith on the conversion of NO_x (monolith catalyst with 62 cells cm^{-2}; three-way formulation with Reprinted with permission from ref. [34], © 1991 Society of Automotive Engineers, Inc. Pt: 1.42 $g\,l^{-1}$, Rh:0.28 $g\,l^{-1}$ after aging on an engine bench 20 h; engine bench test; space velocity 60 000 $Nl\,l^{-1}\,h^{-1}$; exhaust gas temperature 723 K; exhaust gas composition lambda 0.999; dynamic frequency 1 Hz; amplitude 1 A/F). Reprinted with permission from ref. [34], © 1991 Society of Automotive Engineers, Inc.

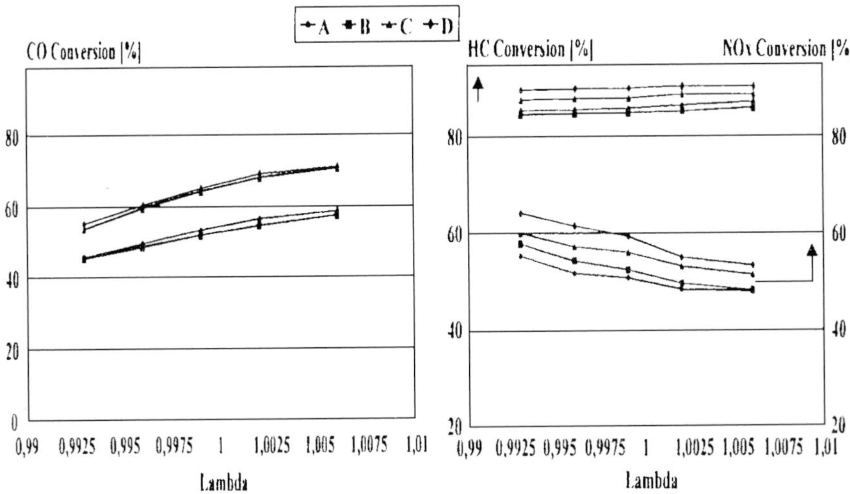

Figure 67. Influence of the washcoat formulation on the conversion of CO, HC and NO_x reached over various engine aged three-way catalysts as a function of the exhaust gas lambda value (monolith catalyst with 62 cells cm^{-2}; three-way formulation with Pt: 1.16 g l^{-1}, Rh: 0.23 g l^{-1}; engine bench test; A/F scan at a space velocity of 60 000 Nl l^{-1} h^{-1}; exhaust gas temperature 723 K; dynamic frequency 1 Hz; amplitude 1 A/F; engine bench high temperature aging cycle for 100 h). Sample A baseline formulation, samples B–D increasing amounts of stabilizers.

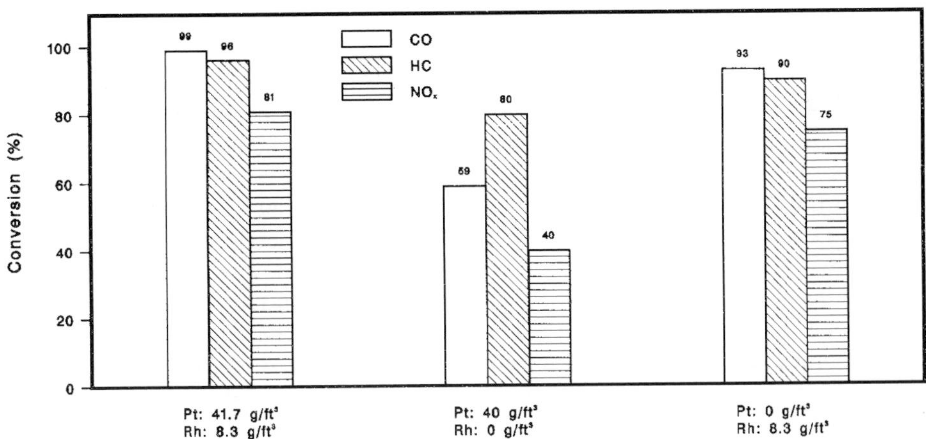

Figure 68. The function of platinum and rhodium on a fully formulated three-way catalyst washcoat in the conversion of CO, HC and NO_x (monolith catalyst with 62 cells cm^{-2}; three-way formulation, aged on an engine bench for 20 h; engine bench test at a space velocity of 60 000 Nl l^{-1} h^{-1}; exhaust gas temperature 673 K; exhaust gas composition lambda 0.999, dynamic frequency 1 Hz; amplitude 1 A/F). Reprinted with permission from ref. [34], © 1991 Society of Automotive Engineers, Inc.

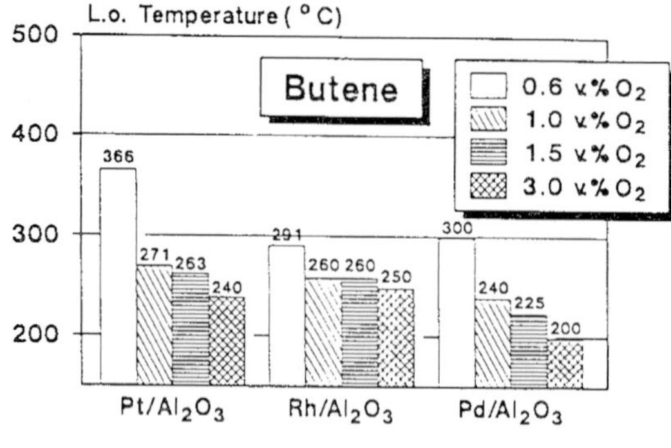

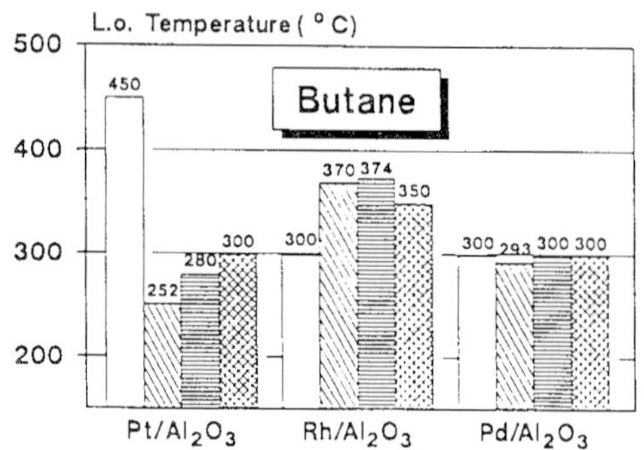

Figure 69. Effect of platinum, palladium and rhodium at an equimolar loading on the temperature needed to reach 50% conversion of butene and butane, as a function of the exhaust gas oxygen current (monolith catalyst with 62 cells cm^{-2}; γ-Al_2O_3 washcoat; fresh precious metal loading 8.8 mmol l^{-1}; model gas light-off test at a space velocity of 60 000 Nl l^{-1} h^{-1}; model gas composition is stoichiometric at 1.0 vol % O_2). Reprinted with permission from ref. [30], © 1994 Society of Automotive Engineers, Inc.

Increasing the total loading of Pt/Rh catalysts, at a fixed platinum to rhodium ratio, directly affects the light-off temperature (Fig. 70). This is consistent with the fact that the catalytic activity is under kinetic control in the temperature range where the light-off occurs. The conversion recorded on a fresh catalyst at temperatures above light-off is practically independent of the total amount of precious metals within the indicated range. This is consistent with the fact that the catalytic activity is under interface mass transfer control under these conditions. However, as shown in Fig. 71, after aging of the catalyst, the reaction kinetics again become rate controlling and, therefore, the total amount of precious metals strongly affects the conversion.

The influence of the total amount of precious metals on the conversion also depends upon the catalyst operation conditions (Fig. 72). After engine aging, the HC light-off under lean conditions is improved by about 40 K upon increasing the Pd loading from 3.5 g l^{-1} to 40 g l^{-1}, whereas the improvement is only about 20 K under stoichiometric conditions.

When multibrick converters with different catalysts are used, the sequence in which the catalysts are mounted in the converter will also affect the overall conversion (Fig. 73).

Each of the precious metals will interact in a different way with the major washcoat constituents, Al_2O_3 and CeO_2. The extent of these interactions is controlled by the catalyst operation conditions, especially the temperature and the net oxidizing

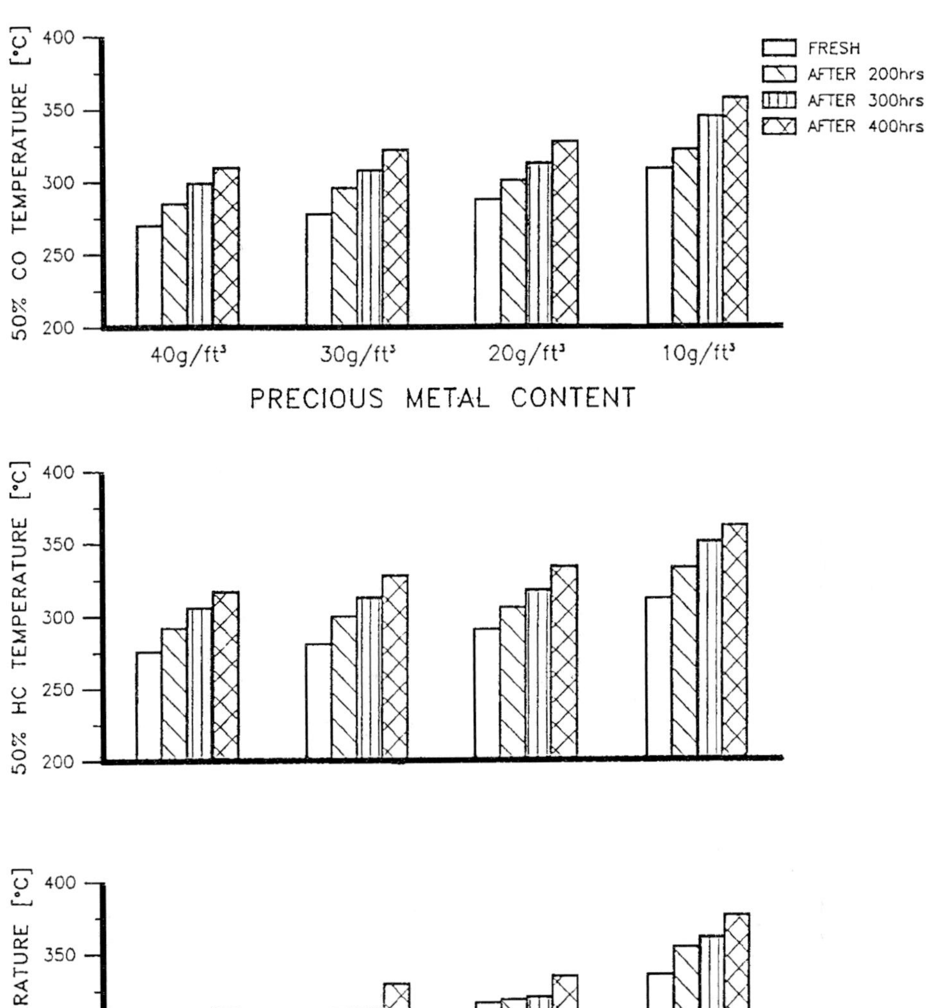

Figure 70. Effect of the total precious metal loading at a fixed Pt:Rh ratio of 5:1 g/g on the temperature needed to reach 50% conversion of CO, HC and NO_x, as a function of the duration of the engine aging (monolith catalyst with 62 cells cm^{-2}, three-way catalyst formulation; engine bench light-off test at a space velocity of 60 000 $Nl\,l^{-1}\,h^{-1}$, exhaust gas lambda 1.02 for CO and HC; lambda 0.984 for NO_x; engine bench aging cycle for various lengths of time). Reprinted with permission from ref. [76], © 1990 Society of Automotive Engineers, Inc.

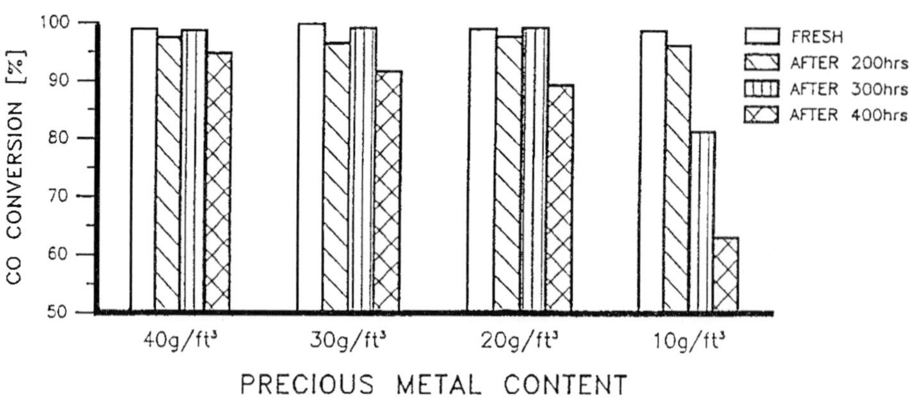

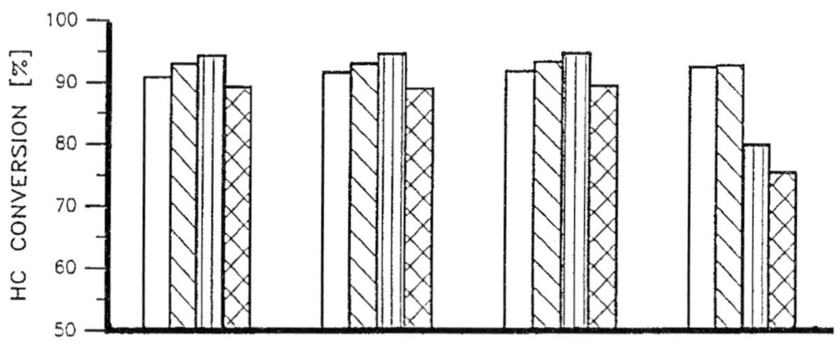

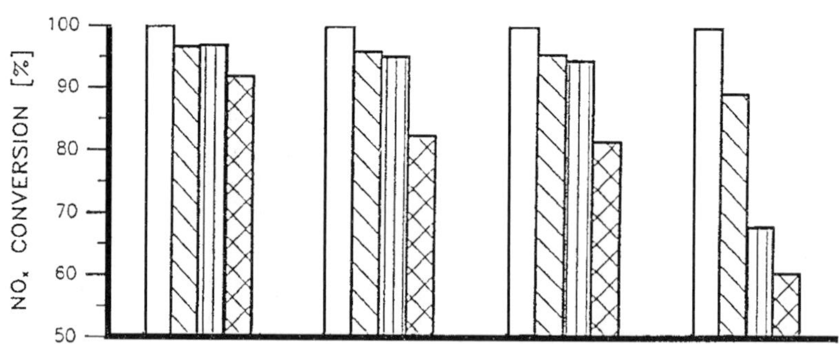

Figure 71. Effect of the total precious metal loading at a fixed Pt:Rh ratio of 5:1 g/g on the conversion of CO, HC and NO_x reached over the catalyst at a stoichiometric exhaust gas composition, as a function of the duration of the engine aging (catalyst as Fig. 70; engine bench activity test at a space velocity of 60 000 Nl l^{-1} h^{-1}; exhaust gas temperature 723 K; exhaust gas composition lambda 0.995; dynamic frequency 1 Hz; amplitude 0.5 A/F; engine bench aging cycle for various lengths of time). Reprinted with permission from ref. [76], © 1990 Society of Automotive Engineers, Inc.

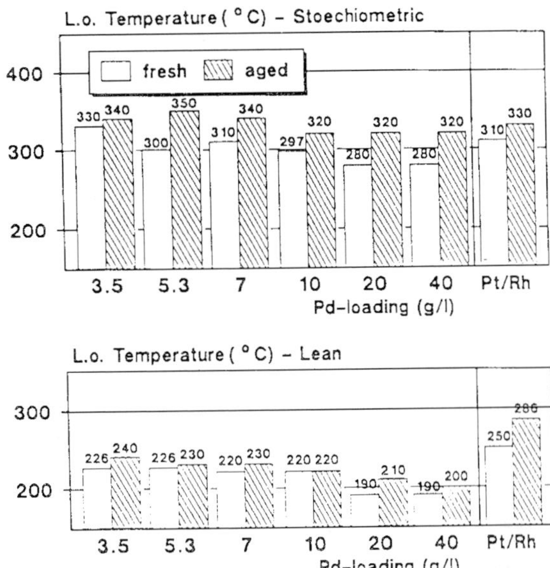

Figure 72. Effect of the palladium loading in "palladium only" three-way catalysts on the temperature needed to reach 50% conversion of the hydrocarbons with a stoichiometric and with a lean exhaust gas, for fresh and for engine aged catalyst (monolith catalyst with 62 cells cm^{-2}; dedicated washcoat for palladium only catalyst; reference catalyst Pt: 1.42 g l^{-1}, Rh: 0.28 g l^{-1}; engine bench light-off test at a space velocity of 60 000 Nl l^{-1} h^{-1}; exhaust gas composition stoichiometric and lean, corresponding to a lambda value of 1.15; engine bench aging cycle). Reprinted with permission from ref. [30], © 1994 Society of Automotive Engineers, Inc.

power of the exhaust gas. These interactions affect the valency state of both the precious metals and of some of the washcoat constituents. Furthermore, the dispersion and the stability of the dispersion of the precious metals is influenced as well as the solid state reactions between the various washcoat constituents. Figure 74 shows the differences in the valency state of Pt and Rh as a function of the type of washcoat oxide on which they are deposited [43, 44].

Similarly, Fig. 75 shows the effect of both the amount and the type of precious metal on the extent of the solid state reaction between aluminum oxide and cerium oxide, leading to the formation of cerium aluminate. The overall effect of these interactions is that the catalytic activity and the durability of the precious metals vary as a function of the type of washcoat oxide they are deposited on [45].

Figure 76 shows the conversion of CO, HC and NO_x reached after engine aging on different combinations between Pt, Pd and Rh at equimolar loading, and on Al_2O_3 and CeO_2.

Figures 77 and 78 demonstrate the effect of the different combinations between the precious metals Pt and Rh and of the washcoat oxides Al_2O_3 and CeO_2 on catalyst resistance against poisoning by SO_2 [46, 47].

1.3.7 Deactivation of Three-way Catalysts

A Introduction

In the application of heterogeneous catalysts to the chemical and petrochemical industries, either all precautions are taken to minimize deactivation of the catalyst, or the process is designed for regular regenera-tion of the catalyst. In contrast, au-

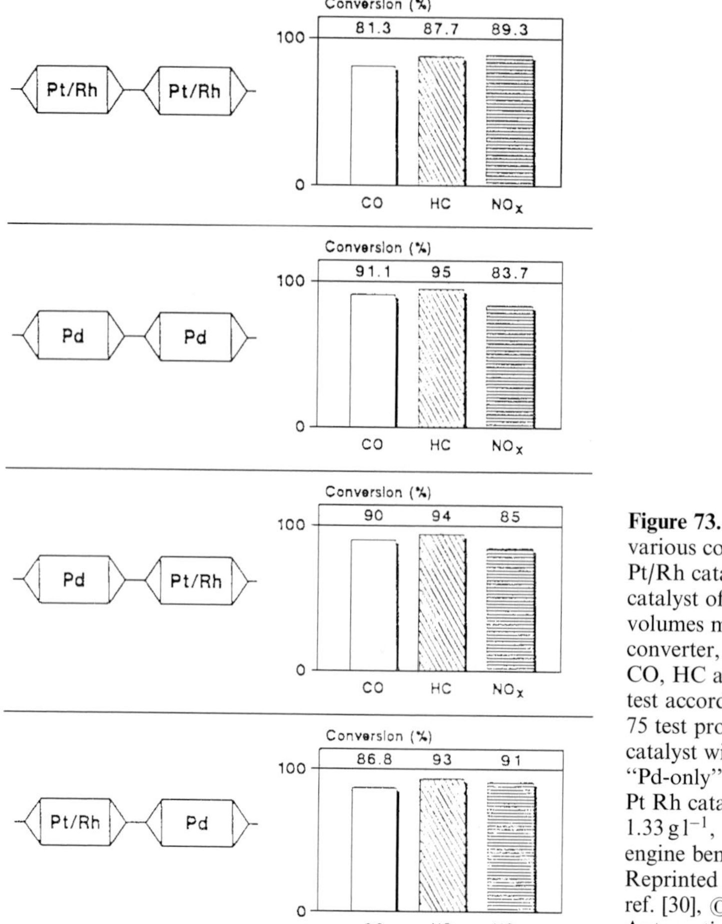

Figure 73. Performance of various combinations between a Pt/Rh catalyst and a "Pd only" catalyst of identical catalyst volumes mounted in one converter, for the conversion of CO, HC and NO_x in a vehicle test according to the US-FTP 75 test procedure (monolith catalyst with 62 cells cm^{-2}; "Pd-only" catalyst 15 g Pd l^{-1}; Pt Rh catalyst with Pt: 1.33 g l^{-1}, Rh: 0.26 g l^{-1}; engine bench aging for 45 h). Reprinted with permission from ref. [30], © 1994 Society of Automotive Engineers, Inc.

tomotive emission control catalysts are a mass application in which the operating conditions cannot be controlled and in which pretreatment of the "feedstock" is not possible. Despite this, legislation calls for catalyst durability to be the same order of magnitude as the vehicle life time. Because of these particularities in the use of automotive emission control catalysts, they experience a great number of deactivation phenomena, some of which are reversible (Fig. 79).

B Durability Measurement

The catalytic converter is only one part in the complex emission control system. Many parameters outside of the catalytic converter influence its durability. Therefore, the most convincing way to prove the durability of a catalyst is to perform a

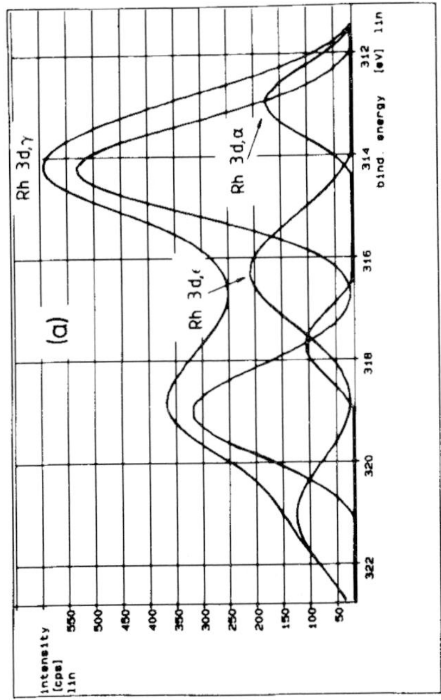

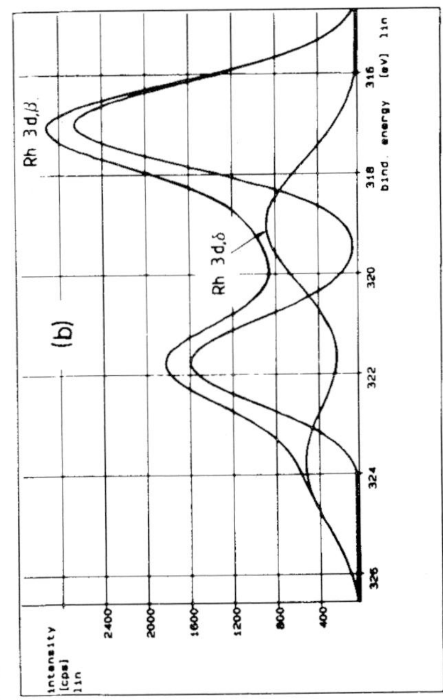

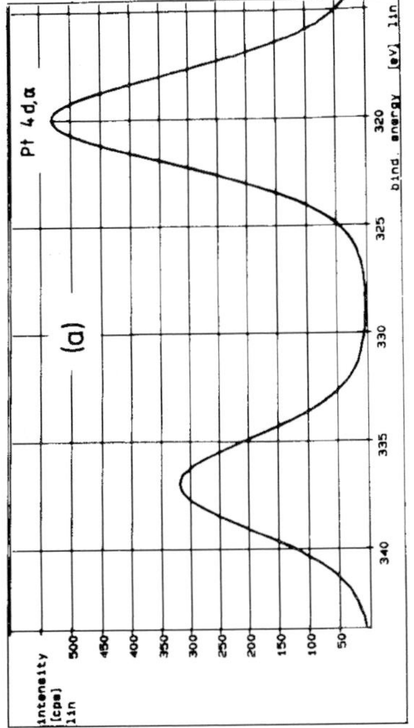

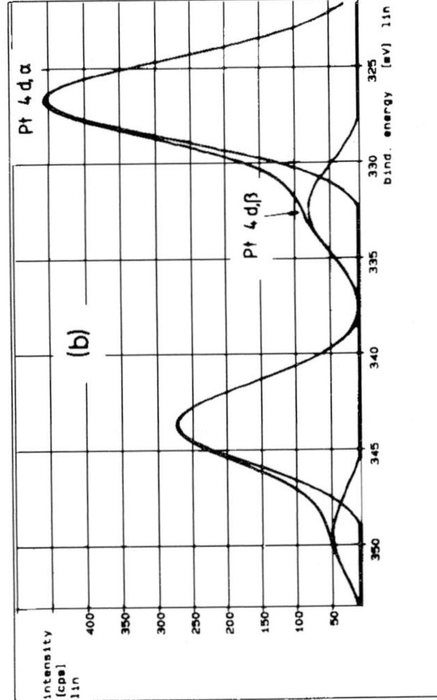

Figure 74. X-ray photoelectron spectra (XPS) showing the valency states of platinum and rhodium as a function of the washcoat composition (model catalyst on monoliths with 62 cells cm^{-2}; washcoat (a) γ-Al_2O_3; washcoat (b) CeO_2; fresh Pt or Rh at 1.76 g l^{-1}). Adapted from [43].

durability test with a fleet of vehicles on the road under uncontrolled driving conditions (Fig. 80). As expected for multivariate systems, the fleet test result obeys the Gaussian distribution [48].

As time and cost constraints make it impossible to use a fleet test as the test procedure in the development of emission control catalysts, several simplified durability tests have been developed. In order of decreasing complexity, these tests are a vehicle driven on the road according to a well defined driving schedule, a vehicle mounted on a vehicle dynamometer driven according to a well defined procedure by robots, tests with an engine mounted on an engine dynamometer, and tests in a laboratory furnace (Table 19).

The durability tests performed in laboratory furnaces mainly aim at studying thermal deactivation effects. They are usually carried out at temperatures between about 1170 K and 1320 K, under an oxidizing (air) or an inert (nitrogen with or without water and carbon dioxide) atmosphere.

The durability test performed on engine dynamometers may differ from one application to the other. As shown in Fig. 81, some tests target high temperature stability, whereas other tests target poisoning effects and are therefore performed at lower temperatures. The common feature of the present day engine durability tests is that a multitude of engine operation conditions are used, in which the composition of the exhaust gas and its temperature are varied. Usually, correlations can be established between the engine and furnace tests, and the vehicle durability tests. In most cases, however, these correlations differ between vehicles.

C Selected Deactivation Phenomena

A very drastic and common deactivation phenomenon with automotive emission control catalysts is the irreversible mechanical destruction of the support during road use: by breakage in the case of ceramic monoliths, by telescoping of the matrix or breakage of the foil in the case of some metallic substrates and, formerly, by attrition in the case of bead catalysts. With ceramic monoliths, sudden temperature changes can cause thermal stresses and consequent breakage.

A broad range of deactivation phenomena are related to the operation temperature of the catalyst. Operation at both low and very high temperatures can induce specific catalyst deactivations (Fig. 82).

Usually, deactivation phenomena occurring at low temperatures are reversible, this means that they are removed by the operation of the catalyst at a higher temperature, eventually in conjunction with a different net oxidizing or reducing power of the exhaust gas. Examples of low temperature deactivation phenomena are the adsorption and/or chemisorption both of reactants and of reaction products such as CO_2, as are the reactions between the various sulfur oxides (SO_2 and SO_3) and the washcoat oxides, and the oxidation of the precious metals. Figure 83 shows an example of the decomposition of the surface carbonates formed on Al_2O_3 and CeO_2, as a function of the temperature. Figure 84 shows the removal of the sulfite and sulfate phases formed on the main washcoat constituents, as a function of the temperature and the composition of the catalyst [49].

Deactivation phenomena occurring at higher temperatures are typically irre-

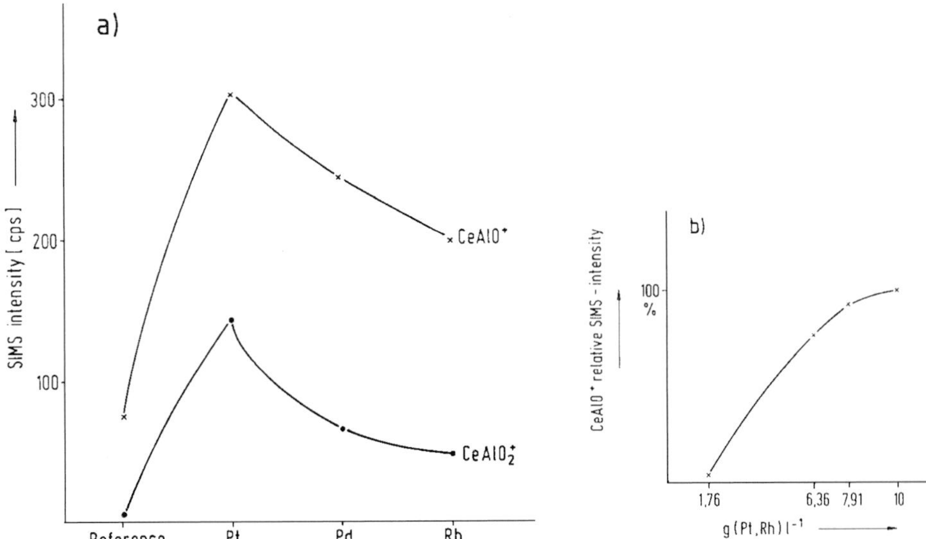

Figure 75. Secondary ion mass spectra (SIMS) showing the extent of the solid state reactions between alumina and ceria in a three-way catalyst washcoat: (a) as a function of the type of precious metal; and (b) as a function of the total precious metal loading (model catalyst on monoliths with 62 cells cm^{-2}; Pt, Pd or Rh loading 1.76 g l^{-1}; various total loadings at fixed Pt:Rh ratio of 5:1 g/g; model washcoat with 70 wt % alumina and 30 wt % ceria, after aging under air at 973 K).

versible. These include solid-state reactions involving the washcoat components, between the precious metals resulting in the formation of nonuniform alloys, and between the different precious metals and the washcoat oxides. A common example of the latter is the migration of Rh^{3+} ions in the lattice of the transitional aluminum oxide, as the crystal structure Rh_2O_3 is isomorphic with the crystal structure of γ-Al_2O_3.

The most important deactivation phenomena occurring at a high catalyst operation temperature are the loss of the internal surface area of the washcoat components, and the loss of the precious metals dispersion. The internal surface area of the washcoat is decreased by the sintering of the washcoat oxides. This is a dramatic process, that also leads to the inclusion of precious metals, as they are supported on these washcoat oxides (Fig. 85). The most important parameter that governs the loss of the washcoat internal surface area is the temperature (Fig. 86). It should be noted that each of the washcoat constituents has a different temperature stability (Fig. 87).

Sintering of the precious metals leads to a loss in precious metals surface area and to the establishment of a broader distribution of the precious metals particle diameter. As shown in Fig. 88, a catalyst in the fresh state features a rather homogenous size of the precious metal particles, whereas after aging, very large precious metal

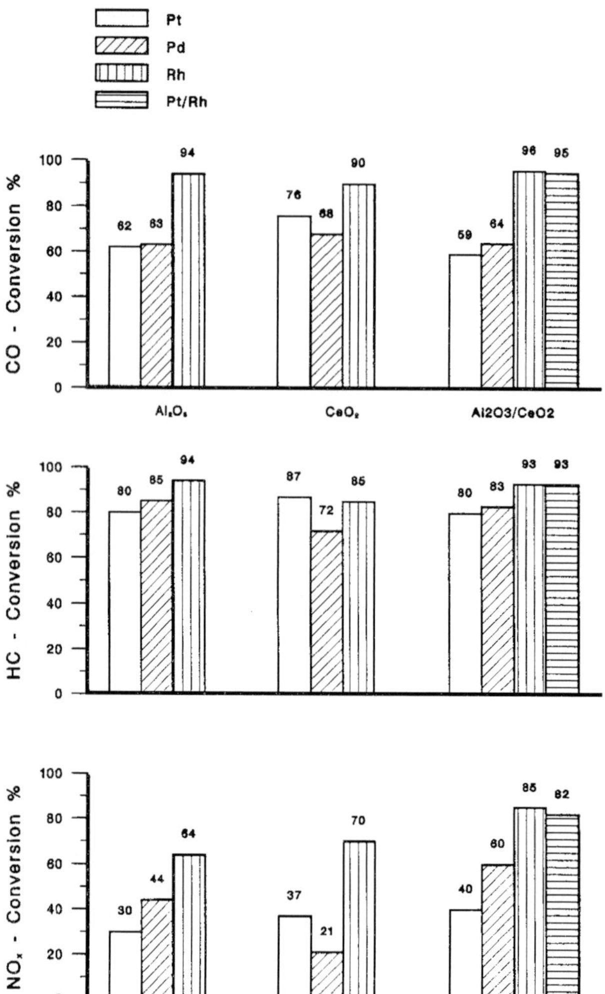

Figure 76. Conversion of CO, HC and NO_x reached at fixed reaction conditions over model catalysts representing the different combinations between the precious metals Pt, Pd and Rh, and the washcoat constituents alumina and ceria, after engine aging (monolith catalyst with 62 cells cm^{-2}; Pt, Pd or Rh at an equimolar loading of $8\,mmol\,l^{-1}$; Pt/Rh catalyst with Pt: $1.42\,g\,l^{-1}$, Rh: $0.28\,g\,l^{-1}$, washcoats are 100% alumina. 100% ceria; 70% alumina 30% ceria; engine bench activity test at a space velocity of 60 000 $Nl\,l^{-1}\,h^{-1}$; exhaust gas temperature 673 K; exhaust gas composition lambda 0.999; dynamic frequency 1 Hz; amplitude 1 A/F; Engine bench aging cycle).

crystallites coexist with small ones. The extent to which precious metals sinter depends upon their initial dispersion, the nature of the interaction between the precious metal and the washcoat, and the type of the precious metal and the net oxidizing power of the exhaust gas. Platinum and rhodium sinter much faster under an oxidizing exhaust gas, whereas palladium sinters faster under a reducing exhaust gas. This behavior is consistent with the oxidation state of these elements under the respective atmospheres, and is explained by the difference in the vapor pressure between the metals and their respective oxides (Table 20) [50, 51].

With spark ignition engines, the solids temperature range during use is broader

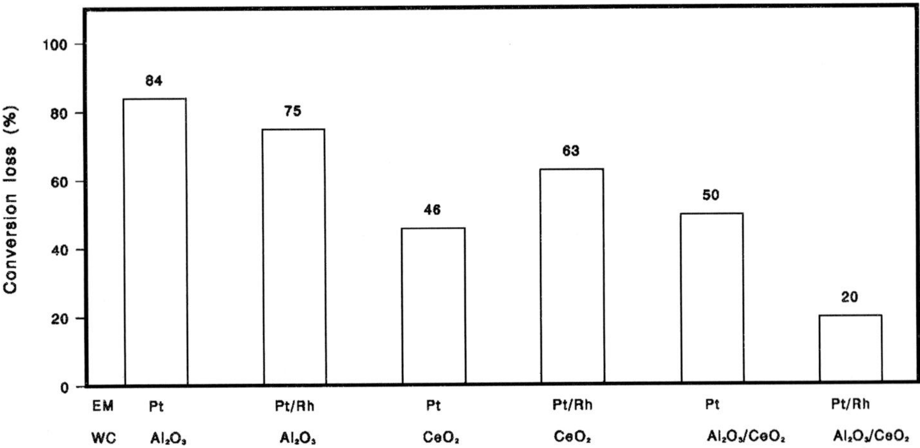

Figure 77. Loss in the CO conversion at a net reducing exhaust gas composition between an activity test using a SO_2 free model gas and an activity test with a model gas containing 20 vol. ppm SO_2, for various combinations between the precious metals Pt and Rh and the washcoat constituents alumina and ceria (monolith catalyst with 62 cells cm^{-2}; washcoats are 100% alumina, 100% ceria; 70% alumina 30% ceria; precious metal content Pt 1.48 g l^{-1}, Rh 0.283 g l^{-1}, fresh; model gas test at a space velocity of 60 000 Nl l^{-1} h^{-1}; exhaust gas temperature 773 K; exhaust gas composition lambda 0.88). Reprinted with permission from ref. [34], © 1991 Society of Automotive Engineers, Inc.

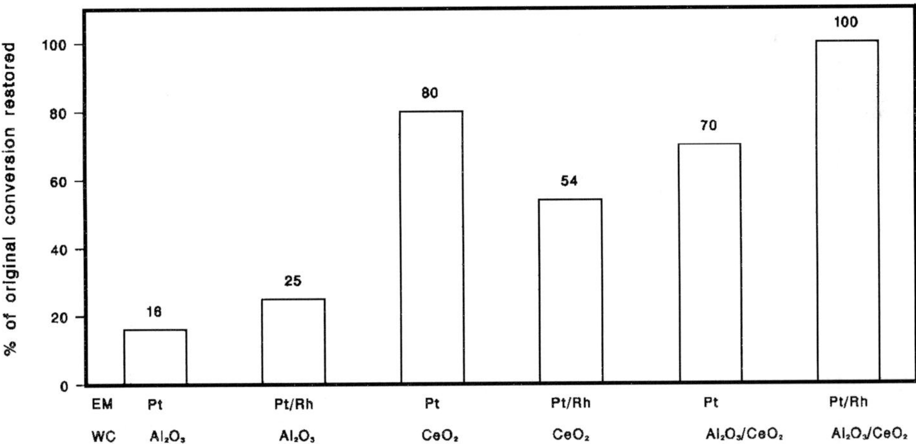

Figure 78. Restoration of the CO conversion at a net reducing exhaust gas composition between an activity test using a model gas with 20 vol ppm SO_2 and an activity test using a SO_2 free model gas, for various combinations between the precious metals Pt and Rh and the washcoat constituents alumina and ceria (experimental conditions as for Fig. 77). Reprinted with permission from ref. [34], © 1991 Society of Automotive Engineers, Inc.

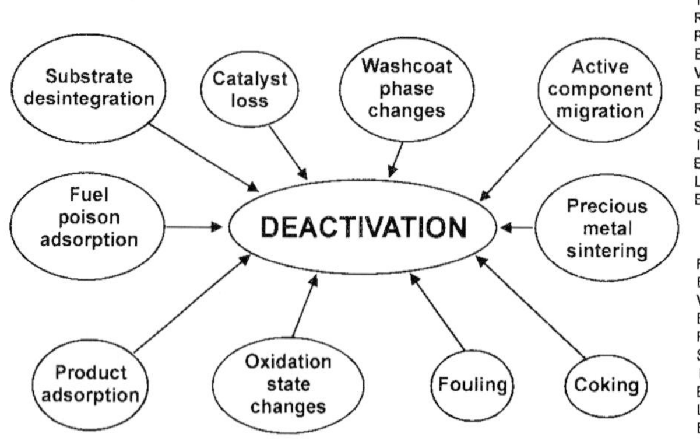

Figure 79. Overview of the reversible and the irreversible deactivation phenomena of three-way catalysts.

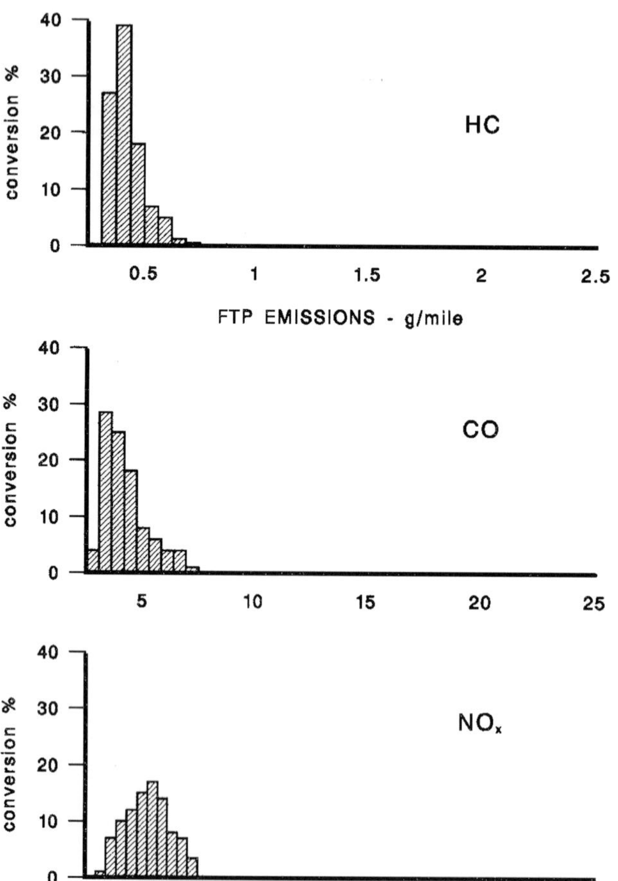

Figure 80. Emission of CO, HC and NO_x from a sample of 532 identical vehicles equipped with three-way catalysts, used on the road since 1985, as measured in the US-FTP 75 vehicle test cycle. Reprinted with permission from ref. [48], © 1988 Society of Automotive Engineers, Inc.

Table 19. Main differences between the procedures used to study the deactivation of three-way catalysts.

Deactivation mechanism	Vehicle tests		Engine tests	Oven tests
	Road	Dynamometer		
Mechanical				
Disintegration	yes	no	no	no
Fouling	yes	yes	yes/no	no
Thermal				
Sintering	yes	yes/no	yes	yes
Phase change	yes	yes/no	yes	yes
Compound formation	yes	yes/no	yes	yes/no
Hot spots	yes	yes	yes/no	no
Component migration	yes	yes/no	yes	no
Chemical				
Poison adsorption	yes	yes/no	yes/no	no
Product adsorption	yes	yes/no	yes/no	no
Oxidation state changes	yes/no	yes/no	yes	no

than the gas temperature range. This is caused by the exothermic nature of the conversion reactions, and by the fact that under ignition failures or vehicle deceleration conditions – for vehicles not equipped with appropriate fuel-cut devices – unburned fuel can reach the catalyst. The unburned fuel will be oxidized by the

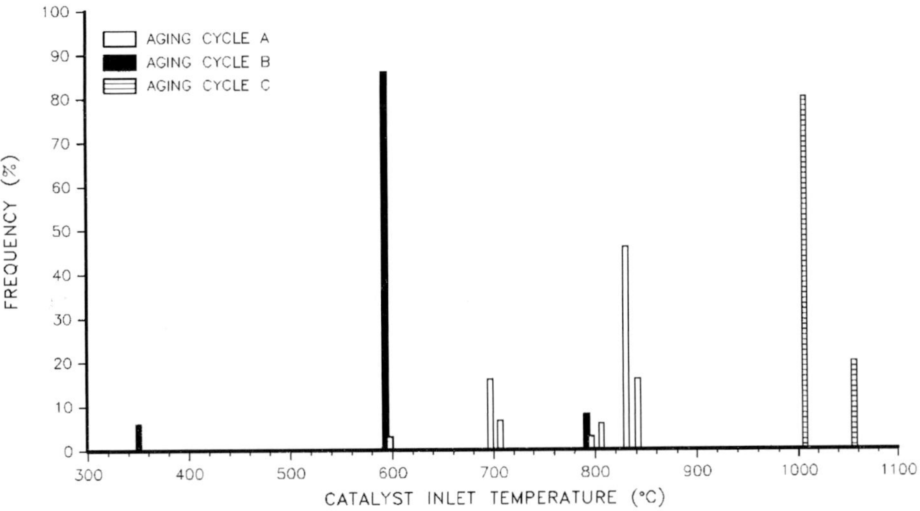

Figure 81. Temperature histogram of three engine bench aging cycles using gasoline spark ignition passenger car engines.

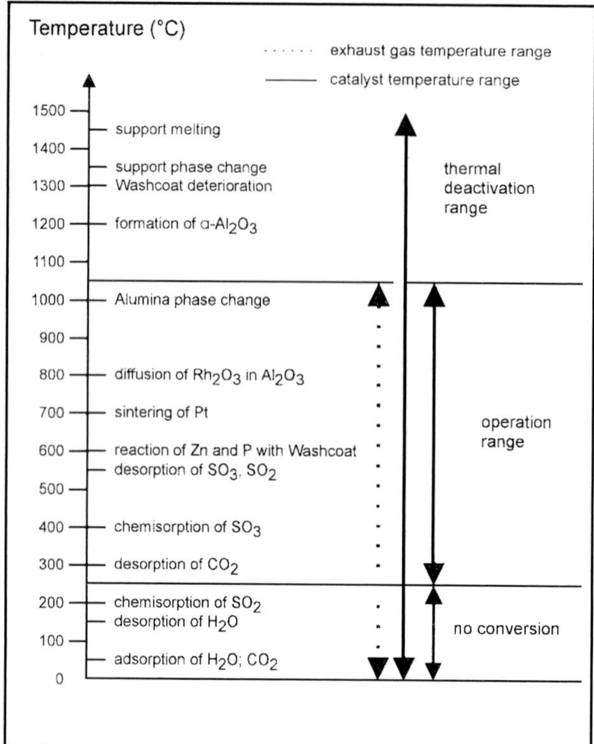

Figure 82. Schematic overview of some deactivation phenomena with three-way catalysts, as a function of catalyst temperature.

catalyst, thereby dramatically raising its temperature. Figure 89 shows as an example the evolution of the temperature of the gas phase and the solid phase at deceleration conditions. Despite the fact that the temperature of the exhaust gas in front of the catalyst decreases rapidly, the temperature of the catalyst itself increases by the exothermic combustion of the unburned fuel. The solids temperature may even exceed 1600 K, causing the melting of the supports [52].

Finally, deactivation of the catalyst by poisoning elements should be mentioned. Precious metal based catalysts are poisoned by sulfur oxides which mainly originate from the combustion of sulfur-containing fuel constituents, by phosphorus and zinc which mainly originate from some additives in the engine lubricating oil, and by silicium which was sometimes present in some engine seals (Table 21). Also, traces of lead, present in the fuel because of contamination of the fuel supply chain, made an important contribution to the deactivation of the catalyst in the past.

The fraction of the poisoning elements retained by the catalyst is distributed nonuniformily over the length of the catalyst and the depth of the washcoat layer. Highest concentrations occur at the entrance to the monolith (Figs. 90 and 91). Each of the poisoning elements interact in a different way with the washcoat con-

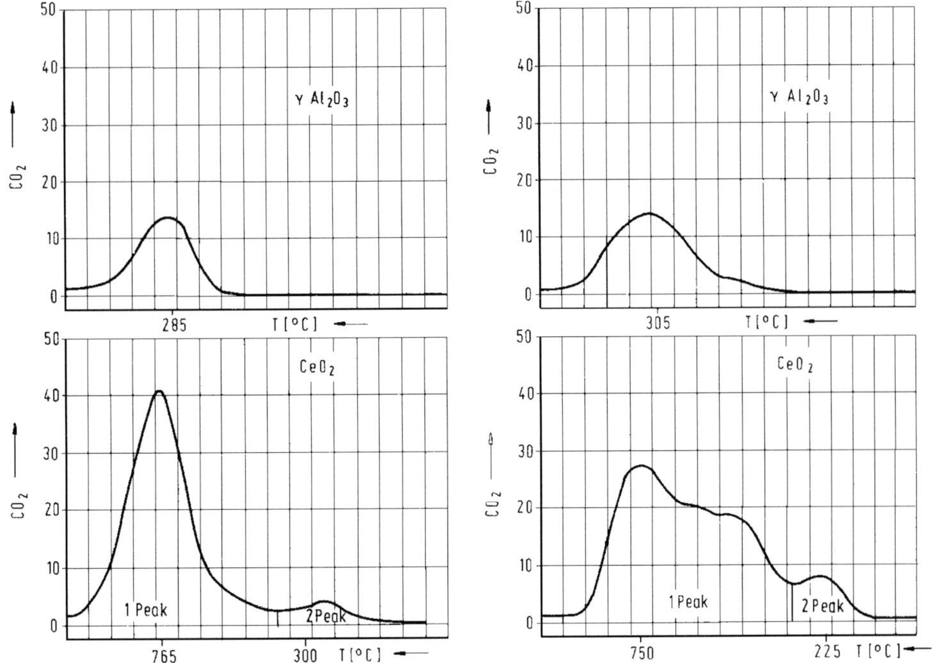

Figure 83. Thermal desorption spectra of carbon dioxide from alumina and ceria, preloaded with carbon dioxide at 298 K (left) and at 723 K (right). Adapted from [44].

stituents and the precious metals. The interaction phenomena are rather complex, and sometimes the simultaneous presence of several poisoning elements causes specific interactions, that would not occur if each of these elements would be present alone. Examples of this are the effects caused by the simultaneous presence of Pb and S or of P and Zn [53].

The type and the extent of the interaction between the poisoning elements and the catalyst constituents depend strongly upon the operation temperature of the catalyst during and after the deposition of the poisoning elements. As a rule of thumb, Pb and Si will mainly affect the function of the precious metals, by the formation of alloys or chemical compounds between the precious metals and these poisoning elements, whereas P, Zn and S will mainly affect the function of the washcoat components [54, 55]. For example, phosphorus will interact with the aluminum oxide of the washcoat and cause of considerable loss in the stability of its internal surface area at a high operating temperatures. Another example is the joint interaction of phosphorus and zinc with the aluminum oxide of the washcoat, leading to a mechanical obstruction of the washcoat pores.

The net result of the chemical and thermal deactivation phenomena is a decrease of catalyst activity for the main reactions that govern the conversion of CO, HC and NO_x, as well as for the side reactions that would cause the formation of sec-

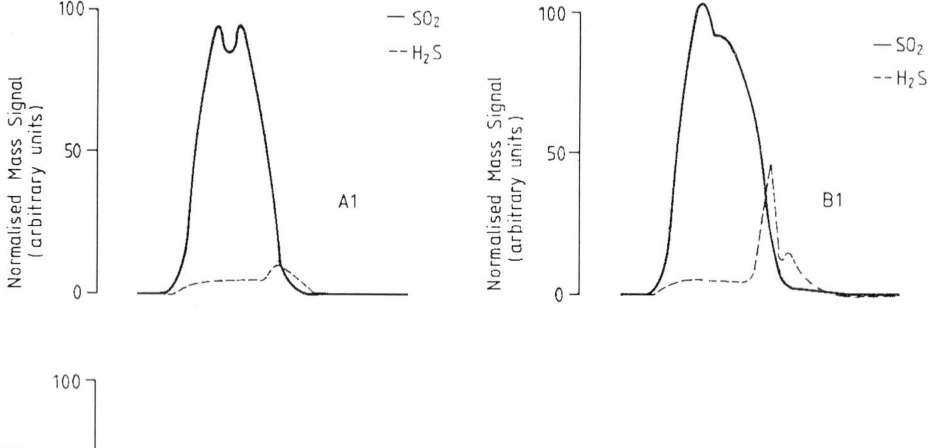

Figure 84. Reactive thermal desorption spectra in a 5 vol.% H_2/95 vol.% N_2 atmosphere: (A1) alumina; (B1) 70% alumina 30% ceria; (C1) Pt/Rh on 70% alumina 30% ceria, all preloaded at 298 K with SO_2). Reprinted with permission from ref. [49], © 1988 Society of Automotive Engineers, Inc.

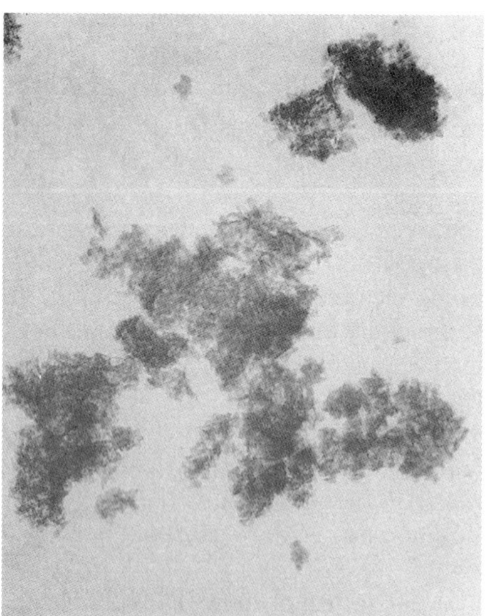

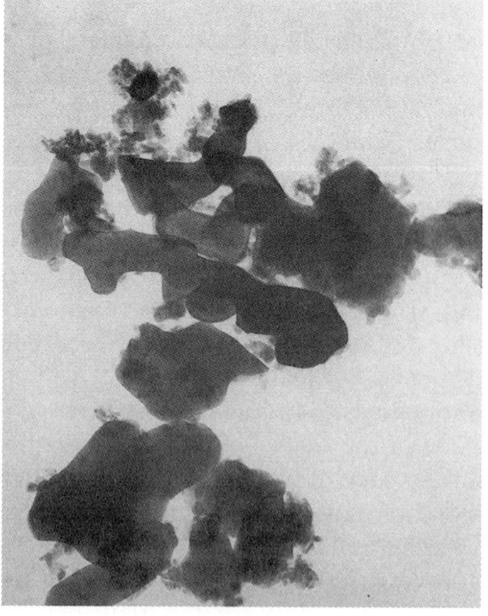

Figure 85. Transmission electron microscope view of a washcoat alumina in the fresh state (left) and after aging in air during 24 h at 1373 K (right). Total magnification 200 000×.

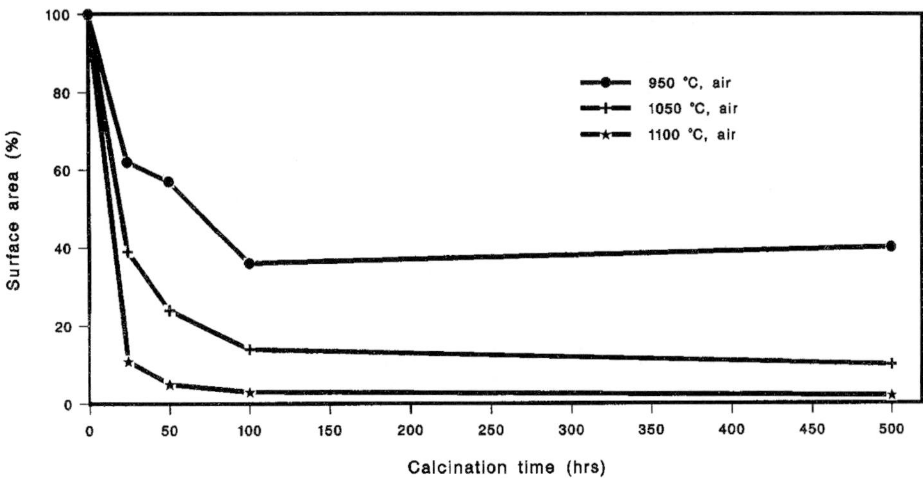

Figure 86. Stability of the internal surface area of a three-way catalyst washcoat as a function of temperature and aging duration in an air atmosphere. Reprinted with permission from ref. [34], © 1991 Society of Automotive Engineers, Inc.

ondary emissions under operation of the catalyst with nonstoichiometric exhaust gas compositions.

As some of the main reactions consist of several sequential steps, the deactivation of the catalyst can cause an enhanced concentration of the intermediate reaction

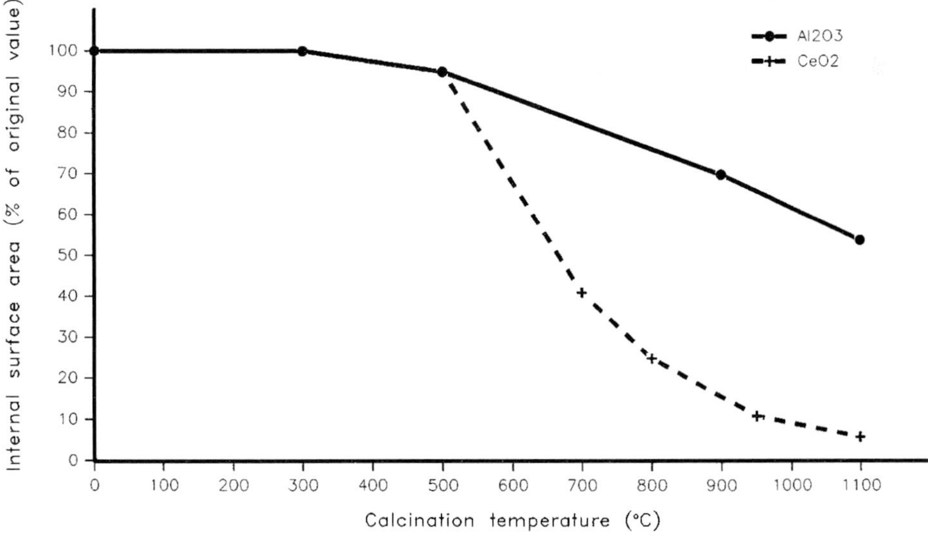

Figure 87. Stability of the internal surface area of the two main washcoat constituents alumina and ceria, as a function of the temperature during aging in air for 24 h.

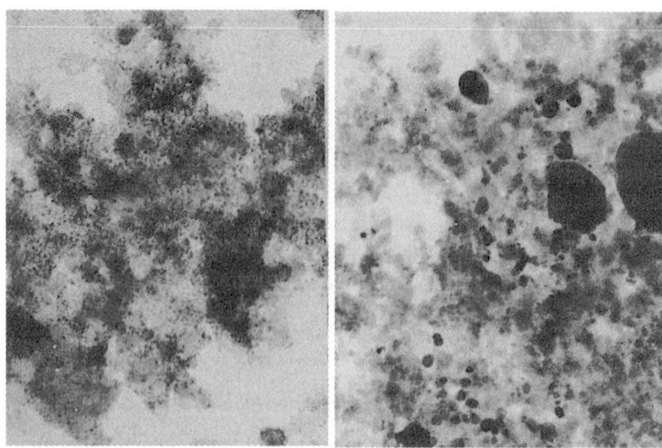

Figure 88. Transmission electron microscopic view of platinum crystallites on alumina, left for a fresh catalyst, right for a catalyst aged at 1323 K in air.

products in the tailpipe exhaust gas. Examples are enhanced concentrations of aldehydes, which can be the intermediate oxidation products of alcohols; and of N_2O which can be the intermediate reduction product of the nitrogen oxides. Again, the extent to which these intermediate reaction products will occur in the tailpipe exhaust gas depends upon the formulation and the operating conditions of the deactivated catalyst.

The decrease of catalytic activity by deactivation has the strongest influence on the conversion of CO, HC and NO_x during those vehicle driving conditions where the exhaust gas temperature is below about 770 K. Figure 92 shows as an example the conversion of CO, HC and NO_x for a fresh and a vehicle aged catalyst in different parts of the European driving cycle. The difference in the conversion reached over the catalyst in the fresh and in the aged state is most apparent during the first phase. This is caused by the relatively low average catalyst operation tem-

Table 20. Sintering behavior of platinum, palladium and rhodium as a function of the aging atmosphere (washcoat La_2O_3-doped Al_2O_3, precious metal content 0.14 wt. %). Reprinted from refs. [50, 51] with kind permission of Elsevier Science.

Parameter	Pt	Pd	Rh
Particle size (Å)			
N_2, 1373 K	210	970	140
Exhaust gas, 1373 K	780	680	880
Air, 1373 K	970	n.a.	n.a.
Vapor pressure at 1073 K (torr)			
Metal	9.1×10^{-17}	1.2×10^{-9}	2.9×10^{-17}
Oxide	1.2×10^{-5}	≈ 0	5.8×10^{-6}

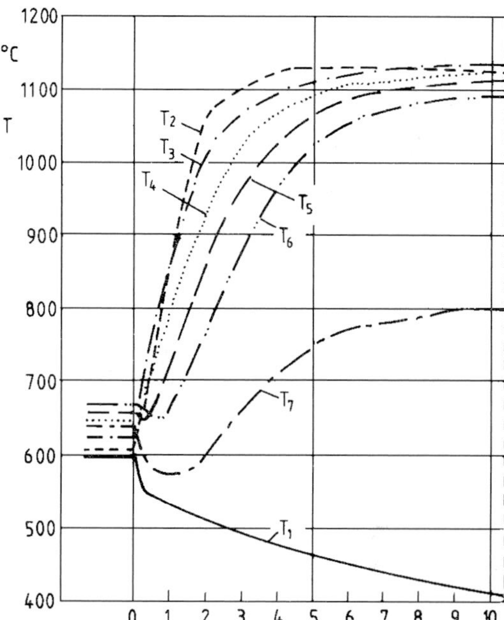

Figure 89. Evolution of the exhaust gas temperature in front of the catalyst (T1) and behind the catalyst (T7), as well as the solids temperature at different positions inside a monolithic three-way catalyst (T2–T6) as a function of the time at deceleration of a vehicle equipped with a gasoline engine without a fuel-cut device. Deceleration stars at time zero. Adapted from [52]

perature during this phase. In the third phase of the driving cycle, only small differences occur between the conversion over the fresh and the engine aged catalyst, because of the relatively high average catalyst operation temperature during this phase.

Table 21. Order of magnitude of the poisoning effects during the aging of three-way catalysts.

Item	Quantity ($g\,l_{cat}^{-1}$)
Catalyst formulation	
Al_2O_3	100–200
CeO_2	40–80
Pt	1–2
Rh	0.1–0.4
Poisons offered during 80 000 km use	
Pb	10–130
S	1600–5000
P	20–50
Zn	80–120

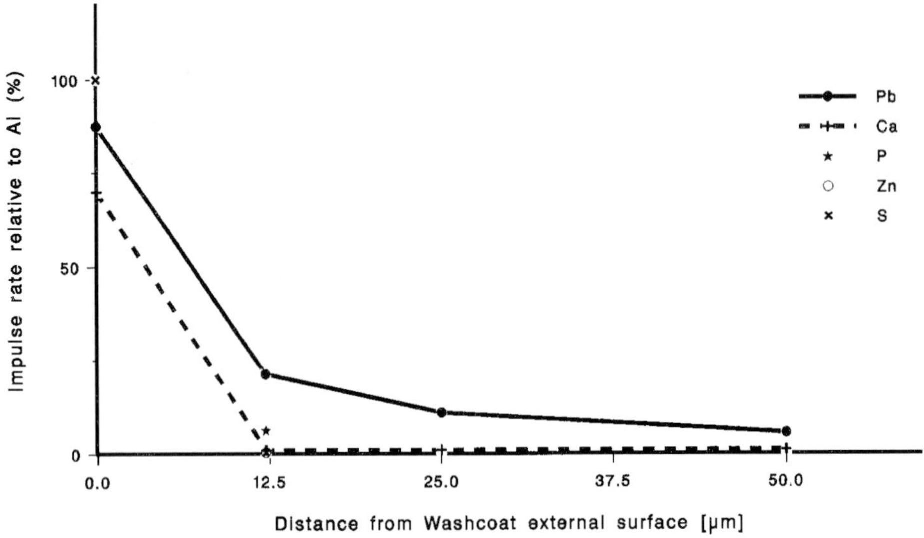

Figure 90. XRF impulse rate of Pb, Ca, P, Zn and S, relative to A1, as a function of the depth in the washcoat layer for a vehicle-aged monolithic three-way catalyst used in the Federal Republic of Germany in 1985–1989.

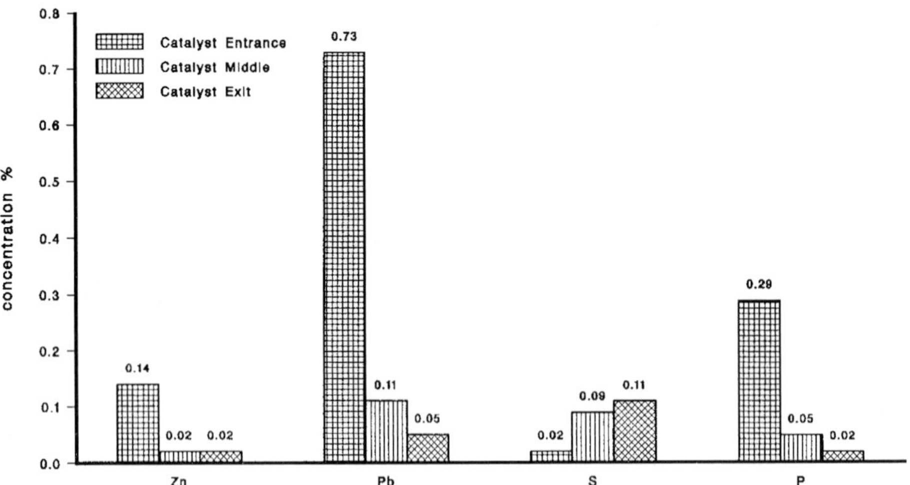

Figure 91. Concentration of Zn, Pb, S and P as a function of the distance from the catalyst inlet for a vehicle aged monolithic three-way catalyst used in the Federal Republic of Germany in 1985–1989.

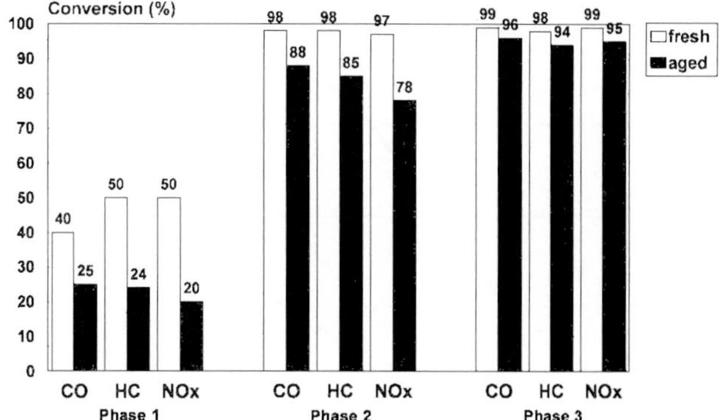

Figure 92. Conversion of CO, HC and NO_x in three different phases of the European vehicle test cycle, reached over a three-way catalyst in the fresh state and after high temperature engine bench aging: (phase 1 first city driving cycle; phase 2 second, third and fourth city driving cycle; phase 3 extra-urban driving cycle).

1.3.8 Future Concepts

A Introduction

As discussed in Section 1.2, legislation has moved towards a further reduction in the tailpipe emissions of CO, HC and NO_x. To meet this requirements, changes in the design of the engine and of the catalytic exhaust gas aftertreatment system are planned, eventually along with a reformulation of the fuel.

The continuous efforts to reduce the fuel consumption of the passenger cars renewed the interest in lean burn engines. These engines require dedicated exhaust after treatment systems.

B Closed-Loop-Controlled Spark Ignition Engines

One important development direction is the further improvement of the performance of the three-way catalytic converter applied to closed-loop-controlled spark ignition engines. This improvement is mainly targeted at increasing the performance of the three-way catalyst during the first minutes following the cranking of the engine.

Numerous technologies are being developed for the enhancement of this so called cold start performance. One approach is to increase the temperature of the exhaust gas in front of the catalyst during the cold start phase. This can be achieved by mounting the catalyst closer to the engine outlet or by the incorporation of a so called light-off-catalyst in the exhaust gas system, upstream of the main catalyst and close to the engine outlet. These measures call for the development of catalysts with excellent stability at high operation temperatures. Another approach is to minimize losses of exhaust gas heat by insulating the exhaust gas pipe. A further approach is to quickly increase the catalyst temperature by the supply of additional heat. This is achieved either by resistively heating the catalyst support or by the installation of a small fuel burner in front of the catalyst.

The technologies that lead to a quicker increase of the catalyst temperature will improve the catalytic performance for the conversion of all three exhaust gas components under consideration. However, some of the new legislation focuses on the

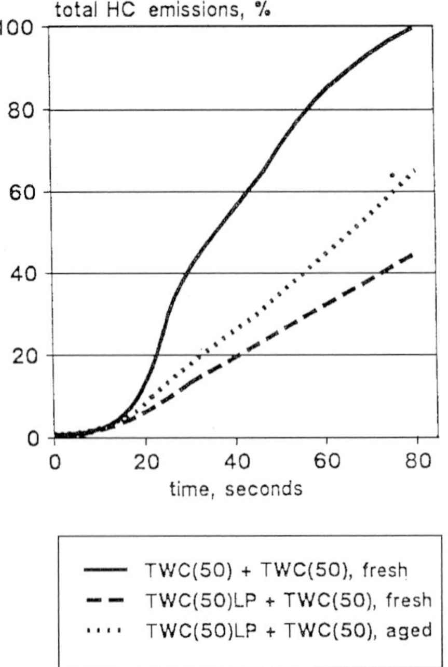

Figure 93. Integrated emission of hydrocarbons as a function of time during the first 80s of the US-FTP 75 vehicle test cycle, for a vehicle equipped with a gasoline spark ignition engine, using a conventional three-way catalyst converter system (TWC (50) + TWC (50)) and using a converter system in which a hydrocarbon adsorber is combined with a three-way catalyst (TWC(50)LP + TWC(50)) (monolith catalyst with 62 cells cm^{-2}; three-way catalyst formulation Pt: 1.42 $g\,l^{-1}$, Rh: 0.28 $g\,l^{-1}$, fresh and aged on an engine bench; vehicle test according to the US-FTP 75 test cycle, using a vehicle with a gasoline spark ignition engine without secondary air supply). Reprinted with permission from ref. [39], © 1993 Society of Automotive Engineers, Inc.

reduction of the tailpipe HC emission, so that catalytic aftertreatment systems are developed that mainly improve the HC conversion. An example is the incorporation of hydrocarbon adsorbers into the aftertreatment system, in addition to the three-way catalyst. The goal of these hydrocarbon adsorbers is to remove the hydrocarbons out of the exhaust gas by adsorption on suitable materials, as long as the catalyst temperature is lower than its light-off temperature, and then to supply these adsorbed hydrocarbons back to the exhaust gas by desorption as soon as the catalyst temperature is high enough to ensure their conversion. Examples of suitable adsorption materials are activated carbon and zeolites.

If adsorption materials are used that withstand the high temperatures of exhaust gases containing oxygen and water vapor, the corresponding hydrocarbon adsorbers can be either incorporated into the three-way catalyst or can be mounted in front of the three-way catalyst. Figure 93 compares the HC emission performance of such a combined adsorber-catalyst system to the performance of the catalyst alone. If the adsorption materials cannot withstand the harsh operating conditions, they can be mounted in a bypass to the exhaust gas system, and the exhaust gas passed through the adsorber only during the cold start phase [56–59].

C Lean Burn Engines

The operation of a spark ignition gasoline engine above the stoichiometric point reduces the fuel consumption and therefore the CO_2 emission of the engine. At the same time, the engine-out emissions of CO, NO_x and up to some defined A/F-ratio,

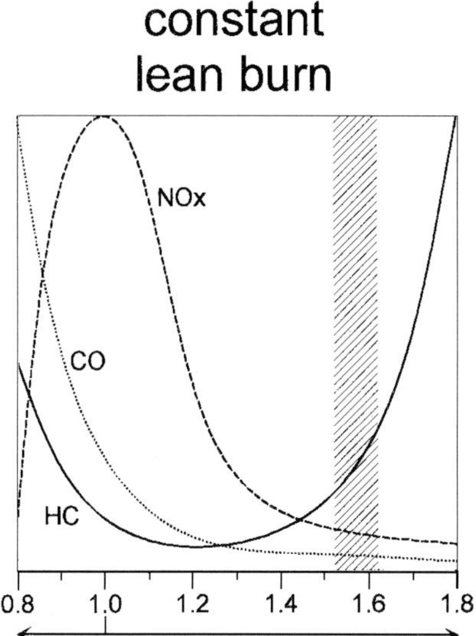

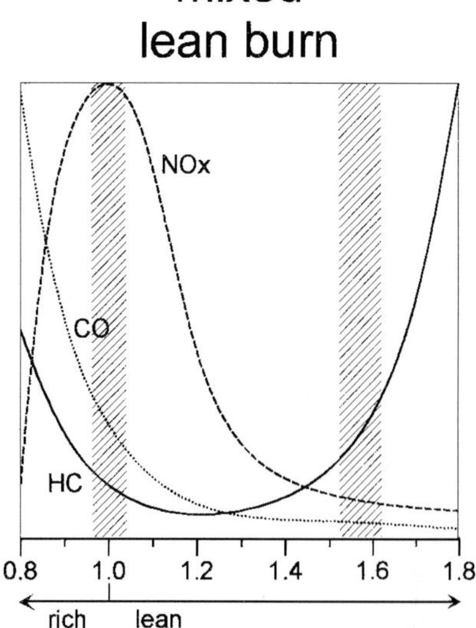

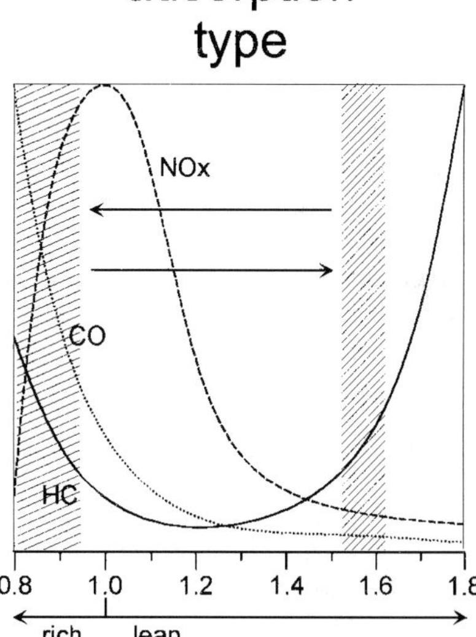

Figure 94. Emission of CO, HC and NO_x at the outlet of a gasoline spark ignition engine as a function of the engine lambda value, and range of the lambda value in which three different catalytic exhaust aftertreatment concepts for lean burn gasoline engines operate. Adapted from [60].

also of HC are decreased. These so called lean burn engines still require exhaust gas aftertreatment, but the conversion level needed to meet the legislative requirements is somewhat lower as compared to engines operated at the stoichiometric point.

As of today, three types of lean burn engines are proposed (Fig. 94) [60, 61]. The first type always operates in the lean burn range, with a lambda value between 1.5 and 1.6. The exhaust gas composition will be net oxidizing at all vehicle driving conditions. The second type operates under some driving conditions in the lean burn mode, and for other driving conditions in the stoichiometric mode. The latter conditions occur, for example, during acceleration. The third type of engine also operates under both lean burn and stoichiometric conditions, the difference to the second type being that the operation mode does not depend upon the vehicle driving conditions, but is forced by the engine management system. Such engines will operate for a few minutes under lean burn conditions, after which the engine management system will adjust to stoichiometric or even rich operation for a few seconds.

Each of these engine types requires a different catalytic aftertreatment system, as will be explained below.

With conventional three-way catalysts, sufficient conversion can be reached for CO and HC under the lean exhaust gas compositions of lean burn engines. However, three-way catalysts are unable to convert NO_x under these net oxidizing exhaust gas compositions. Intensive research programmes have shown that elements such as Co and Cu, supported on specific zeolites, are able to convert the NO_x under these conditions. Corresponding catalysts show some promise, although their durability, especially high temperature stability and resistance against sulfur poisoning, still needs improvement. Recently, also a supported iridium containing catalyst was proposed.

For the first and the second type of lean burn engines, therefore, catalytic aftertreatment systems are considered that include such a Cu/Zeolite catalyst to convert NO_x under the lean burn conditions, and a conventional three-way catalyst to convert CO and HC as well as, with the second engine type, NO_x under stoichiometric operation conditions. Some applications actually use a supported Iridium catalyst instead of the Cu/Zeolite catalyst.

For the third type of lean burn engines, special catalysts are developed, which have the ability to convert NO to NO_2 and to store the NO_2 under the lean burn operation conditions. The stored NO_2 is reduced to N_2 once the engine is operated under stoichiometric and rich conditions. The catalyst proposed in the patent literature for this application contains precious metals such as platinum, on three-way washcoats modified so as to increase their capacity to store and release NO_2.

1.4 Catalysts for Diesel Fueled Compression Ignition Engines

1.4.1 Introduction

Compression ignition engines differ from spark ignition engines in the air and fuel mixing and the ignition of the air and fuel mixture. Furthermore, they use different operating pressures, operating temperatures and air-to-fuel ratios (Table 22). A variety of compression engine designs are currently used. A distinction can be made on the base of both the fuel and the air admission principle. With respect to fuel

Table 22. Main design and operation differences between compression ignition engines and spark ignition engines.

Feature	Compression ignition	Spark ignition
Process type	Internal combustion	Internal combustion
Combustion type	Cyclic	Cyclic
Air–fuel mixing	Heterogeneous	Homogeneous
Ignition type	Auto	External
Operating pressure	30–55 bar	15–25 bar
Temperature at compression	973–1173 K	673–873 K
Lambda during operation	$0 \leq \lambda < \infty$	$0.8 \leq \lambda < 1.2$
Exhaust gas oxygen content	lean	lean to rich

admission, the two most important designs are indirect injection (IDI) engines where the fuel is injected in a precombustion or a swirl chamber connected to the main cylinder, and direct injection (DI) engines where the fuel is injected directly into the cylinder. With respect to air admission, a distinction can be made between natural aspirated (NA) engines in which the air is sucked into the engine, and turbocharged (TC) engines where the air is compressed before it enters the cylinder. The latter is in some cases combined with a cooling of the compressed air (TCI).

Present day diesel engines are in most cases also equipped with an exhaust gas recirculation device (EGR), which causes part of the exhaust gas to be returned to the combustion cylinder. Obviously, the composition and the temperature of the exhaust gas depends upon the design of the engine.

The exhaust gas of diesel engines has a complex composition as gaseous components are present together with liquid and even with solid components (Table 23). The solid exhaust gas components are denoted particulate matter, and defined as any matter that can be collected on a teflon-coated filter paper from diluted exhaust gas at a temperature below 325 K. Scanning and transmission electron microscopic pictures of such particulates are shown in Fig. 95.

The complex composition of the particulate matter as shown in Fig. 96. This composition is the result of a series of chemical and physical processes, that take

Table 23. Overview of the kind of emissions from diesel engines.

Gaseous	Liquid	Solid
N_2	H_2O	Soot
CO_2	H_2SO_4	Metals
CO	Hydrocarbons (C_{15}–C_{40})	Inorganic oxides
H_2	Oxygenates	Sulfates
NO/NO_2	Polyaromatic hydrocarbons	Solid hydrocarbons
SO_2/SO_3		
Hydrocarbons (C_1–C_{15})		
Oxygenates		
Organic nitrogen and sulfur compounds		

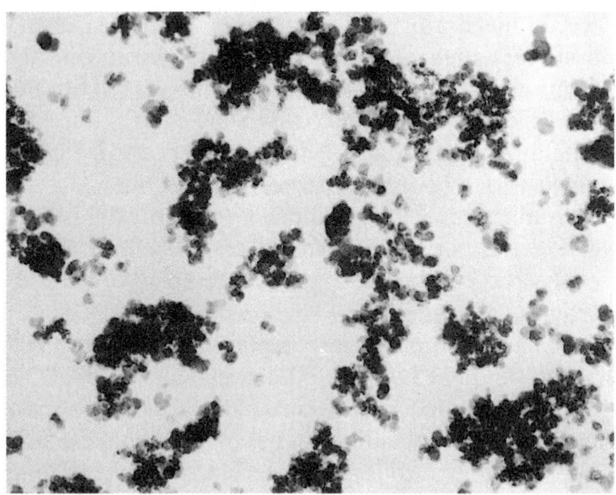

Figure 95. Scanning electron microscopic (top, magnification 20 000×) and transmission electron microscopic picture (bottom, magnification 100 000×) of particulates emitted by diesel engines.

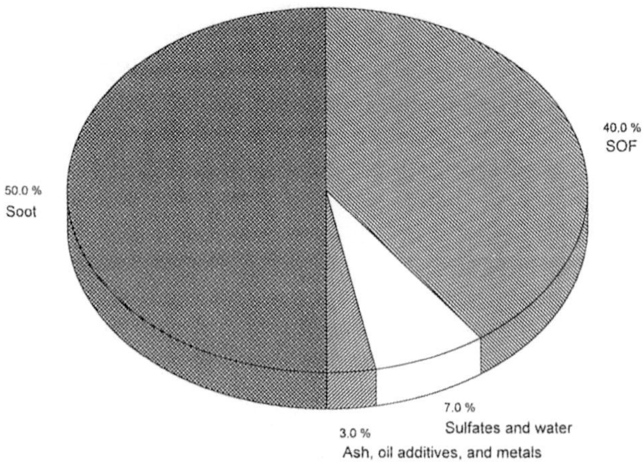

Figure 96. Typical composition of the particulates emitted by a midsize diesel engine at partial load using diesel fuel with 0.05 wt % sulfur.

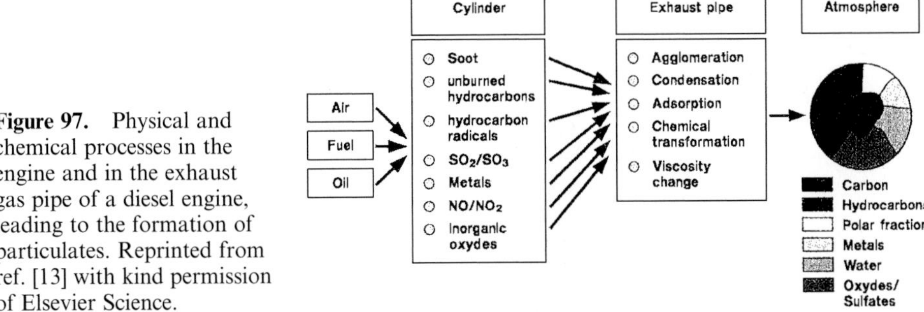

Figure 97. Physical and chemical processes in the engine and in the exhaust gas pipe of a diesel engine, leading to the formation of particulates. Reprinted from ref. [13] with kind permission of Elsevier Science.

place both in the cylinder and in the exhaust gas pipe (Fig. 97). The properties of the carbon nucleus are the results of a series of transformations that find their origin in the incomplete combustion of the diesel fuel molecules, because of the low rate of oxygen diffusion into the diesel fuel droplets injected in the engine (Fig. 98).

It is believed, that the suspected carcinogenic properties of the particulate matter is related to some kind of polynuclear aromatic hydrocarbon (PAH), adsorbed on the carbon nucleus; nitro PAH and the carboxy PAH have been mentioned in this respect.

The composition of the exhaust gas, as well as the physical (e.g. dimensions) and chemical (e.g. composition) properties of the particulate matter depend upon the engine operation conditions (Fig. 99). At all engine operation conditions there is excess oxygen, so that the exhaust gas composition is net oxidizing. Also, because of the high A/F-ratio, the temperature of the exhaust gas is generally lower than for closed-loop-controlled spark ignition engines.

As described in Section 1.2.4.B, several technologies have been considered to reduce the amount of particulate matter in the tailpipe exhaust gas of diesel engines. Of these technologies, only the diesel oxidation catalyst has found widespread application. This technology will be described below.

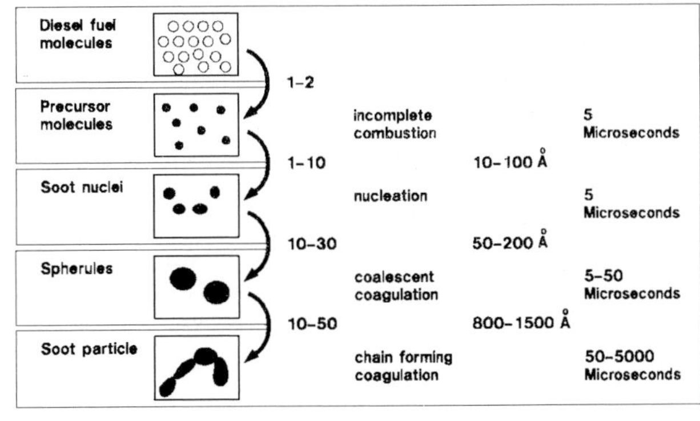

Figure 98. Physical and chemical transformations of diesel fuel molecules in the engine, leading to the formation of the carbon nucleus of a diesel particulate. Reprinted from ref. [13] with kind permission of Elsevier Science.

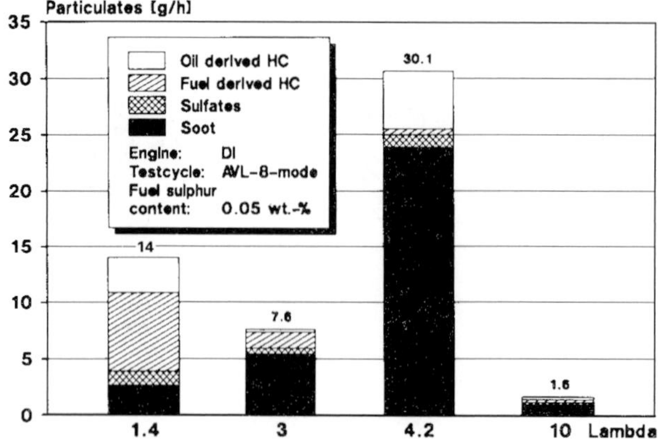

Figure 99. Amount and composition of particulates in the exhaust gas of a heavy duty DI diesel engine at various engine operation conditions in the AVL-8-mode engine test cycle. Reprinted with permission from ref. [68], © 1991 Society of Automotive Engineers, Inc.

A particularity with diesel engines is that the amount of particulate matter emitted is to some extent inversely proportional to the amount of nitrogen oxides emitted. This means, for example, that design modifications that lead to a decrease in the engine out emission of particulate matter, will increase the engine out emission of nitrogen oxides, and vice versa. Therefore, another strategy to simultaneously meet the legislative requirements with respect to the amount of CO, HC, NO_x and particulate matter emitted is to combine an engine design that guarantees a low emission of particulate matter with an exhaust gas catalyst that is able to reduce the amount of nitrogen oxides. Figure 100 gives a schematic representation of this strategy [62–64].

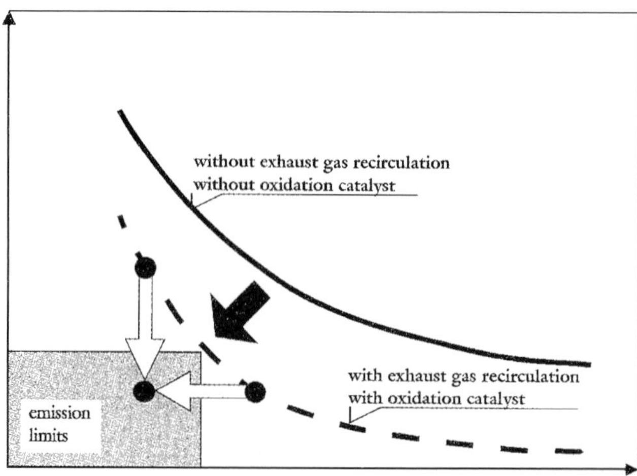

Figure 100. Schematic representation of the relationship between the amount of particulates (x-axis) and the amount of nitrogen oxides (y-axis) emitted by a diesel engine, and visualization of the various strategies to simultaneously meet the corresponding emission limits.

1.4.2 Oxidation Catalysts

Since about 1991, diesel oxidation catalysts have been generally applied to passenger cars in the European Union and to some medium and heavy duty trucks in the USA. Their principle of operation is shown in Fig. 101. The amount of carbon monoxide, hydrocarbons and aldehydes is reduced by oxidation of these components to carbon dioxide and water. The mass of particulate matter emitted is reduced by the oxidation of the liquid hydrocarbons, which are adsorbed on the particulates. These liquid hydrocarbons originate both from the fuel and the lubricating oil, and are commonly denoted as the soluble organic fraction (SOF). The adsorbed polynuclear aromatic hydrocarbons are also removed by oxidation.

The key is to design the catalyst in such a way, that it selectively catalyzes the above mentioned oxidation reactions, and that it neither catalyzes the oxidation of NO to NO_2, nor the oxidation of SO_2 to SO_3. The latter reaction needs to be suppressed because, irrespective of environmental concerns and catalyst durability issues, any SO_3 formed on the catalyst would react with water and condense as sulfuric acid during the sampling of the particulate matter, therefore increasing the mass of particulates, by their definition.

Just as with three-way catalysts, diesel catalysts consist of a ceramic or a metallic monolithic support, on which a special washcoat is deposited, which bears the precious metal component. The supports used have a cell density of typically 62 cells cm^{-2} or lower. Lower cell densities, such as 47 cells cm^{-2} and 31 cells cm^{-2}, are sometimes preferred because their correspondingly wider channels are of advantage in avoiding plugging of the support by particulate matter even during prolonged operation at an exhaust gas temperature below about 420 K. The ratio of catalyst volume to engine displacement is generally somewhat lower as compared to catalysts for spark ignition engines. So, the diesel catalysts are generally operated at a higher space velocity.

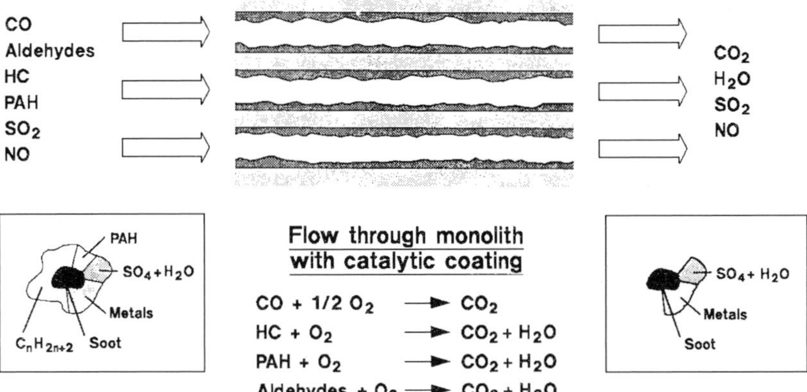

Figure 101. Operation principle of a diesel oxidation catalyst. Reprinted from ref. [13] with kind permission of Elsevier Science.

Table 24. Influence of the operating conditions of a diesel engine on the reaction conditions over a diesel catalysts (IDI/TC diesel engine with four cylinders and an engine displacement of 1.9 l; ceramic monolithic catalyst with 62 cells cm^{-2}).

Operating conditions						
Torque (Nm)	22	36	43	90	93	140
Rotational speed (rpm)	4000	2000	1100	2000	4000	4000
Catalyst inlet temperature (K)	533	448	438	593	773	973
Exhaust gas flow ($kg\,h^{-1}$)	295	134	71	157	303	308
Space velocity ($Nl\,l^{-1}\,h^{-1}$)	277 801	126 187	66 861	147 847	285 334	290 043
Linear velocity ($m\,s^{-1}$)	14.88	5.68	2.94	8.81	22.16	28.36
Residence time (s^{-1})	0.0050	0.0131	0.0254	0.0085	0.0034	0.0026
Gas composition						
CO (vol ppm)	181	132	177	120	149	300
HC (vol ppm)	6.4	9	6.8	5.6	10	2.5
NO_x (vol ppm)	220	170	180	280	410	500
O_2 (vol %)	14.9	15.3	14.2	11.7	8.6	3.4
SO_2 (vol ppm)	28	24	22	41	55	74
CO_2 (vol %)	3.85	3.5	4.4	6.33	8.64	13
λ-value	3.28	3.51	2.95	2.17	1.64	1.18
Dimensionless quantities						
Reynolds	385	197	106	190	309	273
Nusselt	3.23	1.65	0.89	1.60	2.60	2.30

As shown in Table 24, the Reynolds and the Nusselt numbers inside the monolith channels are higher than for spark ignition engines, although laminar flow prevails at all engine operation conditions. As shown in Fig. 102, an increase in the ratio between the catalyst volume and the engine displacement ensures a higher conversion of the gaseous exhaust gas components, but has practically no influence on the conversion of particulate matter.

The washcoat composition is quite different to that used for three-way catalysts. The washcoat oxides used are chosen so as to ensure a minimal adsorption both of the soluble organic fraction and of the sulfur oxides SO_2 and SO_3. The properties of three different oxides in these respects are shown in Figs. 103 and 104. The composition of the washcoat is, together with the choice of precious metal formulation,

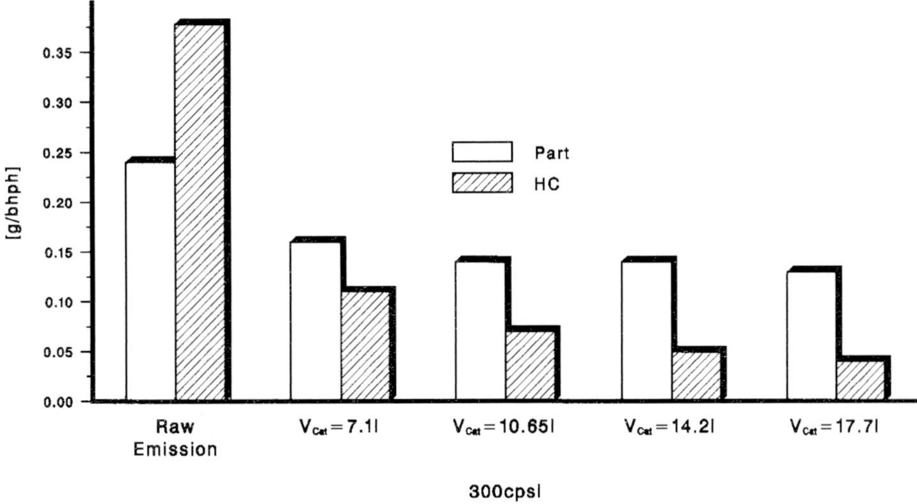

Figure 102. Emission of particulates (part) and gaseous hydrocarbons (HC) from a heavy duty diesel engine in the US transient engine test cycle, as a function of the catalyst volume (monolith catalyst with 46 cells cm^{-2}; diesel oxidation catalyst formulation with platinum at a loading of 1.76 g l^{-1}, fresh).

the key factor in minimizing the oxidation of SO_2. Figure 105, shows the oxidation performance for CO, HC and SO_2 for two catalysts with the same precious metals composition, but with a different washcoat formulation [65–67].

Finally, in typical diesel catalysts platinum or palladium are used. The perfor-

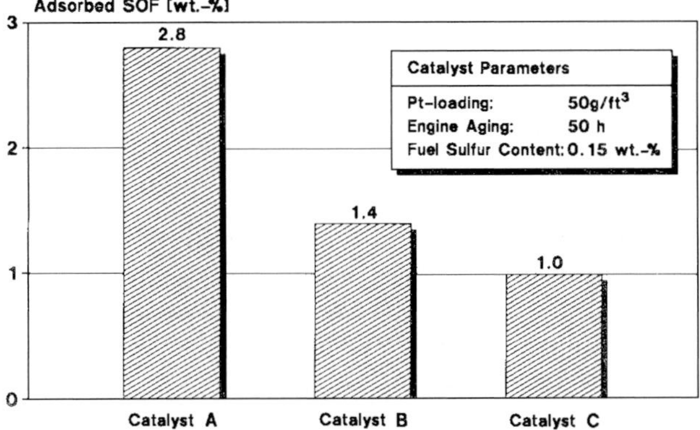

Figure 103. The amount of soluble organic fraction (SOF) adsorbed on an aged diesel oxidation catalyst, as a function of the washcoat formulation (monolith catalyst with 62 cells cm^{-2}, dedicated diesel washcoat formulations with platinum at a loading of 1.76 g l^{-1}; diesel engine bench aging for 50 h; diesel fuel containing 0.15 wt. % sulfur). Reprinted with permission from ref. [68], © 1991 Society of Automotive Engineers, Inc.

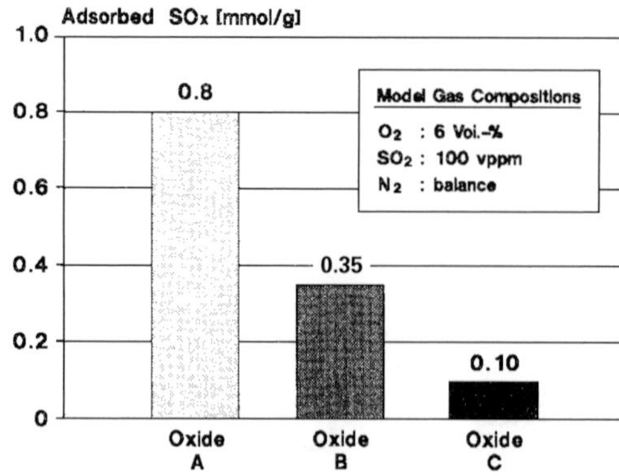

Figure 104. The amount of sulfur oxides adsorbed on diesel oxidation catalysts in a model gas reactor test, as a function of the washcoat formulation (monolith catalyst with 62 cells cm^{-2}; dedicated diesel washcoat formulations with platinum at a loading of 1.76 $g\,l^{-1}$). Reprinted with permission from ref. [68], © 1991 Society of Automotive Engineers, Inc.

mance of Pt, Pd and Rh at equimolar loading, for the conversion of the exhaust gas components CO, HC and SO_2, reveals that Pt is the most active for all three oxidation reactions, followed by Pd and Rh (Fig. 106). Taking the precious metal cost and availability difference into consideration explains why rhodium is rarely used in diesel catalysts [68].

For current legislation the loading of Pt or Pd is between about 0.35 $g\,l^{-1}$ and about 1.76 $g\,l^{-1}$. Again, conversion of the gaseous components is typically more

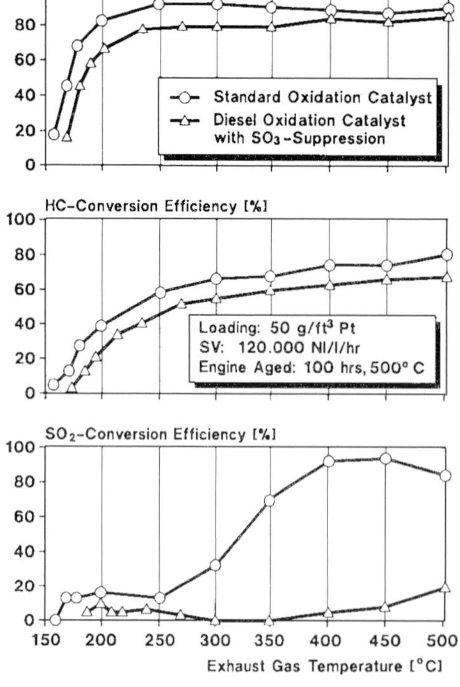

Figure 105. Conversion of carbon monoxide, gaseous hydrocarbons and sulfur dioxide reached over a diesel catalyst with and without measures to suppress the formation of sulfates, as a function of the exhaust gas temperature (monolith catalyst with 62 cells cm^{-2}; dedicated diesel washcoat formulations with platinum at a loading of 1.76 $g\,l^{-1}$; diesel engine test bench; light-off test at a space velocity of 120 000 Nl $l^{-1}\,h^{-1}$; diesel engine bench aging procedure for 100 h at a catalyst inlet temperature of 773 K).

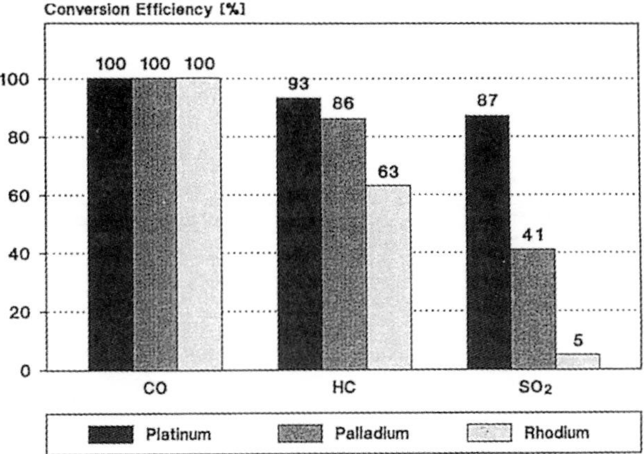

Figure 106. Conversion of carbon monoxide, gaseous hydrocarbons and sulfur dioxide reached over diesel oxidation catalysts as a function of the precious metal formulation at equimolar loading (monolith catalyst with 62 cells cm^{-2}; dedicated diesel washcoat formulations with platinum, palladium and rhodium at an equimolar loading of 8 mmol l^{-1}, fresh; model gas test at a gas temperature at catalyst inlet of 723 K; a space velocity 50 000 Nl l^{-1} h^{-1}; model gas simulates the exhaust gas composition of an IDI passenger car diesel engine at medium load and speed and contains 100 vol. ppm SO_2). Reprinted with permission from ref. [68], © 1991 Society of Automotive Engineers, Inc.

affected by the precious metals content, as is conversion of the soluble organic fraction (Fig. 107).

The performance of diesel catalysts over their lifetime is also subject to changes, caused by thermal and chemical deactivation phenomena. The durability of diesel oxidation catalysts is, however, less affected by high temperature deactivation effects, because the exhaust gas temperature is lower than for spark ignition engines, and because there is practically no exotherm generated during the oxidation reactions. Also, because of the diesel combustion principle, it is practically impossible that large amounts of unburned fuel reach the catalyst, contrary to what can happen with three-way catalysts. So, although the high temperature stability of typical diesel catalyst washcoat oxides is inferior to that of typical three-way washcoat oxides, temperature effects are less important in the diesel catalyst deactivation process (Fig. 108).

A more important deactivation aspect is poisoning. A diesel oxidation catalyst is poisoned by the same elements that poison three-way catalysts, except for lead, which is absent from diesel fuel and the diesel fuel supply chain. Table 25 compares the amount of sulfur, phosphorus and zinc offered to a diesel catalyst and to a three-way catalyst during their lifetime. From this table, it is apparent that a diesel oxidation catalyst has to deal with a considerably higher amount of sulfur during its lifetime than is the case with a three-way catalyst. This is because the diesel fuel specification allows for a higher sulfur content than the gasoline specification. Table 26 gives an overview of the maximum sulfur content in the diesel fuel specification of some selected countries.

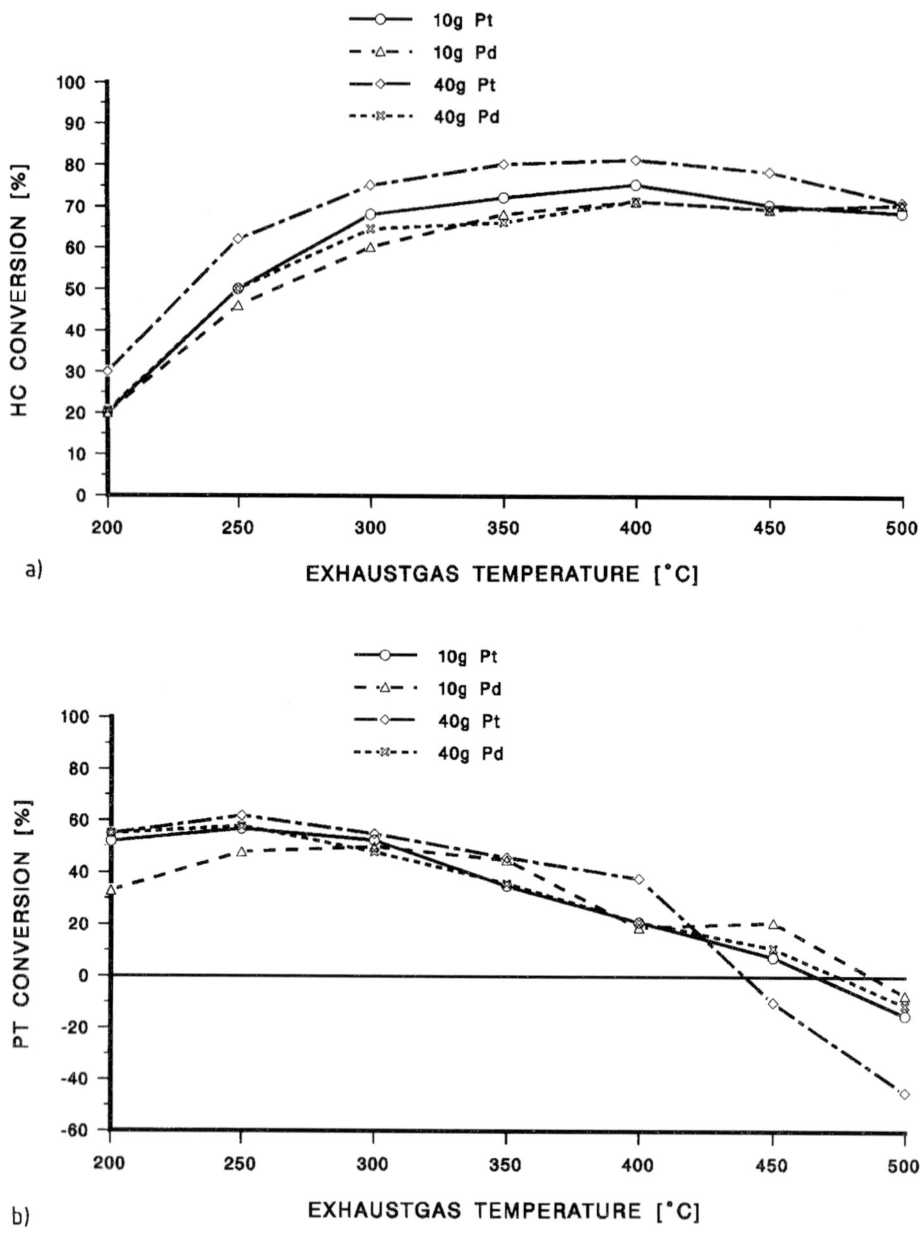

Figure 107. Conversion of (a) gaseous hydrocarbons (b) total particulate (PT) matter and (c) the soluble organic fraction (SOF) as a function of the exhaust gas temperature, reached over diesel catalysts based on platinum or on palladium (monolith catalyst with 62 cells cm^{-2}; dedicated diesel washcoat formulations with platinum and palladium both at a loading of 0.35 g l^{-1} and 1.41 g l^{-1} in the fresh state; heavy duty diesel engine bench light-off test at an average engine speed setting and various engine load settings).

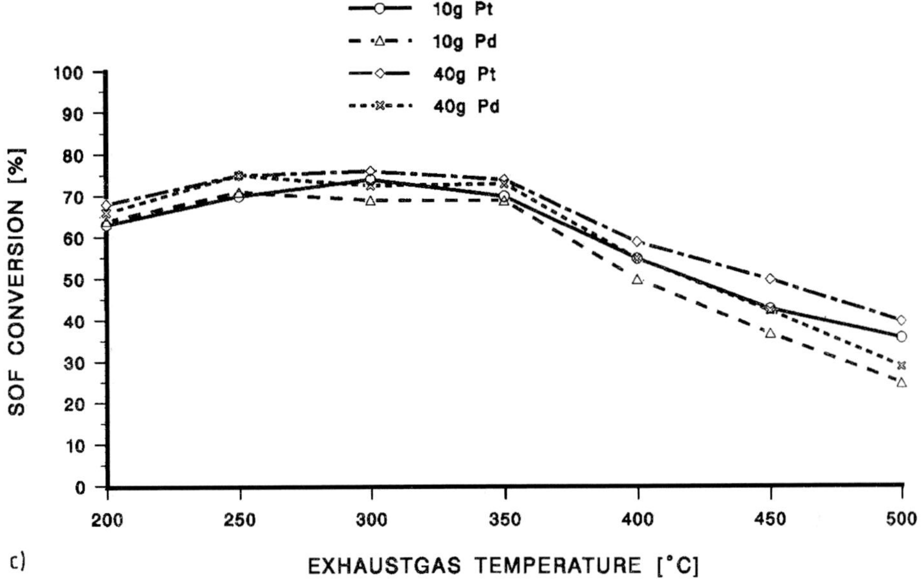

Figure 107 (continued)

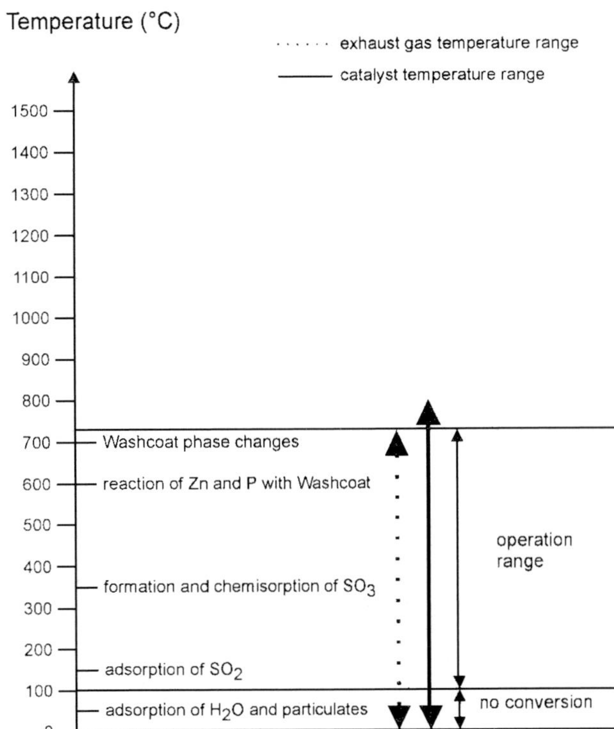

Figure 108. Deactivation phenomena of diesel oxidation catalysts as a function of the catalyst temperature.

Table 25. Order of magnitude of lifetime effects: comparison between a diesel oxidation catalyst and a three-way catalyst.

Component	Diesel catalyst	Three-way catalyst
Components to be converted over 100 000 miles		
Carbon monoxide (kg)	127	1671
Hydrocarbons (kg)	26	210
Nitrogen oxides (kg)	(100)[a]	424
Particulates (kg)	33.5	(2.5)[a]
Catalyst formulation		
Washcoat ($g\,l^{-1}_{engine}$)	60	150
Precious metals ($g\,l^{-1}_{engine}$)	0.9	1.72
Poisons offered over 100 000 miles		
Sulfur ($g\,kg^{-1}_{cat}$)	9600–36000	6400–20000
Phosphorus ($g\,kg^{-1}_{cat}$)	120	80–200
Zinc ($g\,kg^{-1}_{cat}$)	260	320–480
Lead ($g\,kg^{-1}_{cat}$)	0	40–520

[a] No conversion expected.

The presence of SO_2 in the diesel engine exhaust gas already suppresses the activity of a diesel oxidation catalyst at an exhaust gas temperature below about 570 K, especially for the conversion of the gaseous hydrocarbons (Fig. 109). During aging, sulfur is stored on the diesel catalyst. Depending on the catalyst formulation and the engine or vehicle aging conditions, typically up to about 5% of the sulfur offered to the catalyst is found back on the catalyst. The sulfur is almost homogeneously distributed over the length of the catalyst and within the depth of the washcoat layer. XPS studies reveal that several sulfur species coexist on the aged catalyst. Sulfite species, which originate from the reaction between adsorbed SO_2 and water, sulfate and thiosulfate species are detected [69].

The sulfate species originate from adsorbed SO_3, which in turn is formed by the oxidation of SO_2 over the catalyst. The thiosulfate species most probably originate from the reduction of adsorbed sulfur dioxide by adsorbed hydrocarbons or even by adsorbed carbon.

Phosphorus is also found back in the washcoat, to a concentration that corresponds to typically about more than 50% of the amount of phosphorus offered to the catalyst. However, the concentration profile of phosphorus over the length of

Table 26. Overview of the sulfur content (wt.%) of diesel fuel for various countries in 1994; average value and maximum valued allowed by the respective specifications.

Value	Germany (DIN 5 160 1)	Sweden	USA	Australia	Japan	South America
Maximum	0.2	0.3	0.05	0.4	0.2	1.0
Mean	0.15	0.003	0.03	0.18	0.1	0.4

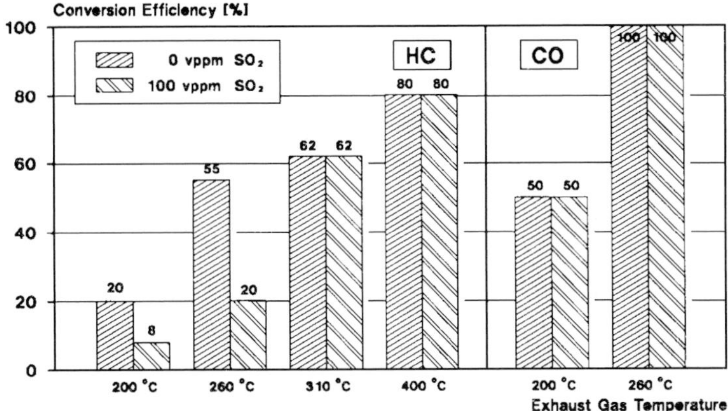

Figure 109. Conversion of carbon monoxide and gaseous hydrocarbons reached over a diesel oxidation catalyst at various settings of the exhaust gas temperature, for a model gas composition with and without SO_2 (monolith catalyst with 62 cells cm^{-2}; dedicated diesel washcoat formulations with platinum loading of 1.76 $g\,l^{-1}$ in the fresh state; model gas light-off test at a space velocity of 50 000 $Nl\,l^{-1}\,h^{-1}$; model gas simulates the exhaust gas composition of an IDI passenger car diesel engine at medium load and speed).

the catalyst and over the depth of the washcoat layer is quite different to that for sulfur. The majority of phosphorus is typically found back at the inlet of the aged catalyst and at the interface between the washcoat layer and the gas phase. The phosphorus is present as phosphate, as identified by XPS. Zinc is also found on the aged catalyst. However, it is not present in the washcoat layer, as is the case with three-way catalysts, but only inside the diesel particulates which are adsorbed on the outer surface of the washcoat layer. Therefore it can be assumed, that zinc has only a limited influence on the deactivation of the catalyst.

Finally, an aged diesel oxidation catalyst always contains carbon, typically about 0.2 wt%. This carbon has to be associated with the fuel and lube oil hydrocarbons adsorbed inside the washcoat, as well as with the diesel particulate matter adsorbed on the top of the washcoat layer. As shown in Table 27, a diesel oxidation catalyst

Table 27. Amount of sulfur, phosphorus and carbon on a diesel catalyst after aging on a diesel engine bench, as a function of the platinum loading, and of the composition of the fuel and the engine lubricating oil. (Reprinted with permission from ref. [69], © 1992 Society of Automotive Engineers, Inc.)

Precious metal content ($g\,ft^{-3}$)	Fuel sulfur content (wt.%)	Sulfur offered ($mg\,m^{-2}$)	Lube oil phosphorus content (wt.%)	Phosphorus offered ($mg\,m^{-2}$)	Catalyst sulfur content ($mg\,m^{-2}$)	Catalyst phosphorus content ($mg\,m^{-2}$)	Catalyst carbon content ($mg\,m^{-2}$)
Pt, 50	0.15	15	0.1	0.08	0.6	0.05	0.21
Pt, 50	0.04	4	0.1	0.08	0.25	0.05	0.21
Pt, 50	0.15	15	0.05	0.04	0.45	0.03	0.21
none	0.15	15	0.1	0.08	0.20	0.05	0.55

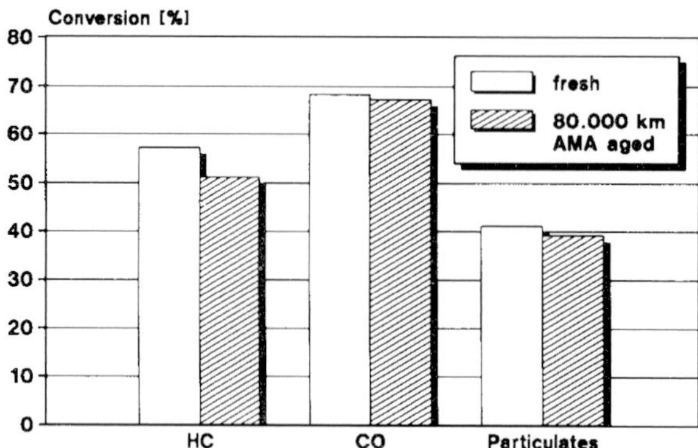

Figure 110. Conversion of carbon monoxide, gaseous hydrocarbons and particulate matter reached over a fresh and a vehicle aged diesel catalyst, in the US-FTP 75 vehicle test procedure (monolith catalyst with 62 cells cm^{-2}; dedicated diesel catalyst formulations with a platinum loading of 1.76 g l^{-1} fresh and vehicle aged; vehicle dynamometer test according to the US-FTP 75 test procedure, with a four cylinder IDI/TC diesel engine of displacement 1.9 l; vehicle aging for 80 000 km according to the AMA cycle with a passenger car equipped with a four cylinder IDI/TC diesel engine, of displacement 1.9 l; sulfur content of the diesel fuel used in aging and testing 0.15 wt. %). Reprinted with permission from ref. [69], © 1992 Society of Automotive Engineers, Inc.

without precious metals exhibits a much higher carbon content than one with precious metals.

The carbon-containing species do not have a chemical poisoning effect as such, but they can cause an enhancement of the intraphase diffusion resistance, for example by narrowing the washcoat pores. The data presented above explain that the performance of diesel oxidation catalysts is generally less affected by deactivation phenomena than the performance of three-way catalysts. Figure 110 shows the conversion for CO, HC and particulate matter reached for a fresh and a 80 000 km vehicle aged diesel oxidation catalyst mounted on a passenger car, equipped with a 1.9 l IDI/TC diesel engine [69]. Table 28 gives a typical breakdown of the conversion in the different phases of the legislated passenger car test cycles, for fresh and engine aged catalysts.

In addition to the exhaust gas components legislated for, the diesel oxidation catalyst also reduces the emissions of exhaust gas components such as aldehydes and polynuclear aromatic hydrocarbons, see Fig. 111 [70].

Likewise, Fig. 112 shows the conversion efficiency for diesel particulate matter reached in the US transient cycle with a diesel catalyst mounted on a heavy duty engine. Table 29 shows the conversion for CO, HC and particulate matter in the European 13-mode test performed with a heavy duty engine for a fresh and for an engine-aged catalyst. The catalyst performance at operating conditions, reflecting the lowest and the highest exhaust gas temperature in this 13-mode test, is also reported.

Table 28. Emission of carbon monoxide, gaseous hydrocarbons, nitrogen oxides and particulate matter from a passenger car equipped with an IDI/NA diesel engine, and conversion over a diesel catalyst in the fresh and the engine aged state, in the different phases of the US-FTP 75 vehicle test procedure and of the European MVEG-A vehicle test procedure.

Parameter	Phase 1	Phase 2	Phase 3	Total
US-FTP 75				
Raw emission	0.65	0.56	0.42	0.54
CO ($g\,mile^{-1}$)	0.12	0.10	0.06	0.09
HC ($g\,mile^{-1}$)	0.78	0.65	0.64	0.67
NO_x ($g\,mile^{-1}$)	0.15	0.16	0.14	0.15
Particulates ($g\,mile^{-1}$)				
Fresh catalyst conversion				
CO (%)	69	87	70	80
HC (%)	60	68	59	62
Particulates (%)	46	30	8	28
Aged catalyst conversion				
CO (%)	68	87	71	79
HC (%)	59	64	60	62
Particulates (%)	20	31	18	26
MVEG-A				
Raw emission				
CO ($g\,km^{-1}$)	1.42	0.83	0.24	0.52
HC ($g\,km^{-1}$)	0.31	0.17	0.06	0.11
NO_x ($g\,km^{-1}$)	1.04	1.01	0.60	0.74
Particulates ($g\,km^{-1}$)	0.08	0.07	0.06	0.06
Fresh catalyst conversion				
CO (%)	20	88	88	46
HC (%)	44	33	33	40
Particulates (%)	35	38	38	37
Aged catalyst conversion				
CO (%)	20	87	87	45
HC (%)	24	42	42	31
Particulates (%)	47	40	40	38

1.4.3 NO_x-Reduction Catalysts

Whereas the diesel oxidation catalyst became the accepted technology to limit the emission of particulate matter, carbon monoxide and gaseous hydrocarbons, the major remaining challenge is the removal of NO_x.

One possibility is to lower the engine-out emission of NO_x using exhaust gas recirculation (EGR) systems; however, these can increase the engine-out emission of particulate matter.

Conventional diesel oxidation catalysts or three-way catalysts are unable to convert NO_x in the net oxidizing exhaust gas of diesel engines. One way to reduce NO_x emission by catalytic aftertreatment is to use the selective catalytic reduction (SCR)

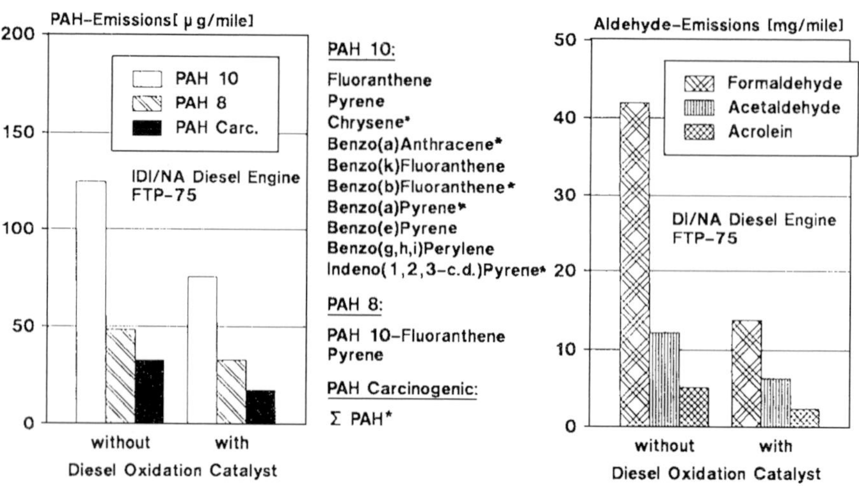

Figure 111. Emission of aldehydes, acrolein and various polynuclear aromatic hydrocarbons of two passenger cars equipped with an IDI/NA and with a DI/NA diesel engine, once without and once with a diesel oxidation catalyst, in the US-FTP 75 vehicle test cycle (monolith catalyst with 62 cells cm^{-2}; dedicated diesel washcoat formulation with a platinum loading of 1.76 g l^{-1} in the fresh state; vehicle dynamometer tests according to the US-FTP 75 vehicle test procedure, with passenger cars equipped with a DI/NA and with an IDI/NA diesel engine of displacement 2.0 l). Reprinted with permission from ref. [70], © 1990 Society of Automotive Engineers, Inc.

process. However, as this process needs ammonia or a substance that forms in situ ammonia as the reduction agent, it is not likely to be applied to on-road vehicles, especially passenger cars. Thermodynamics show that nitrogen oxide is unstable with respect to N_2 and O_2 in the exhaust gas composition and at the exhaust gas temperatures that occur at typical diesel engine operating conditions.

Both the direct decomposition of NO,

$$NO \rightleftharpoons 0.5N_2 + 0.5O_2 \tag{44}$$

as well as the reduction by hydrocarbons and alcohols,

$$(6m + 2)NO + 2C_mH_{2m+2} \rightleftharpoons (3m + 1)N_2 + 2mCO_2 + (2m + 2)H_2O \tag{45}$$

and

$$6mNO + 2C_mH_{2m+1}OH \rightleftharpoons 3mN_2 + 2mCO_2 + (2m + 2)H_2O \tag{46}$$

are thermodynamically feasible.

To date, no heterogeneous catalyst has been found that can catalyze the direct decomposition within the boundary conditions (space velocity, durability) that apply to on-road vehicles. Therefore, the majority of development effort has been devoted to the catalysis of reactions between NO_x and hydrocarbons and/or alcohols. As the concentration of hydrocarbons and alcohols in the engine-out exhaust gas of diesel engines is too low to fulfill the stoichiometry of eqs 45 and 46, appropriate components need to be added to the exhaust gas. Promising results have been

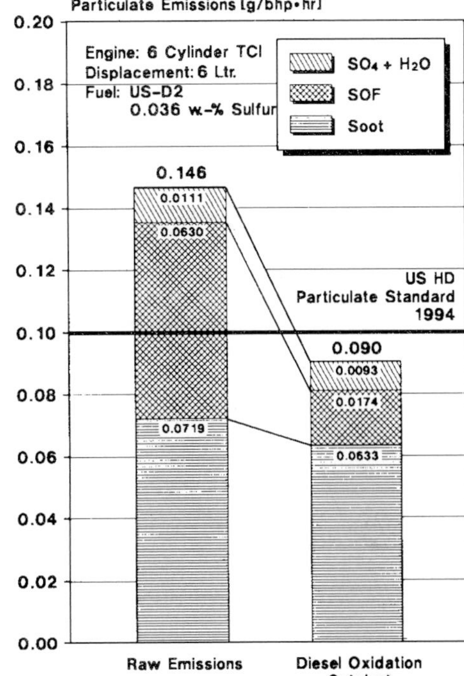

Figure 112. Emission of particulate matter of a heavy duty diesel engine without and with a diesel catalyst, in the US transient engine test cycle (monolith catalyst with 62 cells cm^{-2}; dedicated diesel catalyst formulation, in the fresh state; heavy duty DI/TCI diesel engine with six cylinders and an engine displacement of 6 l, in the US transient engine cycle on an engine dynamometer test with the US-D2 diesel fuel containing 0.036 wt % sulfur).

obtained with catalysts based upon zeolites, together with elements from Groups Ib and VIII. Figure 113 shows the HC and NO_x conversion obtained in model gas reactor experiments with a series of air-aged catalysts.

Table 29. Emission of carbon monoxide, gaseous hydrocarbons, nitrogen oxides and particulate matter of a heavy duty diesel engine, and conversion over a fresh and an engine aged diesel oxidation catalyst, in the European 13-mode engine test procedure.

	Total in test	Mode 3	Mode 6
Raw emission			
CO	0.56 g kWh^{-1}	18.5 g h^{-1}	35.4 g h^{-1}
HC	0.09 g kWh^{-1}	4.9 g h^{-1}	4.3 g h^{-1}
NO_x	0.57 g kWh^{-1}	12.7 g h^{-1}	41.3 g h^{-1}
Particulates	0.147 g kWh^{-1}	3.5 g h^{-1}	11.1 g h^{-1}
Fresh catalyst conversion			
CO (%)	80	86	87
HC (%)	66	69	74
Particulates (%)	25	51	0
Aged catalyst conversion			
CO (%)	68	80	78
HC (%)	66	73	83
Particulates (%)	20	55	0

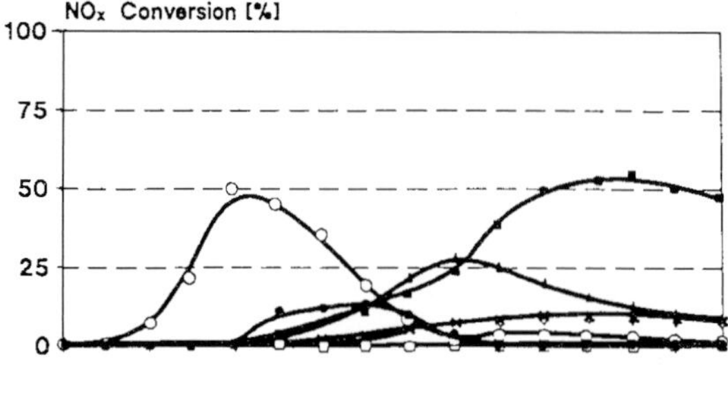

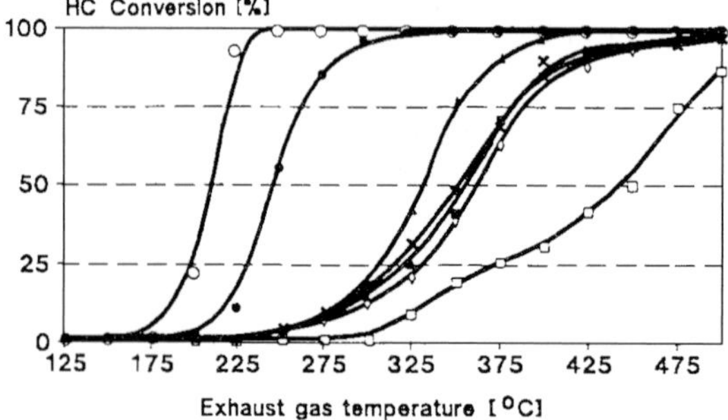

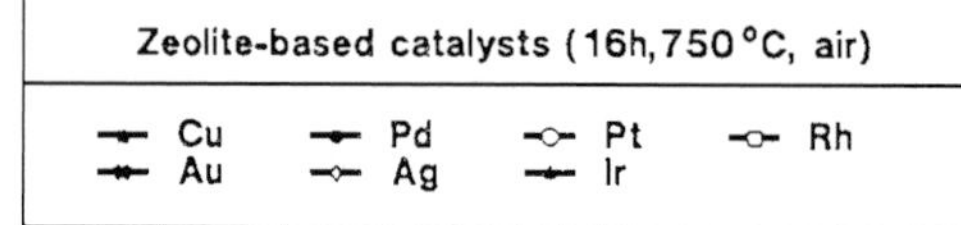

Figure 113. Conversion of nitrogen oxides and gaseous hydrocarbons reached over different NO_x-reduction catalyst formulations, as a function of the exhaust gas temperature (monolith catalyst with 62 cells cm^{-2}; dedicated NO_x-reduction catalyst formulations with zeolites and with different active components, after laboratory oven-aging in air at a temperature of 1023 K for 16 hours; light-off test in a model gas reactor at a space velocity of 50 000 Nl l^{-1} h^{-1}; model gas simulates the exhaust gas composition of an IDI/NA passenger car diesel engine at medium load and speed, except for the hydrocarbon concentration, which was increased to reach $y_{HC} : y_{NO_x}$ 3:1 (mol : mol)).

About 50% NO_x conversion is obtained both with a platinum-based catalyst at temperatures below about 570 K, and with a copper-based catalyst at temperatures above about 570 K. These temperatures coincide with the temperature range in which these two catalysts convert hydrocarbons. However, whereas the platinum-containing catalyst is able to convert carbon monoxide efficiently, the copper-containing catalyst generates carbon monoxide, probably by the incomplete oxidation of the added hydrocarbons [71].

Platinum based catalysts to reduce NO_x in the exhaust gas of diesel engines with hydrocarbon addition have been proved successfull under the dynamic conditions that occur during an activity test with a vehicle. As shown in Fig. 114, the degree of NO_x conversion depends primarily on the exhaust gas temperature during the different phases of the vehicle test. In addition to the exhaust gas temperature, the space velocity (Fig. 115), the NO_x concentration (Fig. 116), the hydrocarbon concentration (Fig. 117), and the type of hydrocarbon (Fig. 118) have a strong impact on the overall NO_x conversion [71, 72].

The detailed conversion mechanism is still not very well understood. Initial selectivity studies point to nitrogen as the primary product of the reaction and both N_2O

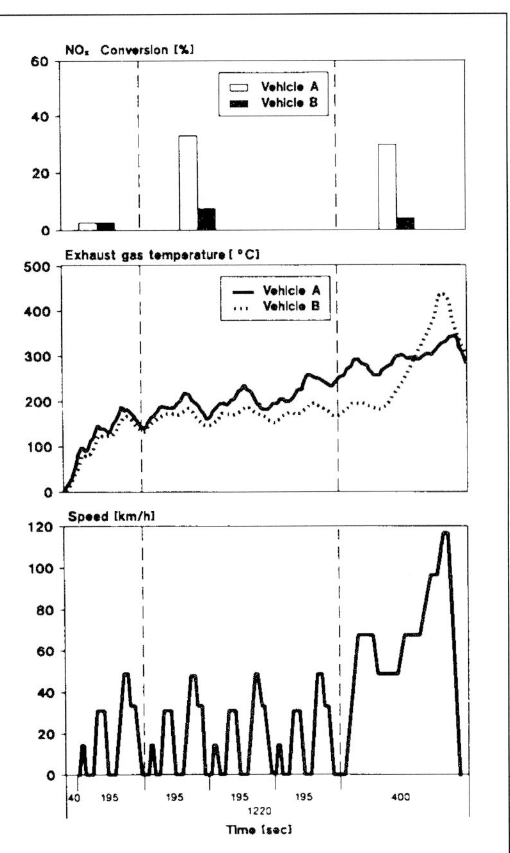

Figure 114. Conversion of nitrogen oxides reached over a NO_x-reduction catalyst mounted on two different passenger cars equipped with an IDI/NA and a DI/TC diesel engine in three different phases of the European vehicle test procedure. The exhaust gas temperature in front of the catalyst is also shown as a function of the time in the vehicle test procedure (monolith catalyst with 62 cells cm^{-2}; dedicated NO_x-reduction catalyst formulation containing platinum, in the fresh state; vehicle dynamometer test according to the European vehicle test procedure with passenger cars equipped either with a DI/TC diesel engine with four cylinders and an engine displacement of 2.5 l (vehicle A) or with an IDI/NA diesel engine with four cylinders and an engine displacement of 1.9 l (vehicle B). In both cases, a 2:1 molar mixture of 1-butene : *n*-butane was added to raise the concentration of gaseous hydrocarbons in the exhaust gas to 800 vol ppm). Reprinted with permission from ref. [71], © 1993 Society of Automotive Engineers, Inc.

and NO_2 as secondary products (Fig. 119). Together with information gathered from temporal analysis of products (TAP) experiments and from in situ catalyst characterization, a dual site reaction mechanism is proposed, a simplification of which is shown in Fig. 120 [72–74].

The first results on the durability of the diesel NO_x reduction catalyst show that engine aging decreases the maximum NO_x conversion and shifts the onset of the conversion to higher temperatures. This deactivation was found to be caused by phenomena similar to those identified for the deactivation of diesel oxidation catalysts.

1.5 Conclusion

The application of catalytic aftertreatment devices to passenger cars and trucks both with spark ignition and with compression ignition engines is an established

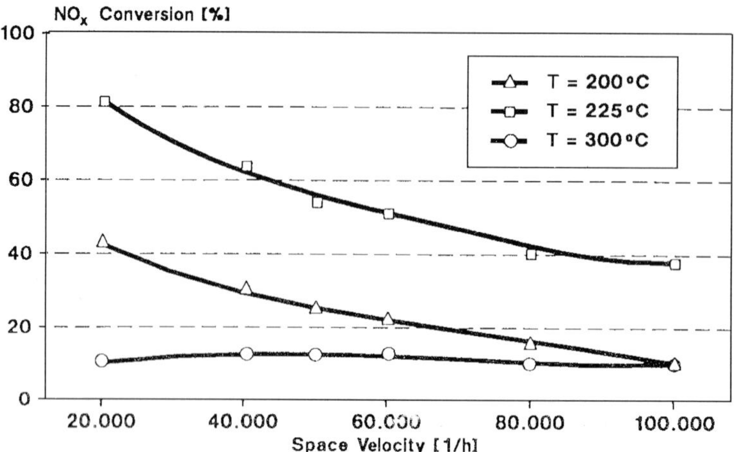

Figure 115. Conversion of nitrogen oxides reached over a NO_x reduction catalyst as a function of the space velocity, at three different settings of the exhaust gas temperature (monolith catalyst with 62 cells cm^{-2}, dedicated NO_x reduction catalyst formulation containing platinum, in the fresh state; light-off test in a model gas reactor, space velocity settings obtained by adjusting the length of the catalyst sample. Model gas simulates the composition of the exhaust gas of an IDI/NA passenger car diesel engine at medium load and speed, except for the hydrocarbon concentration which was raised to reach $y_{HC}:y_{NO_x}$ 3:1 (mol:mol)). Reprinted with permission from ref. [71], © 1993 Society of Automotive Engineers, Inc.

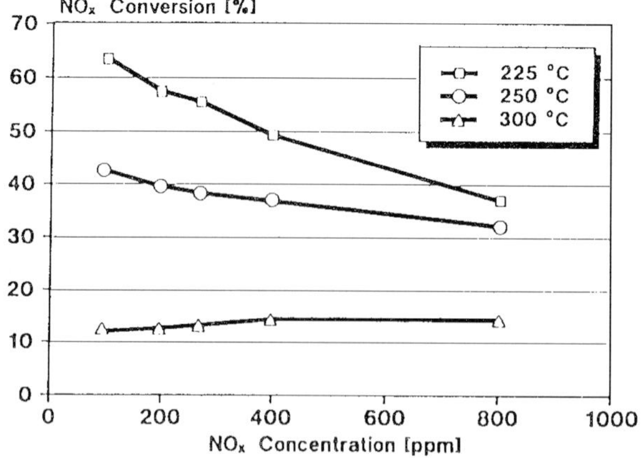

Figure 116. Conversion of nitrogen oxides reached over a NO_x reduction catalyst at three different settings of the exhaust gas temperature, as a function of the concentration of the nitrogen oxides in the exhaust gas (light-off test in a model gas reactor at a space velocity of 50 000 Nl l^{-1} h^{-1}. Model gas composition based upon the composition of the exhaust gas of an IDI/NA passenger car diesel engine at medium speed and load, but modified by adding *n*-hexadecane to reach a hydrocarbon concentration of 3200 vol.ppm C_1 and by varying the concentration of NO). Reprinted from ref. [72] with kind permission of Elsevier Science.

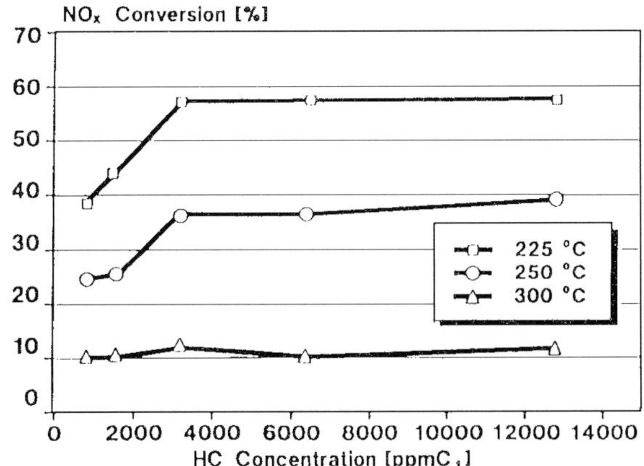

Figure 117. Conversion of nitrogen oxides over a NO_x reduction catalyst at a fixed inlet concentration of NO_x and at three different settings of the exhaust gas temperature, as a function of the concentration of the hydrocarbon component (catalyst and test as for Fig. 116 except that the NO_x concentration is fixed at 270 vol.ppm and that the concentration of *n*-hexadecane is varied). Reprinted from ref. [72] with kind permission of Elsevier Science.

method of achieving the increasingly demanding legislative emission limits for carbon monoxide, hydrocarbons, nitrogen oxides and particulate matter. Dedicated research and development effort by the automotive industry and by the catalyst industry has led, since the 1970s, to the rapid introduction of a broad variety of new and/or considerably improved catalysts to the market.

As a result of this effort, automotive catalysis has become one of the most inno-

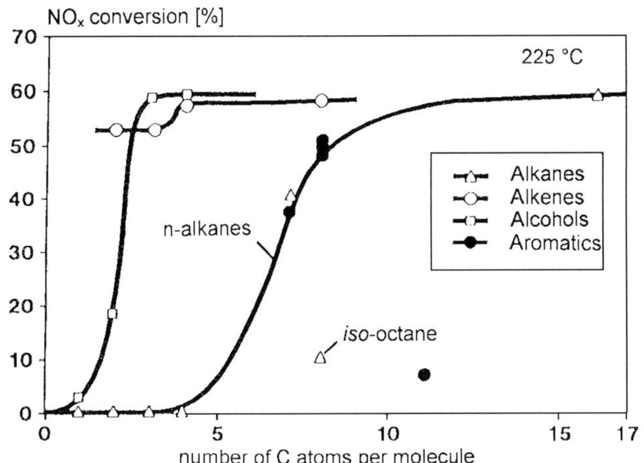

Figure 118. Conversion of nitrogen oxides reached over a NO_x reduction catalyst at 498 K for different classes of hydrocarbons and oxygenates added to the exhaust gas, as a function of the number of carbon atoms in the molecule of the organic component (catalyst and test as for Fig. 116 except that the NO_x concentration is fixed at 270 vol.ppm and that various hydrocarbons or oxygenates are added to a fixed concentration of 3200 vol.ppm C_1). Reprinted from ref. [72] with kind permission of Elsevier Science.

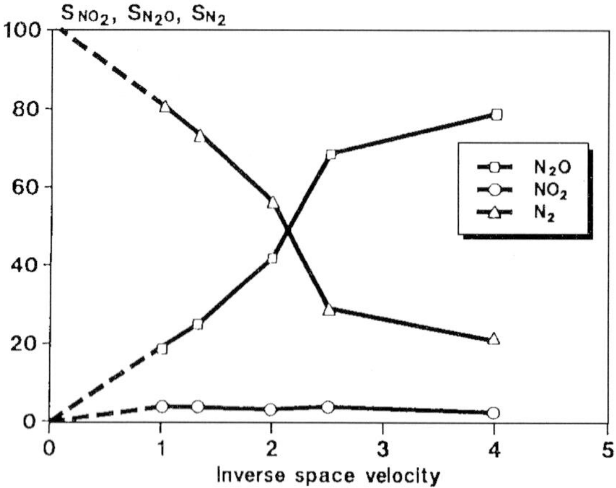

Figure 119. Selectivities for nitrogen dioxide S_{NO_2}, dinitrogen oxide S_{N_2O} and nitrogen S_{N_2}, reached over a NO_x reduction catalyst as a function of the inverse of the space velocity (monolith catalyst with 62 cells cm^{-2}; dedicated NO_x reduction formulation containing platinum, in the fresh state; test as for Fig. 116 except that the NO_x concentration is fixed at 270 vol.ppm. The various settings of the space velocity are reached by varying the length of the catalyst; the selectivity is defined as mol of the component formed per 100 mol of NO_x converted). Reprinted from ref. [72] with kind permission of Elsevier Science.

vative fields of catalytic science and one of the most important applications of heterogeneous catalysts. Despite its complexity and harsh environment, automotive catalysis has been shown to be a robust technology, that can perform over a broad range of operation conditions and applications [40, 75]. Finally, an increasing understanding of catalyst operation and deactivation is leading to new developments that have the potential for successfull application to new engine designs, or to further emission reductions for conventional engine designs.

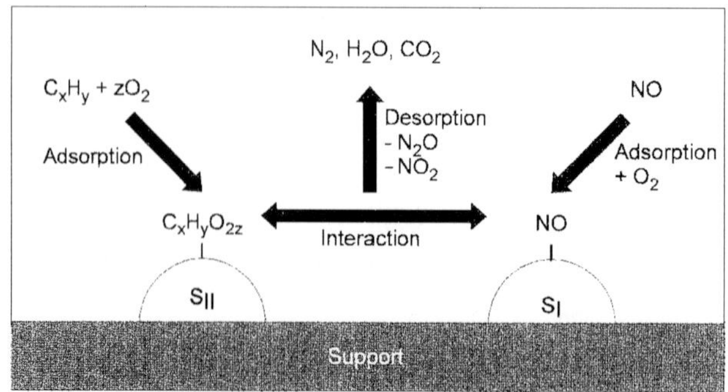

Figure 120. Proposed reaction mechanism for the reduction of nitrogen oxides in the exhaust gas of diesel engines over an NO_x reduction catalyst. Reprinted from ref. [72] with kind permission of Elsevier Science.

Acknowledgment

The authors wish to express their gratitude to all their collegues and coworkers at Degussa, who contributed to the vast amount of data presented here. Special recognition is due to Dr R. Domesle, Dr J. Leyrer, Dr D. Lindner, Dr A. Schäfer-Sindlinger, Dr R. van Yperen and Dr W. Strehlau for the invaluable scientific work and discussions, to Ms B. Schäfer and Dr J. Gieshoff for the editing of the manuscript, and Ms A. van den Bergh, Mr J. Emge and Mr T. Bickert for preparing the many figures.

Elsevier Science Publishers B.V., Sara Burgerhart straat 25, NL-1055 KV Amsterdam, and the Society of Automotive Engineers, Inc., are acknowledged for their kind permission to reproduce the figures quoted.

References

1. G. A. Ahrens et al., *Verkehrsbedingte Luft- und Lärmbelastungen*, Umweltbundesamt, Berlin, **1991**.
2. J. J. MacKenzie, M. P. Walsh, *Driving Forces*, World Resources Institute, Washington, **1990**.
3. A. Lowi, W. P. L. Carter, *Technical Paper Series 900710*, Society of Automotive Engineers, Warrendale, **1990**.
4. Association des Constructeurs Europeens d'Automobiles (ACEA), *Future Exhaust Emission Standards of Passenger Cars*, AE/71/91/VE/ACEA/2, Brussels, **1991**.
5. D. M. Heaton, R. C. Rijkeboer, P. van Sloten, *Proceedings Symposium Traffic Induced Air Pollution*, Graz, Austria, **1992**.
6. J. S. Mc Arragher et al., *Concawe Report 93/51*, CONCAWE, Brussels, **1993**.
7. J. T. Kummer, *Prog. Energy Combust. Sci.*, **1979**, *6*, 177–199.
8. J. E. McEvoy, *Catalysts for the Control of Automotive Pollutants*, American Chemical Society, Washington, **1975**.
9. K. C. Taylor, *Catalysis and Automotive Pollution Control* (Eds: A. Crucq, A. Frennet), Elsevier, Amsterdam, **1987**, p. 97,
10. F. Schäfer, R. van Basshuysen, *Die Verbrennungskraftmaschine* (Eds.: H. List, A. Pischinger), Springer Verlag, Wien, **1993**, vol. 7.
11. H. J. Foerster, *Automobiltechnische Zeitschrift*, **1991**, *93*, 342–352.
12. D. Hundertmark, *Automobil-Industrie*, **1990**, *1*.
13. E. S. Lox, B. H. Engler, E. Koberstein, *Catalysis and Automotive Pollution Control II* (Ed.: A. Crucq), Elsevier, Amsterdam, **1991**, p. 291.
14. J. Brettschneider, *Bosch Technische Berichte*, **1979**, *6*, 177–186.
15. A. J. L. Nievergeld, *Institute for Continuing Education*, Eindhoven University of Technology, **1994**.
16. S. H. Oh, J. C. Cavendish, L. L. Hegedus, *Chem. Eng. Prog.* **1980**, *26*, 935–943.
17. S. H. Oh, J. C. Cavendish, *AIChE J.* **1985**, *31*, 935–949.
18. A. Renken, *Int. Chem. Eng.* **1994**, *33*, 61–71.
19. C. N. Montreuil, S. C. Williams, A. A. Adamczyk, *Technical Paper Series 920096*, Society of Automotive Engineers, Warrendale, **1992**.
20. L. L. Hegedus, J. J. Gumbleton, *Chemtech* **1980**, *10*, 630–642.
21. P. Nortier, M. Soustelle in *Catalysis and Automotive Pollution Control* (Eds: A. Crucq and A. Frennet), Elsevier, Amsterdam, **1987**, p. 275.
22. J. C. Summers, L. Hegedus, US Patent 4152301, **1979**.
23. G. Kim, M. V. Ernest, S. R. Montgomery, *Industrial and Engineering Chemistry Product Research and Development*, **1984**, *23*, 525–531.

24. J. S. Howitt in *Catalysis and Automotive Pollution Control* (Eds: A. Crucq, A. Frennet), Elsevier, Amsterdam, **1987**, p. 301.
25. J. P. Day, L. S. Socha, *Technical Paper Series 881590*, Society of Automotive Engineers, Warrendale, **1988**.
26. J. Yamamoto, K. Kato, J. Kitagawa, M. Machida, *Technical Paper Series 910611*, Society of Automotive Engineers, Warrendale, **1991**.
27. W. Maus, H. Bode, A. Reck, *Technische Akademie Wupperthal*, **1989**.
28. M. Nonnenmann, *Automobiltechnische Zeitschrift* **1989**, *91*, 4.
29. W. B. Kolb, A. A. Papadimitriou, R. L. Cerro, D. D. Leavitt, J. C. Summers, *Chem. Eng. Prog.* **1993**, *2*, 61–67.
30. B. H. Engler, E. S. Lox, K. Ostgathe, T. Ohata, K. Tsuchitani, S. Ichihara, H. Onoda, G. T. Garr, D. Psaras, *Technical Paper Series 940928*, Society of Automotive Engineers, Warrendale, **1994**.
31. M. C. F. Steel in *Catalysis and Automotive Pollution Control II* (Ed: A. Crucq), Elsevier, Amsterdam, **1991**, p. 105.
32. C. Hagelücken, *Proceedings Second European Precious Metals Conference*, Lisboa, **1995**.
33. W. Kuemmerle, R. Duesmann, *Motortechnische Zeitschrift*, **1994**, *55*, 708–713.
34. E. S. Lox, B. H. Engler, E. Koberstein, *Technical Paper Series 910841*, Society of Automotive Engineers, Warrendale, **1991.**
35. A. Cybulski, J. A. Moulijn, *Chem. Eng. Sci.* **1994**, *49*, 19–27.
36. E. Koberstein, G. Wannemacher in *Catalysis and Automotive Pollution Control* (Ed: A. Crucq, A. Frennet), Elsevier, Amsterdam, **1987**, p. 155.
37. D. Schweich in *Catalysis and Automotive Pollution Control II*, (Ed: A. Crucq), Elsevier, Amsterdam, **1991**, p. 437.
38. J. P. Leclerc, D. Schweich, J. Villermaux in *Catalysis and Automotive Pollution Control II* (Ed.: A. Crucq), Elsevier, Amsterdam, **1991**, p. 465.
39. B. H. Engler, D. Lindner, E. S. Lox, K. Ostgathe, A. Schäfer-Sindlinger, W. Müller, *Technical Paper Series 930738*, Society of Automotive Engineers, Warrendale, **1993**.
40. E. Schwizer, D. Burch, P. Riedwyl, *10 Jahre Katalysatorautos*, Touring Club der Schweiz, Emmen, Switzerland, **1994**.
41. D. W. Wendland, W. R. Matthes, *Technical Paper Series 861554*, Society of Automotive Engineers, Warrendale, **1986**.
42. D. W. Wendland, P. L. Sovrell, J. E. Kreucher, *Technical Paper Series 912372*, Society of Automotive Engineers, Warrendale, **1991**.
43. B. Engler, E. Koberstein, P. Schubert, *Appl. Catal.* **1989**, *48*, 71–92.
44. E. Koberstein, *Dechema Monographs*, *120*, VCH, Weinheim, **1990**, p. 67.
45. P. Albers, B. Engler, J. Leyrer, E. S. Lox, G. Prescher, K. Seibold, *Chem. Eng. Tech.* **1994**, *17*, 161–168.
46. B. H. Engler, E. Koberstein, D. Lindner, E. S. Lox in *Catalysis and Automotive Pollution Control II* (Ed: A. Crucq), Elsevier, Amsterdam, **1991**, p. 641.
47. B. H. Engler, D. Lindner, E. S. Lox, P. Albers, *Proceedings of the 10th International Congress on Catalysis* (Eds: L. Guczi, F. Solymosi, P. Tetenyi), Akademiai Kiado, Budapest, **1993**, p. 2701–2704.
48. H. H. Haskew, J. J. Gumbleton, *Technical Paper Series 881682*, Society of Automotive Engineers, Warrendale, **1988**.
49. E. S. Lox, B. H. Engler, E. Koberstein, *Technical Paper Series 881682*, Society of Automotive Engineers, Warrendale, **1988**.
50. H. Skinjoh, H. Muraki, Y. Fujitani in *Catalysis and Automotive Pollution Control II*, (Ed.: A. Crucq) Elsevier, Amsterdam, **1991**, p. 617.
51. G. Mabilon, D. Durand, M. Prigent in *Catalysis and Automotive Pollution Control II*, (Ed.: A. Crucq) Amsterdam, **1991**, p. 569.
52. Forschungsvereinigung Verbrennungskraftmaschinen, *Report 419*, Frankfurt, **1988**.
53. M. Shelef, K. Otto, N. C. Otto *Adv. Catal.* **1978**, *27*, 311.
54. W. B. Williamson, H. K. Stephien, W. L. H. Watkeins, H. S. Gandhi, *Environ. Sci. Technol.* **1979**, *13*, 1109.

55. K. C. Taylor, *Automobile Catalytic Converters, in Catalysis, Science and Technology* (Eds: J. R. Anderson, M. Boudart), Springer Verlag, Berlin, **1984**, p. 5.
56. K. Kollmann, J. Abthoff, W. Zahn, *Auto. Eng.* **1994**, *10*, 17–22.
57. M. Theissen, P. Langer, J. Mallog, R. Zielinski, *Proceedings 4. Aachener Kolloqium Fahrzeug- und Motorentechnik*, Aachen, **1993**.
58. E. Otto, W. Held, A. Donnerstag, P. Kueper, B. Pfalzgraf, A. Wirth, *Proceedings 16. Internationales Wiener Motorensymposium*, Vienna, **1995**.
59. P. Hofmann, F. Indra, *Proceedings 16. Internationales Wiener Motorensymposium*, Vienna, **1995**.
60. T. Kreuzer, E. S. Lox, D. Lindner, J. Leyrer in *Proceedings Second Japan – EC Joint Workshop on the Frontiers of Catalytic Science and Technology for Energy, Environment and Risks Prevention*, Catalysis Today, 29, **1996**, 17–27.
61. M. Miyoshi, T. Tanizawa, S. Takeshima, N. Takahashi, K. Kasahara, *Toyota Technical Review*, **1995**, *40*, 21–26.
62. W. Held, A. König, T. Richter, L. Puppe, *Technical Paper Series 900496*, Society of Automotive Engineers, Warrendale, **1990**.
63. J. R. Needham, D. M. Doyle, S. A. Faulkner, H. D. Freeman, *Technical Paper Series 891949*, Society of Automotive Engineers, Warrendale, **1989**.
64. L. Burgler, P. L. Herzog, P. Zelenka, *Proceedings Worldwide Engine Emission Standards and How to Meet Them*, Institution of Mechanical Engineers, London, **1991**, p. 57.
65. J. Leyrer, E. S. Lox, B. H. Engler, *Proceedings Symposium Diesel engines, Technical Academy Esslingen* (Eds: U. Esser, K. H. Prescher), Esslingen, **1993**, 55–78.
66. M. Horiuchi, K. Saito, S. Ichihara, *Technical Paper Series 900600*, Society of Automotive Engineers, Warrendale, **1990**.
67. U. Graf, P. Zelenka, *Proceedings 15. Internationales Wiener Motoren-symposium*, Vienna, **1994**.
68. B. H. Engler, E. S. Lox, K. Ostgathe, W. Cartellieri, P. Zelenka, *Technical Paper Series 910607*, Society of Automotive Engineers, Warrendale, **1991**.
69. R. Beckmann, W. Engeler, E. Müller, B. H. Engler, J. Leyrer, E. S. Lox, K. Ostgathe, *Technical Paper Series 922330*, Society of Automotive Engineers, Warrendale, **1992**.
70. P. Zelenka, K. Ostgathe, E. S. Lox, *Technical Paper Series 902111*, Society of Automotive Engineers, Warrendale, **1990**.
71. B. H. Engler, J. Leyrer, E. S. Lox, K. Ostgathe, *Technical Paper Series 930735*, Society of Automotive Engineers, Warrendale, **1993**.
72. B. H. Engler, J. Leyrer, E. S. Lox, K. Ostgathe in *Catalysis and Automotive Pollution Control III* (Ed: A. Frennet, J. M. Bastin), Elsevier, Amsterdam, **1995**, p. 529.
73. J. Leyrer, E. S. Lox, *Proceedings Dechema Jahrestagung (Preprints)*, Frankfurt, **1995**.
74. Ricardo Consulting Engineers *Automotive Diesel Engines and the Future*, Ricardo, Shoreham-by-Sea, England, **1994**.
75. R. C. Rijkeboer, M. F. van den Haagen, *TNO-Report 730210064*, TNO Industrial Research, Delft, Netherlands, **1992**.
76. B. Engler, E. Koberstein, E. Lox, Technical Paper Series 900271, Society of Automotive Engineers, Warrendale, **1990**.

2 Environmental Catalysis – Stationary Sources

F. J. JANSSEN, Arnhem Institutions of the Dutch Electricity Utilities,
N. V. KEMA, R & D Division, Department of Chemical Research,
P.O. Box 9035, 6800 ET Arnhem/The Netherlands

2.1 General Introduction

Environmental catalysis is a rather new and broad field and not really explored. There is no comprehensive treatise available, yet. However, it is the author's hope that the chemist seeking information in some particular field of environmental catalysis might find all he or she desired to know by consulting this modest review.

Environmental catalysis is required for cleaner air, soil, and water. Various catalysts are in use to improve and/or protect our environment. Catalysts are used in environmental technologies to convert environmentally hazardous materials into harmless compounds. Deactivation of the environmental catalysts occurs as a result of thermal aging, physical and chemical poisoning, and masking mechanisms. Regeneration procedures, which include thermal, physical, and chemical treatment have been developed in order to extend catalyst life.

Examples of environmental catalytic issues are: selective catalytic reduction (SCR) of NO_x; removal of volatile organic compounds (VOC); decomposition of NO; oxidation of CO from exhaust gas. SCR is now in a mature state. Several reviews have been published recently.

In the review by Armor [1] a variety of pollutants are discussed with a focus on commercially applied processes using catalysis as a solution. Issues such as the removal of NO_x, SO_x, chlorofluorohydrocarbons (CFC), VOC, carbon monoxide, auto exhaust emission, ozone, nitrous oxide, byproducts from chemicals production, odor control, and toxic gas removal are discussed. In another review Armor [2] discusses specific topics such as monolith technology, new catalytic materials, and specific processes. Additionally, key suggestions for future research effort are given.

Spivey [3] reviewed the catalytic oxidation of volatile organic compounds. A number of reported applications of catalytic oxidation are included.

An extensive review on NO_x removal has been published by Bosch and Janssen [4], which may serve as a quick introductory guide and which gives a comprehensive picture of SCR up to 1986.

Bond and Flamerz Tahir [5] reviewed the literature on the preparation, structure and catalytic properties of vanadium oxide monolayer catalysts. This catalyst plays a key role in the selective catalytic reduction of NO_x.

Recently, the proceedings of the congress on "Catalysts in Environmental Technology" were published in an issue of *Catalysis Today* [6]. This issue is devoted to the conversion/removal of NO_x and SO_x from exhaust gases. Additionally, two issues of *Catalysis Today* were published concerning catalytic processes for environmental protection in Japan [7] and environmental industrial catalysis [8].

Pollutants are emitted daily with the effluents produced by stationary and non-stationary sources. The first type includes industrial processes, energy production. waste incineration, and domestic burning, whereas the second type includes automotive emissions and emissions from aeroplanes. Road and air traffic make a substantial contribution to NO_x and particulate emissions.

Fossil fuels such as coal, oil, gas, and all types of biomass and domestic and industrial wastes are burnt or gasified for the conversion of energy. In Western European countries, Japan, and the United States activities are carried out for decreasing emissions of hydrocarbons, CO, NO_x, N_2O, NH_3, SO_x, dioxins, and CO_2 from the combustion and gasification of fossil fuels and the burning of biomass and wastes.

Emissions of NO_x, SO_x and NH_3 cause acidification of the environment. Greenhouse gases such as CO_2, N_2O, and CH_4 give rise to global warming.

The need to remove NO_x from stationary sources such as industrial boilers, power plants, waste and biomass incinerators and gasifiers, engines, and gas turbines was emphasized in the 1980s.

Japan was the first country to do this, starting with SCR for the removal of NO_x from gas, coal and oil-fired power plants [7]. Utility companies in Japan introduced the catalytic SCR process as early as 1972. At present many countries utilize SCR, with Germany and Japan being the leading countries. Although priority is currently given to combustion modifications for NO_x control in European countries, extended application of catalytic SCR is expected in the near future.

Coal gasification (Section 3) is a relatively new technique for the production of electric energy. During gasification a variety of compounds are formed. These are CO, H_2, CO_2, HCN, NH_3, H_2S, COS, HCl, and HF. The coal gas is burnt in the gas turbine. Before combustion takes place all compounds containing sulfur and nitrogen have to be removed in order to avoid formation of SO_x and NO_x.

The reduction of nitrogen oxides by hydrocarbons in the presence of excess oxygen has been applied to the exhaust gases of diesel engines. Biomass and solid waste are used as fuel to produce electricity. The energy in biomass may be obtained by upgrading it by thermochemical conversion. Application of catalysts in the fields of gasification, pyrolysis, and liquefaction of biomass are reviewed by Bridgewater [9].

This section on environmental catalysis highlights the development of catalysts for stationary sources. These are coal-, oil-, and gas-fired power plants, including gas turbines, waste incinerators, and gasifiers.

Selective catalytic reduction using monolith technology of NO_x is emphasized. A huge amount of scientific work was done in the past twenty years. Both mechanistic work and catalyst preparation is reviewed here.

The decomposition of NO into its elements has not been realized yet. Another important issue of environmental catalysis is the removal of volatile organic compounds such as chlorohydrocarbons and dioxins. The simultaneous removal of NO_x and SO_x is another important subject for catalytic research.

In gas turbine technology catalytic combustion (Section 3) is an important method to lower NO_x emissions. Section 2.2 summarizes briefly aspects of the catalytic decomposition of NO, whereas Section 2.3 is concerned with SCR catalysts. Industrial experiences with respect to power plants, nitric acid plants, gas turbines and new concepts are described in Section 2.4. Section 2.5 presents the removal and conversion of volatile organic compounds. Finally, future trends are given in Section 2.6. Because of the limited space in this contribution only a selection of the vast literature (after 1988) relating to environmental catalysis is given. No reference is made to the application of clean processes using catalysts such as catalytic combustion or energy-saving catalytic-driven processes.

2.2 Decomposition of NO

Nitrogen oxides are formed in practical combustion systems by high-temperature thermal fixation of atmospheric nitrogen and by oxidation. These types are called thermal and fuel NO_x, respectively. According to thermodynamics NO is an unstable compound even at very high temperatures. However, the decomposition rate is very low and, hence, it is advantageous to apply a catalyst. A large number of experimental studies have been carried out on suitable catalysts for the decomposition. However, it has been found that most types of catalysts are not resistant against oxygen poisoning.

NO decomposition has been studied over a variety of catalysts such as oxides of various metals, zeolites, perovskites, and noble metals. Pt on Al_2O_3, and unsupported oxides such as Co_3O_4, CuO, NiO, Fe_2O_3, and ZrO_2 are suitable catalysts [10]. For all catalysts the reaction is first order with respect to NO. Oxygen strongly inhibits the decomposition reaction. The inhibition by oxygen is ascribed to the chemisorption of oxygen on sites on which nitric oxide chemisorption takes place.

Zeolites are also able to decompose N_2O. Li and Armor [11] prepared 25 zeolite samples by exchanging the cation Na^+. These catalysts were tested for the catalytic decomposition of N_2O. The Co- and Cu-exchanged zeolites are active at temperatures ranging from 623 K to 673 K. Rhodium and ruthenium supported on ZSM-5 are active catalysts between 523 and 573 K.

In contrast to Co-ZSM-5, Cu-ZSM-5 and perovskites show very high catalytic activities for the decomposition of NO.

According to Li and Hall [12] the selectivity to N_2 on Cu-ZSM-5 is not 1. An amount of undecomposed NO and O_2 react homogeneously at low temperatures downstream from the reactor. Based on their results the rate equation for the NO decomposition reaction can be written in Langmuir–Hinshelwood form as in eq 1.

$$r = \frac{\mathrm{d}[N_2]}{\mathrm{d}\tau} = \frac{k[NO]}{1 + K\sqrt{[O_2]}} \tag{1}$$

where r is the turnover rate in s^{-1} $site^{-1}$; k is the rate constant in s^{-1} $site^{-1}$ mol^{-1} L, and K is the adsorption equilibrium constant for oxygen in $(mol\,L^{-1})^{-1/2}$. The NO decomposition was first order in NO and was inhibited by O_2 being half order.

Valyon and Hall [13] employed isotopically labeled molecules such as $^{18}O_2$ and $^{15}N^{18}O$ to study the very interesting issue of the release of molecular oxygen during the reduction step. It was shown in their qualitative study that during the decomposition of NO the oxygen atom of the NO molecule is attached to the surface of the zeolite. Molecular oxygen leaves the surface and consists of the chemisorbed oxygen and lattice oxygen. Moreover, it was shown that nitrogen dioxide-containing species are present on the catalyst surface.

According to Iwamoto and co-workers [14], Cu–zeolites are the most suitable for the decomposition of NO. Infrared spectroscopy together with an isotopic tracer method were used to study the adsorption of NO over copper ion-exchanged ZSM-5 zeolites and a reaction mechanism for the catalytic decomposition of NO was presented [14]. A mixture of NO and helium was passed over the catalyst at room temperature. The evolution of nitrogen started after 12 min. The amount of nitrogen formed went through a maximum and dropped to zero after 60 min. After 45 min N_2O started to pass through. The amount of N_2O formed showed a maximum after 100 min and then decreased gradually. After 350 min no reactions occurred. The catalyst becomes inactive as a result of the adsorption of oxygen. Above 573 K the active sites, Cu^+ ions, are regenerated by the desorption of oxygen. Three NO species were detected. These are $NO^{\delta+}$, $NO^{\delta-}$, and $(NO_2)^{\delta-}$. The first species is formed on copper(I) ions, whereas the negative ions occur on copper(II) ions. The anions are intermediates for the decomposition of NO. The amount of anion decreases with reaction time while the amount of $NO^{\delta+}$ at the surface increases.

The superconducting material $Ba_2YCu_3O_{7-\delta}$ is catalytically active for the decomposition of NO. Tabata [15] found that NO was selectively decomposed over this catalyst even in the presence of excess oxygen. The author also suggested that N_2O was formed in addition to N_2 and O_2. NO chemisorbs on $YBa_2Cu_3O_7$ forming, nitride, nitrate, nitrite, and nitrito groups at room temperature evolving N_2, N_2O and NO_2. The various nitrogen species were observed at differing temperatures. Nitride species and predominant nitrate ions were observed at temperatures ranging from 300 to 540 K. The nitrite ion becomes predominant at high temperatures. The nitrito group was found to be present for the entire temperature range.

Hardee and Hightower [16] studied the decomposition of NO and N_2O over rhodium on γ-Al_2O_3. At temperatures in the range from 573 K to 673 K the reaction is initially fast, but it slows down after 20 min. N_2 and N_2O were identified as reaction products. Gas phase O_2 or NO_2 were not formed. It was suggested that the catalyst becomes poisoned due to oxygen according to the reaction

$$6NO + 2Rh \rightarrow Rh_2O_3 + 3N_2O$$

Nitrous oxide decomposition is extremely fast at 593 K and both O_2 and N_2 are formed as products.

Recently, Yang and Chen [17] used the heteropoly compound $H_3PW_{12}O_{40} \cdot 6H_2O$ for the decomposition of NO by means of a two-stage process. Because of the very low concentrations of NO in flue gas and consequently the low decomposition rate, NO is first concentrated on the sorbent at 423 K. NO_x is chemisorbed forming nitrates and nitrites. After saturation, the partial pressure of NO_x in the catalyst is 1 MPa, the catalyst is heated rapidly in the presence of water to a temperature of 723 K and N_2 is formed. CO_2 and SO_2 do not affect the reaction rate. The rate equation for absorption is of the form shown in eq 2.

$$r = k[\mathrm{NO}]^{0.5}[\mathrm{O_2}] \tag{2}$$

Conclusively, nitric oxide can be decomposed by a two step approach. The absorption reaction is

$$H_3PW_{12}O_{40} \cdot 6H_2O + 3NO = H_3PW_{12}O_{40} \cdot 3NO + 6H_2O$$

The heteropoly compound remains unchanged and H_2O is exchanged by NO. The decomposition reaction that occurs after rapid heating is

$$H_3PW_{12}O_{40} \cdot 3NO = H_3PW_{12}O_{40} \cdot 6H_2O + 1.5N_2$$

In the last equation, three oxygen atoms are missing. The authors state in their article that the role of oxygen is still unclear. However, this compound is a promising option for NO decomposition.

2.3 SCR Catalysts

Both titania and titania/silica-supported vanadia, molybdena, tungsta, and chromia have been applied as SCR catalysts. Low-temperature and high-temperature catalysts have been developed. The vanadia on titania catalysts have received most attention.

A variety of reducing agents such as CH_4, CO, H_2, and NH_3 have been used. Ammonia is widely used as reductant in the selective catalytic reduction.

Applications of SCR are found in power plants, waste incinerators, and gas turbines.

Flue gas produced by a coal fired burner is passed to the SCR reactor together with ammonia. The catalytic SCR reactor is usually placed between the economizer and the air preheater (Fig. 1). Downstream the air preheater fly ash is collected by means of the electrostatic precipitators. Via a heat exchanger and flue gas desulphurization the flue gas is passed to the stack. The SCR catalyst used in this way is in the so-called high dust mode. This means that the resistance of the catalyst against attrition should be very high. That is one the reasons for using titania as a support.

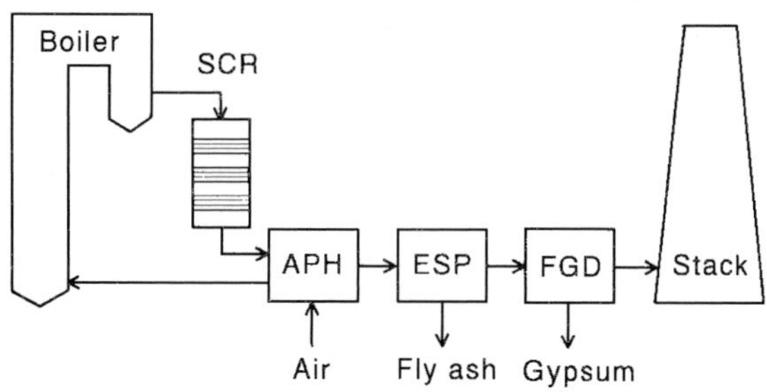

Figure 1. A scheme of a power plant equipped with both selective catalytic reduction (SCR) and flue gas desulfurization (FGD): APH, air preheater; ESP, electrostatic precipitator.

Table 1. Reaction heats of various reactions (673 K) [18].

No	Reaction	$-\Delta H$ (kJ)
1	$4NO + 4NH_3 + O_2 = 4N_2 + 6H_2O$	1625
2	$4NO + 4NH_3 + 3O_2 = 4N_2O + 6H_2O$	1304
3	$6NO + 4NH_3 = 5H_2 + 6H_2O$	1806
4	$4NH_3 + 3O_2 = 2N_2 + 6H_2O$	1263
5	$2NH_3 + 2O_2 = N_2O + 6H_2O$	553
6	$4NH_3 + 5O_2 = 4NO + 6H_2O$	907

The overall reactions over highly selective supported vanadia catalysts may be described by the stoichiometry

$$4NO + 4NH_3 + O_2 = 4N_2 + 6H_2O$$

$$6NO_2 + 8NH_3 = 7N_2 + 12H_2O$$

The first reaction predominates if the flue gas contains more NO. The high activity of NH_3 combined with the enhanced reaction rate in the presence of oxygen, makes NH_3 the preferred reducing agent. Apart from this main reaction other reactions are feasible. The variety of reactions occurring are given in Table 1 [18].

In the following sections issues such as vanadium-based catalysts, noble metal catalysts, zeolites, carbon-supported catalysts, and promoters are discussed.

2.3.1 Preparation and Characterization

In this section preparation and characterization of catalytic materials are briefly reviewed with respect to their applications in environmental catalysis. A number of techniques for the preparation of the supports and catalysts are emphasized. Techniques such as impregnation, homogeneous deposition precipitation, grafting, hydrolysis, sol-gel, and laser-activated pyrolysis are used for the preparation of catalysts for fundamental studies.

The preparation and characterization of titania-supported vanadia catalysts have been reviewed by Bond and Flamerz Tahir and provide a guide to the literature on preparation, structure and catalytic properties of vanadium oxide monolayer catalysts [5]. Preparative methods such as grafting, heating mechanical mixtures, or coprecipitation are also discussed.

It is well known that the activity of SCR catalysts depends on the amount of vanadia present on the support [4]. Dispersion of vanadia is necessary in order to increase the number of catalytically active species. For instance, four layers of vanadia on titania exhibit an increase of two orders of magnitude in reaction rate. Titania shows a strong interaction with vanadia. In order to decrease the influence of titania, silica is added to the support. It was found that vanadia on silica/titania catalysts are far more active catalysts than vanadia on silica and less active than vanadia on titania materials (see Table 4).

One of the simplest ways of preparing SCR catalysts is wet impregnation of the carrier with an acidified aqueous solution of ammonium metavanadate, followed by drying and calcination.

Bond et al. [19] prepared both WO_x and MoO_x on low-area anatase catalysts by means of wet impregnation with aqueous solutions of molybdate and tungstate salts. MoO_x on TiO_2 containing 50% of a theoretical monolayer of MoO_x was prepared by means of a grafting technique. A solution of $MoOCl_4$ in dry CCl_4 was added to the support. After refluxing the solid was filtered, dried and subsequently hydrolyzed with moist air. The catalysts were characterized by means of temperature programmed reduction (TPR), laser Raman spectroscopy, and X-ray photoelectron spectroscopy (XPS). Monolayer capacity (wt%) was calculated and corresponded to 0.16 and 0.25 for MoO_3 and WO_3, respectively. Monolayer capacities derived in various ways were compared. The peak ratios of the XPS signals of I_{Mo}/I_{Ti} or I_W/I_{Ti} linearly increase with increasing metal oxide (MoO_3 or WO_3) content up to the monolayer loading [19].

Vanadia on titania catalysts prepared by wet impregnation (ammonium metavanadate) and monolayer catalysts prepared by grafting using vanadyl acetylacetonate were compared [20]. It was demonstrated that monolayer catalysts show better activities at similar vanadium loadings than those of commercial catalysts.

Wet impregnation is not feasible for all catalyst types. For instance, vanadia on silica cannot be prepared by means of wet impregnation. With homogeneous deposition precipitation (HDP), however, it is possible to prepare vanadia on silica catalysts [21]. The principle of this method is to use a lower valence of the metal, which may be produced by cathodic reduction of the respective metal ion. The reason for using lower valence state metal ions is the lower acidity compared to that of the higher valence state and the higher solubility of the metal ions. This technique was used for the preparation of silica-supported vanadia, titania, and molybdena catalysts [22, 23].

Grafting has been used for the preparation of vanadia on titania catalysts [24]. Vanadyl triisobutoxide was used as precursor. The vanadia species were well dispersed. It was observed by electron microscopy that silica was present as an amorphous phase. Moreover, the amount of titania present in the catalysts influences the specific surface area and the pore volume. Lower titania concentrations give rise to both higher specific surface areas and pore volumes. The most stable catalyst contained equimolar titania–silica or pure titania.

It was shown that the amount of vanadia per m^2 increased almost linearly with the number of subsequent grafting steps [25]. After five grafting treatments the amount of V^{5+} present was $7\,\mu mol\,m^{-2}$. Maximum turnover frequencies were obtained for vanadia concentrations of $3\,\mu mol\,m^{-2}$ and above.

Vanadia on titania samples were prepared by wet impregnation and by grafting using three types of vanadyl sources [26]. These were NH_4VO_3/oxalic acid solution, $VOCl_3$, and $VO(O\text{-}i\text{-}Bu)_3$. Again, the XPS intensity ratio V:Ti increases linearly with the amount of V_2O_5 in the catalyst. It was observed that catalysts obtained by grafting showed the same TPR spectrum as those obtained by wet impregnation.

Handy et al. [27] prepared TiO_2/SiO_2 mixed gels as supports for vanadia. The supports were grafted with vanadyl triisopropoxide. The following surface reaction occurs

$$(TiO_2/SiO_2){-}OH + VO(O{-}C_3H_7)_3 \\ \rightarrow (TiO_2/SiO_2){-}O{-}VO(O{-}C_3H_7)_2 + HO{-}C_3H_7$$

The vanadia-site densities after one, two and, three graftings were 0.6, 1.6, and 2.6 V^{5+} nm^{-2}, respectively. The last figure corresponds to 0.4 monolayer of vanadium oxide. No crystalline aggregates of V_2O_5 were formed by the grafting method.

With Raman spectroscopy it was possible to distinguish between vanadia adsorbed on silica or vanadia adsorbed on titania. It was concluded that a very narrow $\nu_{V=O}$ stretchings at 1040 cm^{-1} and at 1030 cm^{-1} could be assigned to silica- and titania-supported vanadia, respectively. During preparation vanadium first covers the titania faces followed by interaction with silica.

Vanadyl triisopropoxide was used for the preparation of vanadia submonolayers on TiO_2, SiO_2, Al_2O_3, and ZrO_2, and physical mixtures such as TiO_2/SiO_2, TiO_2/Al_2O_3, and TiO_2/ZrO_2 [28]. The physical mixtures were covered by chemical vapor deposition of the vanadium compound.

Mariscal et al. [29] prepared a series of titania-silica honeycombs by impregnation and homogeneous deposition precipitation. Only small amounts of titania are deposited on the silica matrix by homogeneous deposition precipitation. It was shown that incipient wetness impregnation was the only method to control both the amount of deposited titania and the concentration profile across the wall of the honeycombs [29].

Went et al. [30] prepared the anatase phase of TiO_2 by hydrolysis of titanium isopropoxide in a water/isopropanol mixture. After calcination at 773 K the anatase contained 5% brookite and no rutile.

The sol-gel technique may be used for the preparation of the supports and catalyst systems [31, 32].

The preparation of silica spheres may be carried out by introducing a solution of sodium silicate, nitric acid, and hexamethylene tetramine via a vibrating capillary into paraffin of silicone oil at a temperature of about 250 K [31].

Laser-activated pyrolysis has been applied for the synthesis of ultrafine TiO_2 with titanium isopropoxide or ethoxide as precursors [33]. A specific surface area of 128 m^2 g^{-1} was obtained.

Brent et al. [32] produced gels from the hydrolysis of vanadyl triisopropoxide, titanium tetraisopropoxide, and silica tetraethoxide.

Several traditional techniques have been used to determine the composition, structure and texture of the catalysts. These include X-ray fluorescence, X-ray diffraction, specific surface-area measurements, mercury porosimetry, and electron microscopy. The application of each technique is straightforward and will not be discussed here. For descriptions of these techniques see the respective sections in Part A of this volume.

The following techniques for the characterization of supported vanadia catalysts and vanadia will briefly reviewed: temperature programmed reduction (TPR) [26, 30], oxidation (TPO) [30] and desorption (TPD) [33, 36, 41, 44]; laser Raman spectroscopy (LSR) [5, 25, 35, 36, 39, 44]; Fourier-transform infrared spectroscopy

Table 2. Catalysts used for reduction experiments and the positions of the peak maxima for the strong peaks only in the NH_3-TPR and H_2-TPR spectra [34].

Reduced species	V_2O_5 (annealed)		V_2O_5 (decomposed)	
	NH_3	H_2	NH_3	H_2
V_6O_{13}	678	945	700	945
VO_2	755	985	767	977

(FTIR) [33, 37, 44]; temperature programmed surface reaction (TPSR) [41]; X-ray photoelectron spectroscopy (XPS) [26, 33, 36, 39, 44, 45].

Janssen [34] studied the reduction of unsupported vanadia and molybdena by means of NH_3-TPR and H_2-TPR. It was found that when the reduction is carried out with NH_3 instead of H_2 the temperatures for the peak maxima shift to lower temperatures. The peak maxima for two types of vanadia, namely well annealed, V_2O_5(a), and prepared by thermal decomposition, V_2O_5(d), of ammonium metavanadate at 673 K for 16 h are given in Table 2. Above 873 K ammonia is decomposed into nitrogen and hydrogen and NH_3-TPR changes into H_2-TPR. At 1073 K the V_2O_5 is completely reduced to V_2O_3 [34, 35].

Oszkan et al. [36] carried out similar experiments and found comparable data. The V_2O_5(d) compound showed peaks at 933, 963 and, 1123 K, whereas the second type compound showed peaks at 703, 943, 969, and 1143 K. Using XRD analysis the corresponding species were established to be V_6O_{13}, VO_2 and V_2O_3, and $V_2O_5 + V_6O_{13}$, $V_6VO_{13} + VO_2$, and $VO_2 + V_8O_{15}$, respectively.

During TPR water is formed and desorbs from the surface [35]. It is observed that the hydrogen consumption starts at 350 K, whereas water is formed at temperatures above 425 K. The monomeric vanadyls are restored after one cycle of reduction and oxidation. For polymeric vanadates terminal vanadyl oxygen atoms are removed leaving the structure unchanged. Upon oxidation the polymeric vanadate splits into a dimeric form and a monomeric vanadyl (Fig. 2). Oszkan et al. [36] concluded from their TPD studies that two types of ammonia species were present at the surface. One type was ammonia coordinatively bonded to surface vanadyls (Lewis sites) and a second type of ammonia species formed by the reaction of NH_3 and surface hydroxyl groups [36].

Little information is available on the WO_3 on titania system. It was observed by Ramis et al. [37] that WO_3 on TiO_2 contains Lewis sites and Brønsted sites. WO_3 in the catalyst stabilizes the morphology of the support and inhibits the SO_2 into SO_3

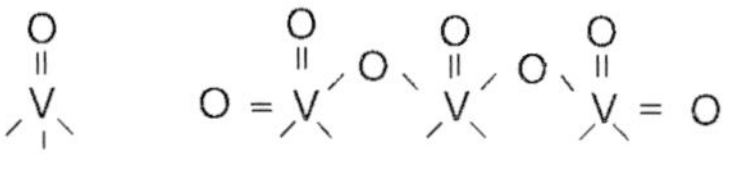

Figure 2. Monomeric vanadyl (left) and polymeric vanadates (right) (adapted from Ref. 35).

oxidation reaction. The redox properties of the WO_3 on titania catalyst shows redox properties associated with W^{6+} ions. For V_2O_5 on titania, however, the redox character as compared to the acidic character is more prominent than in WO_3 on titania [37].

The infrared spectrum of 4 wt% vanadia on titania shows a band at 940 cm^{-1}, which may assigned to the V=O stretching [38]. Going from 4 wt% to 15 wt% vanadia in the catalyst this peaks shifts to 1020 cm^{-1}. This peak was assigned to segregated vanadia crystallites [38]. FTIR was used to study the type and distribution of acid Lewis and Brønsted sites of TiO_2/SiO_2 and TiO_2 supports with pyridine as a probe molecule. TiO_2/SiO_2 supports exhibit both Lewis and Brønsted acidity [39]. TiO_2 contains mainly Lewis acid sites; there are formed by condensation of adjacent hydroxyl groups. Raman features were found at 1010, 990, and 852 cm^{-1} for vanadia on titania which were attributed to octahedrally coordinated surface species. Dehydrated samples to accommodate isolated and surface vanadate and polymeric metavanadate species.

Evidence for polymeric species of the metavanadate type in submonolayer vanadia on titania catalysts is obtained by the appearance of a $\nu_{V=O}$ stretching mode at 870 cm^{-1} [40]. The stretching mode of V=O for isolated vanadyls is found at 1030 cm^{-1}.

Lietti and Forzatti [41] have shown by means of transient techniques such as TPD, TPSR, TPR and SSR (steady state reaction experiments) that isolated vanadyls and polymeric metavanadate species are present on the surface of vanadia on titania catalysts with V_2O_5 loadings of up to 3.56 wt%. Polyvanadate species are more reactive than isolated vanadyls due to the presence of more weakly bonded oxygen atoms. It has been shown that titania-supported vanadia materials comprise of a distribution of monomeric vanadyl, polymeric vanadates, and crystalline vanadia, the amount of which is dependent on the vanadia content.

Went et al. [35] used in situ laser Raman spectroscopy and temperature programmed reduction and temperature programmed oxidation to establish the presence of monomeric vanadyls, polymeric vanadates and crystallites of V_2O_5 in titania supported catalysts. The distribution of vanadia species as a function of V_2O_5 loading was determined. At low V_2O_5 concentrations a monomeric vanadyl species is present at 1030 cm^{-1} assigned to the V=O stretching mode. At increasing V_2O_5 content in the catalyst the peak at 1030 cm^{-1} broadens significantly and the intensity of the bands located in the region 700–1000 cm^{-1} assigned to polymeric vanadates increases. The band at 840 cm^{-1} is assigned to V–O–V bending vibrations. The average polymer size varies from two to three depending on the vanadia load. For vanadia amounts of 6 wt% and higher a narrow band at 960 cm^{-1} appears [35]. The effect of the loading of vanadia on the fraction of the various species is shown in Fig. 3.

The Raman feature located at 1030 cm^{-1} was assigned to V=O in V_2O_5 on the TiO_2/SiO_2 catalyst. For lower loadings of vanadia an intensity in the range 920–950 cm^{-1} appears, which is attributed to a tetrahedral metavanadate structure. This observation was supported by means of ^{51}V NMR measurements. A sixfold coordinated species is found at higher vanadium loadings. Small amounts of crystallites of V_2O_5 were present for loadings of 4 wt% and above, which appear from the sharp feature at 991–996 cm^{-1} [35].

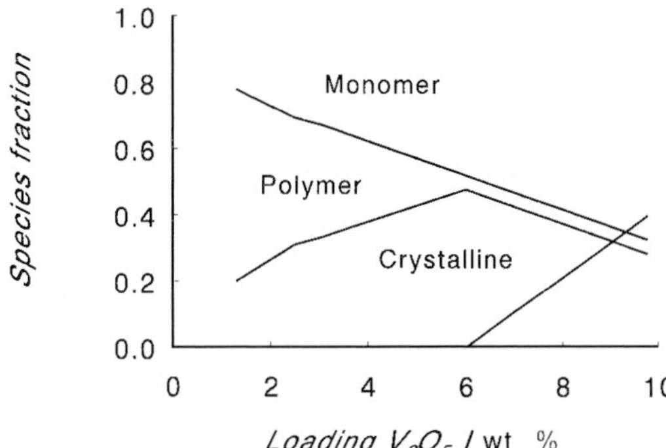

Figure 3. The distribution of monomeric, polymeric, and crystalline vanadia species as a function of the weight loading of vanadia (adapted from Ref. 35).

By means of laser Raman spectroscopy two characteristic band were found at 996 and 285 cm^{-1} for vanadia on titania catalysts which were assigned to V=O groups [42]. A band at 1030 cm^{-1} from a singly grafted catalyst was assigned to small vanadia clusters [25].

WO_3 on titania containing a WO_3 loading higher than 2 wt% exhibits Raman bands at 992 and 1008 cm^{-1}, assigned to a species containing two W=O groups. WO_3 on a high-surface-area titania (55 $m^2\,g^{-1}$) shows a broad band in the region from 946 to 992 cm^{-1} [5].

Nickl et al. [43] characterized submonolayers of vanadia on a variety of supports by means of Raman spectroscopy; TiO_2, SiO_2, Al_2O_3, ZrO_2 and mixtures thereof were used as support.

Raman features and their assignments are shown in Table 3. The Raman spectra of both the supported catalyst and its corresponding physical mixture differ not so much. Moreover, it was observed that well dispersed V_2O_5 species tend to agglomerate.

Table 3. Raman bands of vanadia on a variety of supports and their assignment [43].

Support	Loading[a] (μmol V m^{-2})	band (cm^{-1})	Assignment
TiO_2	4.14	970	decavanadate
		920	square pyramidal coordination of V^{5+} centers
SiO_2	3.65	1030	monooxovanadyl
		966	$\nu_{V=O}$
Al_2O_3	–	nd[b]	no vanadia species
ZrO_2	5.27	985	$\nu_{V=O}$
SiO_2/TiO_2	2.95	1025	two-dimensional layers
		1000 sh	partially hydrated domains
Al_2O_3/TiO_2	1.93	1025–990	small clusters
ZrO_2/TiO_2	0.81	940	dehydrated metatvandate species

[a] All vanadia loadings are below the theoretical monolyaer capacity, namely 15 μmol V m^{-2}.
[b] not detected.

Oszkan et al. [36] showed that the Raman band position at 995 cm^{-1} is similar for V_2O_5 samples which were prepared by decomposition and melting followed by recrystallization.

Ramis et al. [44] studied the effect of dopants and additives on the state of surface vanadyl species of vanadia on titania catalysts by means of FTIR spectroscopy. Additives such as alkali and alkali-earth metal cations (typically Cs, K, Na, Li and Mg), oxoanions (such as sulphates and arsenates), and other species (such as Al^{3+}, MoO^{4+}, and WO^{4+}), influence the position of V=O stretching frequencies. The position of $\nu_{V=O}$ for a 3 wt% V_2O_5 on titania was observed at 1035 $^{-1}$. Two percent W or Mo did not show any shift of the stretching frequency of V=O, whereas Cs lowered the band position by 45 cm^{-1}. This was explained in terms of the formation of strong basic sites and the exchange of Ti in O=V–O–Ti by O=V–O–Cs. The elements Al, S, and As shift the position of $\nu_{V=O}$ to higher frequencies. Oxoanions are coordinatively bond to vanadyl centers [44].

Nickl et al. [43] carried out XPS measurements on model catalysts such as VO_x on Au, VO_x on SiO_2, VO_x on Al_2O_3, and VO_x on TiO_2. These were prepared by thermal evaporation and subsequent oxidation in air at room temperature. A high fraction of V^{3+} ions was observed in the VO_x on TiO_2 material after reduction in H_2 at 573 K. The V $2p_{3/2}$ binding energies of the compounds mentioned were 516.76, 517.2, 517.2, and 517.0 eV, respectively [43]. Bond and Flamerz [26] found a similar value for the binding energy of $2p_{3/2}$ in VO_x on TiO_2 which was prepared by grafting or impregnation.

It is known from the literature that the peak ratio I_V/I_{Ti} increases with increasing vanadia content up to the monolayer loading [19]. At higher coverages the slope of the plot of I_V/I_{Ti} versus vanadia load decreases. This has been attributed to the formation of other vanadia phases, probably crystalline V_2O_5 [19]. The V 2p line in the XPS spectrum broadens up to 2.6 eV (FWHM) if the amount of vanadia increases to 30 wt% [33].

TiO_2 is not active in the SCR reaction; however, if TiO_2 is sulfated, which means treated with SO_2 and O_2, it is very active – even as active as V_2O_5 on TiO_2 (Section 1.2.3.3) [45].

The binding energies for Ti $2p_{1/2}$ and Ti $2p_{3/2}$ were 464.3 and 458.6 eV, respectively, and were the same for the sulfated and unsulfated material.

The sulfated samples show a feature at 168.5 eV which was assigned to S 2p of S^{6+}. Catalysts which were subjected to the SCR reaction show two peaks for N 1s in the XPS spectrum, namely at 399.9 and 401.7 eV. These peaks were assigned to the ammonia chemisorbed on Lewis sites and ammonia chemisorbed on Brønsted sites, respectively. This was confirmed by means of an intensity at 1401 cm^{-1} in the FTIR spectrum which was assigned to chemisorbed ammonia on Brønsted sites.

From the infrared spectra it was concluded that the sulfated titania is bridged bidentate, $Ti_2SO_3(OH)$ with C_{2v} symmetry [45].

Bjorklund et al. [46] showed by means of electrical conductance measurements that vanadia on SiO_2/TiO_2 was highly dispersed on the support. Moreover, it was observed by FTIR and XRD that an amorphous vanadate phase was present.

2.3.2 Catalyst Types and Kinetics

Most of the work reviewed in this section is related to the selective reduction of NO over various catalyst types and using a variety of reductants.

Many catalyst compositions are in use or under study for the SCR reaction [4]. Vanadia-type catalysts show high activities for the reduction of NO with NH_3. For zeolites CO, H_2, or hydrocarbons are used as the reductant.

A Supported Vanadia Catalysts

Supported vanadia catalysts and physical mixtures of V_2O_5 and TiO_2 are described, regarding mechanisms and kinetics. Kinetic data for a variety of supported vanadia catalysts are given in Table 4 [28]. Catalysts with TiO_2 as support have the highest activity.

V_2O_5 catalysts are structure sensitive for various reactions. The V=O group located in the (010) plane plays a key role in many reactions. Another aspect of these catalysts is the nature of the active site that lead to the SCR reaction or the oxidation of ammonia when ammonia is used as reductant. In this section the following catalyst types will be discussed: V_2O_5, V_2O_5 on TiO_2, V_2O_5 on TiO_2/SiO_2, noble metals, zeolites, and a variety of other types such as metal oxides and carbon-supported materials.

Ozkan et al. [47] studied the morphological aspects of unsupported vanadia catalysts. Two types of vanadia were prepared. The preparation procedures described above resulted in thick particles (V_2O_5(a)) and thin, sheet-like particles (decomposition of ammonium metavanadate V_2O_5(d)). V_2O_5(a) contains relatively more (010) planes than V_2O_5(d). The selectivity to N_2 for V_2O_5(d) is higher than that for V_2O_5(a), whereas over the latter more N_2O is formed. The (010) planes promote the ammonia oxidation into N_2O more than the reduction of NO into N_2 [47].

The role of ammonia oxidation was studied more comprehensively [36]. From the experimental results it was suggested that ammonia adsorption takes place at two types of sites located on the (010) and (100) planes. V–O–V groups were found to

Table 4. Physicochemical properties of supported vanadia catalysts [28].

Support	S_{BET} ($m^2\,g^{-1}$)	Loading ($\mu mol\,V\,m^{-2}$)	Monolayer fraction	E_A[a] ($kJ\,mol^{-1}$)	ln(TOF)[b]
TiO_2	50	4.14	0.28	65.5	970
SiO_2/TiO_2	108	3	0.20	65.5	500
ZrO_2/TiO_2	42	2.17	0.10	61.7	420
ZrO_2	37	5.27	0.35	61.5	190
Al_2O_3	81	1	0.07	32.9	63
SiO_2	187	3.65	0.24	–	16
Al_2O_3/TiO_2	58	3.59	0.24	65.1	4

[a] Apparent activation energy.
[b] Turnover frequencies in $mmol(NO)\,mol(V)^{-1}\,ks^{-1}$ at 473 K.

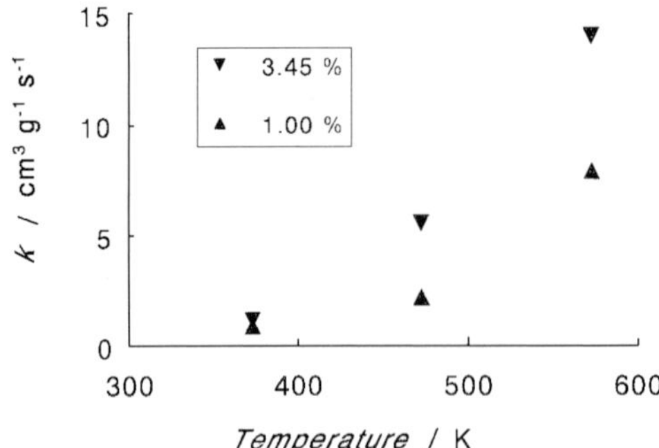

Figure 4. First-order rate constants, k_{NO} ($m^3\,g^{-1}\,s^{-1}$), as a function of temperature for two vanadia-on-titania catalysts: $[NO]_i = [NH_3]_i = 1000$ ppm, $[O_2] = 2\%$, balance N_2, GHSV = 15 000 (adapted from Ref. 49).

be responsible for the SCR reaction, whereas the V=O sites promote the direct oxidation of ammonia to NO and the formation of N_2O from the reaction of NO and ammonia. However, it is stressed here that these studies were carried out with unsupported materials.

Janssen [34] carried out TPR experiments of MoO_3 and V_2O_5 by ammonia and hydrogen. It was concluded that ammonia reduces the oxides forming almost exclusively water and nitrogen. Both catalysts are reduced below 873 K. Above this temperature ammonia decomposes into N_2 and H_2.

The reaction rate of the reaction of NH_3, NO and O_2 over vanadia on titania catalysts at 365 K was 1.9×10^{-18} ($mol\,g^{-1}\,s^{-1}$) and the activation energy 37 kJ mol^{-1} [48]. These data were obtained by using ^{13}NO, a positron emitter, at very low concentrations (5×10^{-9} ppm). Moreover, it was confirmed that the nitrogen atom of ammonia combines with the nitrogen atom of nitric oxide.

Buzanowski and Yang [49] studied the kinetics over both unpoisoned V_2O_5 on TiO_2 and alkali poison-doped catalysts. The results for two unpoisoned vanadia-on-titania catalysts are given in Fig. 4. It was observed that the higher the vanadia load the higher the conversion. Beeckman and Hegedus [50] found a preexponential factor k_0 of $8.64 \times 10^3\,cm\,s^{-1}$ and an activation energy of 80 kJ mol^{-1}. Arrhenius plots of NO conversion rates on powdered material and nine-channel monolith are shown in Fig. 5. (An example of a monolith is shown in Fig. 16 below). Figure 5 clearly shows the strong effect of pore diffusion on the observed rates for the nine-channel monolith.

The technique of preparing the support influences the activity of the catalyst.

Over vanadia on laser-synthesized titania catalysts, large amounts of N_2O were produced [33]; the amount of N_2O increased for vanadia loadings higher than 4 wt%; the overall reaction was

$$4NO + 4NH_3 + 3O_2 = 4N_2O + 6H_2O$$

Kotter et al. [51] prepared V- and Ti-containing catalysts by dissolving appropriate amounts of ammonium vanadate in a titanium oxychloride solution followed by hydrolysis and evaporation. The thus formed 20 wt% V_2O_5 catalyst showed a decrease in BET specific surface area from 85 to 20 $m^2\,g^{-1}$ after heating the sample at 713 K for 100 h. It was suggested that crystalline V_2O_5 was formed during the heat

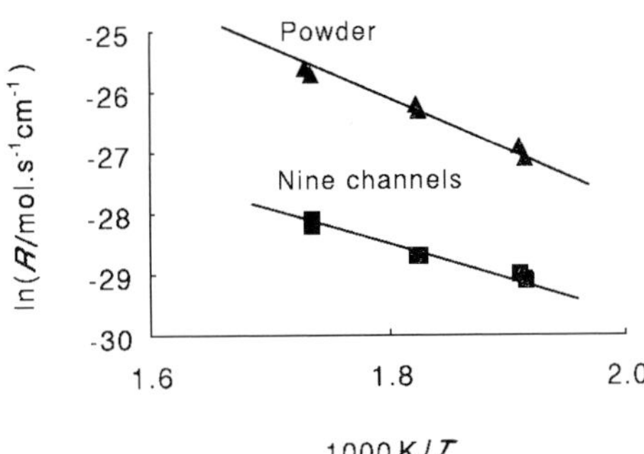

Figure 5. Measured (points) and predicted (lines) NO conversion rates over a vanadia on titania catalyst: $[NO]_i = [NH_3]_i = 1000$ ppm, $[O_2] = 4$ vol. %, balance nitrogen (adapted from Ref. 50).

treatment. Performing the SCR reaction over this pretreated catalyst large amounts of N_2O were formed which, indeed, indicates the presence of crystalline V_2O_5 [18]. The catalytic activity of the catalyst could be restored by washing the catalyst with ammonia [51]. Lintz and Turek [52] determined the intrinsic rates of the following three linearly independent overall reactions over 20 wt% V_2O_5 on TiO_2 in a recirculation system for temperatures ranging from 423 K to 523 K:

$$4NH_3 + 4NO + O_2 \rightarrow 4N_2 + 6H_2O$$

$$4NH_3 + 3O_2 \rightarrow 2N_2 + 6H_2O$$

$$4NH_3 + 4NO + 3O_2 \rightarrow 4N_2O + 6H_2O$$

The catalyst consisted of an alumina plate coated with V_2O_5/TiO_2 mixtures on both sides. The mass specific reaction rate decreased with increasing thickness of the catalyst layer. Intrinsic kinetics were obtained for catalyst layers with a thickness lower than 20 μm. A power law dependence of the main reaction rate on the concentration of NO was found in the absence of water in the feed. Water reduces the activity slightly at temperatures lower than 663 K; however, it increases the selectivity with respect to nitrogen.

The authors presented the rate expression given in eq 3.

$$r_{m,1} = k_{m,1}[NO]^b \left(\frac{[NH_3]}{1 + a[NH_3]}\right) \left(\frac{1}{1 + c[H_2O]}\right) \tag{3}$$

The main reaction rate is valid for NH_3 and NO concentrations ranging from 10^{-6} to 10^{-4} mol L^{-1}, and for water concentrations ranging from zero to 10 vol. % [52]. The values for a, b, c, and $k_{m,1}$ are given in Table 5 [52]. The influence of water on the reactivity of vanadia on titania is hardly described.

Turco et al. [53] studied the influence of water on the kinetics of the SCR reaction over a vanadia on titania catalyst in more detail and found that water inhibits the reaction. The influence of water on the SCR reaction is largest at low temperatures (523–573 K) and low at 623 K. They considered a power rate law (eq 4).

$$r = k[NO]^a[NH_3]^b[H_2O]^c \tag{4}$$

The reaction orders a, b, and c are summarized for three temperatures in Table 6. It

Table 5. Kinetic parameters for the SCR reaction on V_2O_5 on TiO_2 on alumina plate [52].

Temperature (K)	$k_{m,1}$ ($L^{1-b}\,mol^{-b}\,g^{-1}\,s^{-1}$)	a ($L\,mol^{-1}$)	b	c ($L\,mol^{-1}$)
623	11.7×10^3	1.93×10^5	0.7	271
523	7.03×10^3	21.7×10^5	0.59	–

can be seen that the reaction order of ammonia varies slightly with temperature. From the results it was concluded that the sites for ammonia are able to irreversibly adsorb water. The irreversible adsorption of water follows a Temkin isotherm [53]. Topsøe et al. [54] found that water adsorbs on the surface more weakly than ammonia.

Figure 6 shows the effect of water on the NO reduction and the formation of N_2O. Water hydroxylates the surface producing more Brønsted acid sites. Conclusively, water slightly inhibits the catalytic activity at temperatures lower than 663 K and significantly blocks the formation of N_2O at the surface of the catalyst [54]. These results are comparable with those of Odenbrand et al. [55]. The presence of 1 vol.% of water suppressed the formation of N_2O below 775 K, whereas the maximum of the NO conversion shifted to higher temperatures. Goudriaan et al. [56] described a vanadia on titania/silica catalyst in a parallel passage reactor for low temperatures ranging from 400 K to 473 K. The main features of this low-temperature catalyst are:

(i) both Ti and V are highly dispersed and intimately mixed with SiO_2;
(ii) part of the titanium is present in a four-fold coordination with oxygen;
(iii) part of the V is present as V^{4+} species which are stabilized by titanium.

The reaction order in NO is close to unity, whereas the order in NH_3 is lower.

The reduction of NO_2 over V_2O_5 on SiO_2/TiO_2 catalysts was studied by Odenbrand et al. [57] at temperatures ranging from 350 to 700 K. At temperatures lower than 425 K the following reaction occurs:

$$2NO_2 + 2NH_3 \rightarrow N_2 + H_2O + NH_4NO_3$$

In the temperature range 475–625 K the overall reaction is

$$6NO_2 + 8NH_3 \rightarrow 7N_2 + 12H_2O$$

Table 6. Reaction orders of NO(a), NH_3(b), and H_2O(c). [NO] = 100–1000 ppm, [NH_3] = 100–3000 ppm, [H_2O] = 200–3000 ppm, [O_2] = 2.7 vol.% [53].

Reaction order	523 K	573 K	623 K
a	0.82	0.75	0.80
b	0.10	0.12	0.21
c	–0.14	–0.14	–0.10

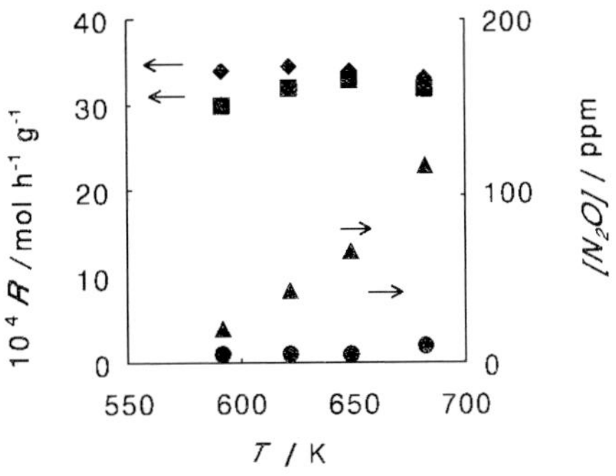

Figure 6. The rate of NO reduction and the amount of N_2O formed as a function of temperature. Diamonds and triangles, and squares and circles correspond to experiments without water and with water, respectively (adapted from Ref. 54).

At temperatures above 595 K ammonia is oxidized homogeneously or according to

$$5NO_2 + 2NH_3 \rightarrow 7NO + 3H_2O$$

Lapina et al. [58] observed by ^{1}H MAS NMR that V_2O_5 on SiO_2/TiO_2 interacts both with SiO_2 and TiO_2. From data obtained from ^{51}V NMR it was found that the structure of vanadium complexes at the surface depends on the sequence of V and Ti deposition. However, the effect of the preparation technique on the SCR reaction was not determined.

Two tetrahedral complexes and one octahedral complex of vanadia species were found when vanadium and titanium were deposited simultaneously. Tetrahedral complexes are less active in redox reactions. All complexes differ from those found on the V_2O_5 on SiO_2 and the V_2O_5 on TiO_2 catalysts.

Zirconia-supported vanadia catalysts are also active in SCR [59]. Most attention was paid to the behavior of submonolayer materials. The catalyst were active in the temperature range 423–623 K. The rate was independent of the partial pressure of O_2 as long as the ratio of the partial pressures $O_2 : NO$ was greater than about nine.

It was concluded that the active site requires two vanadia species in close proximity to one another for the dissociation of ammonia [59].

B Noble Metal Catalysts

Platinum catalysts are subject to many kinds of deactivation, such as:

(i) poisoning by heavy metals, phosphorous, or arsenic;
(ii) deactivation by sulfur oxides, halogen compounds;
(iii) fouling by fly ash.

Thus, platinum as a catalyst in SCR application is limited to nitric acid plants.

Hepburn et al. [60] prepared eggshell and egg-white Rh on γ-Al_2O_3 honeycomb catalysts for the reaction of NO and H_2. In an eggshell-type supported catalyst the active material is deposited at the outer surface of the support, whereas with an egg-white supported catalyst the active material is present in subsurface layers. During the reaction, both N_2 and N_2O are formed with equal selectivities. No ammonia was detected in the product stream. The kinetics fit a simplified Langmuir–Hinshelwood kinetic model in which the rate-limiting step was the surface reaction of an adsorbed NO molecule with an adsorbed H atom. At high concentrations of NO the egg-white catalysts were found to be most active, whereas at low NO concentrations the eggshell catalysts were superior. Both catalyst types are deactivated by SO_2 at nearly identical rates. It was found that fluorine present in the alumina support inhibits the adsorption of SO_2 [60].

Kaspar et al. [61] studied the effect of the rhodium dispersion in rhodium on γ-Al_2O_3 catalysts on the SCR of NO with CO. It was observed that with increasing particle size of rhodium (>1.2 nm) the reaction rate increases.

Platinum foil surfaces have been used to study the SCR reaction. It is found that nitric oxide inhibits the reaction of ammonia and oxygen, while oxygen accelerates the reaction of nitric oxide and ammonia in the temperature range 423–598 K [62]. The reaction order with respect to oxygen was found to be 0.5 over the temperature range. The orders of both NO and ammonia depend on the temperature range. In the low temperature range, 423–493 K, the rate of NO reduction shows zeroth reaction order with respect to NO and NH_3 and the apparent activation energy was $13\,\mathrm{kJ\,mol^{-1}}$. In the high temperature range, 523–598 K, first and half orders with respect to NO and ammonia were found, respectively.

Van Tol et al. [63] studied the kinetics of the reaction of NO and NH_3 using platinum foil and vanadia on platinum in the temperature range 450–560 K by means of Auger electron spectroscopy and mass spectrometry. For platinum the following rate equation was obtained

$$r_{NO} = 5 \times 10^7 [\mathrm{NO}] \mathrm{e}^{-13.6/T}\ (\mathrm{mol\,s^{-1}\,cm^{-2}}) \tag{5}$$

For vanadia on platinum the activation energy is comparable to that of platinum; however, the preexponential factor is a factor of 3000 lower. The active species was found to be V_2O_4 because of the absence of oxygen.

A significant amount of work has been carried out using single crystals of platinum [64]. The activity and selectivity for the reactions of NO and H_2, and NO and NH_3 were performed over a Pt(100) single crystal and the changes in the surface structure and the gas-phase composition were measured by means of low energy electron diffraction (LEED) and mass spectrometry, respectively. Kinetic oscillations (Section A.5.2.7) were found in both systems at a temperature of about 550 K. These oscillations were explained as a very fast periodic depletion of adsorbate coverage followed by refilling of the adsorbate layer [64].

Oscillatory behavior of the reactions NO + NH_3 and NO + H_2 over Pt and Rh single crystals were observed by Van Tol et al. [63] and Janssen et al. [65]. It was shown that the process was structure sensitive [65]. Platinum on $AlPO_4$ and on $B_2O_3/SiO_2/Al_2O_3$ catalysts were studied in the NO reduction with propene as reductant in the presence of SO_2 [66]. The activity of these two systems were compared with that of Pd, Ir, and Ru on $B_2O_3/SiO_2/Al_2O_3$. The maximum conversion over all catalysts was found at a temperature of 473 K. The platinum-containing catalysts showed the highest conversion. Large amounts of N_2O were produced. It was speculated that the role of oxygen is to activate the hydrocarbon as well as to convert NO into NO_2. Sulfur dioxide enhances the activity and decreases the formation of N_2O. From these results it was suggested that the selective reduction of

NO proceeds via an NO_2 intermediate. This species was then reduced by an oxygen-activated hydrocarbon.

C Zeolite Catalysts

Zeolites are often applied for the reduction of nitric oxide in off-gases of diesel or lean-burn gasoline engines.

Centi et al. [67] found that the activity of the Cu-ZSM-5 in SCR was higher than that of V_2O_5 on TiO_2. The activity over Cu-ZSM-5 catalysts depends on the Si : Al ratio [68]. Decreasing the Si : Al atomic ratio increases the activity. The active site consists of two adjacent copper ions according to Moretti [68]. It is well known that the activity of copper-exchanged ZSM-5 catalysts decreases in the presence of water at high temperatures. Moreover, NO_2 formation is necessary but not a sufficient condition for the SCR of NO over Cu-ZSM-5 [69].

In most studies on zeolites, alkanes and alkenes are used as the reductant; alkanes are less active than alkenes. Gopalakrishnan et al. [70] studied the reduction of NO by propane over copper-exchanged zeolites, namely mordenite, X-type and Y-type, at temperatures ranging from 473 to 873 K. The following overall reaction occurs:

$$C_3H_8 + 10NO \rightarrow 5N_2 + 3CO_2 + 4H_2O$$

The activity appeared to increase in the order:

$$\text{Cu-X} < \text{Cu-Y} < \text{Cu-mordenite} < \text{Cu-ZSM-5}.$$

With the catalyst Cu-ZSM-5 high conversions of NO (>90%) were obtained. The molar ratio of propane and NO was 2 and the reaction was carried out at 674 K, and at a space velocity of about 100 000 h^{-1}. An oxygen concentration of about 1% promotes the reaction; however, water shows a detrimental effect on the activity. Above 673 K the oxidation of propane competes with the reduction of NO. One of the disadvantages of NO reduction with propane is the incomplete conversion of propane. Propane is mainly converted to CO_2 and to a small amount to CO [70].

Radtke et al. [71] found that hydrogen cyanide and N_2O were formed during the SCR reaction over Cu-ZSM-5 catalysts with propene or ethene as reductant and in the presence of excess oxygen at temperatures below 700 K and above 450 K. With propene as reductant a maximum amount of HCN corresponding to 3% of the inlet concentration of NO was formed at 500 K, whereas with ethene this temperature shifted to 600 K. The maximum amount of N_2O was produced at 650 K with both reducing agents.

Shelef et al. [69] specified the role of the NO_2 intermediate when the catalyst is Cu-ZSM-5 and the reductant is a hydrocarbon. It was shown that in the absence of oxygen NO_2 decomposes into NO and O_2, whereas under strongly oxidizing conditions NO_2 and NO are reduced to N_2. These authors compared Cu-ZSM-5, H-ZSM-5, and CuO on γ-Al_2O_3; again, Cu-ZSM-5 showed the highest activity.

In the absence of oxygen and in the presence of excess propene, NO_2 is converted into NO which implies that Cu^{2+} plays a prominent role in the selective reduction to N_2. With propane or propene as reductant 80% and 55% of NO_2, respectively, is converted into N_2 at a very high space velocity (500 000 h^{-1}). Water inhibits the reaction of NO_2/NO over Cu-ZSM-5 decreasing the activity by about 60% at 683 K. At a temperature of 800 K this percentage was about 25% [69].

Petunchi et al. [72, 73] have shown that isobutane is an effective reductant in excess oxygen over Cu-ZSM-5 zeolite at temperatures ranging from 573 to 773 K. Oxygen is necessary to form NO_2 and for maintaining a favorable Cu^{2+}/Cu^{+} balance. NO_2 reacts with i-C_4H_8 to CO_2, N_2 and H_2O.

Recently, Komatsu et al. [74] published a study on the SCR of NO with ammonia in the presence of oxygen over Cu-ZSM-5 catalysts. From this study it was concluded that up to temperatures of 573 K the reaction is selective for N_2. Above that temperature, however, the oxidation of ammonia by oxygen from the gas phase occurs. Moreover, it was stated that the specific activity of Cu^{2+} in Cu-mordenite is comparable to that of Cu^{2+} in Cu-ZSM-5. The reaction order of oxygen was 0.6 and first order and zeroth order with respect to the NO and ammonia concentrations, respectively. The activation energy appeared to be 49 kJ mol^{-1}. A reaction mechanism, different from that for V_2O_5 was suggested. Ammonia adsorbs on the copper ions. NO and dissociated oxygen react with the bridging oxygen of two copper ions forming a nitrato complex. This step in the mechanism is supposed to be the rate-determining step. NO attacks the nitrato group forming two nitrito groups. These groups react with the adsorbed ammonia molecules forming N_2 and H_2O [74].

Burch and Scire [75] studied the reaction of NO and H_2 over Pt-, Rh-, Co-, and Cu-ZSM-5. On these catalysts the reaction starts at 323 K, 423 K, 523 K, and 623 K, respectively. The reactivity of the catalysts in decreasing order is: Pt > Rh > Co > Cu. Over the platinum catalyst ammonia was formed which resulted in a decrease of the activity at temperatures ranging from 373 K to 423 K. Hydrogen was more reactive than ethane or methane. A redox mechanism was suggested. NO is adsorbed and decomposes on active sites leading to the formation of N_2 and adsorbed oxygen. Oxygen is removed by the reducing agent, thus restoring the active site.

Li and Armor [76] used metal-exchanged zeolites for the reduction of nitric oxide with methane in the presence of excess oxygen. They used ZSM-5 and zeolite Y exchanged with Co, Cu, Fe, Cr, and Na. Na-ZSM-5 is completely inactive for the reduction of NO with methane. The activity increases in the order Cr < Fe < Cu < Co. In the absence of oxygen the nitric oxide conversion was 17%. The selectivity of CH_4 over a Co-ZSM-5 catalyst decreased with increasing concentration of oxygen in the feed. The presence of oxygen significantly enhances the NO reduction activity. The following overall reactions were suggested:

$$2NO + CH_4 + O_2 = N_2 + 2H_2O + CO_2$$

$$CH_4 + 2O_2 = 2H_2O + CO_2$$

The apparent reaction order of NO and CH_4 over the Co-ZSM-5 catalyst were 0.44 and 0.56, respectively.

Co-ZSM-5 is very appropriate for the simultaneous removal of NO and N_2O at 723 K [77]. It was found that in the absence of oxygen nitrous oxide serves as a source of oxygen. The results of the simultaneous removal are shown in Fig. 7.

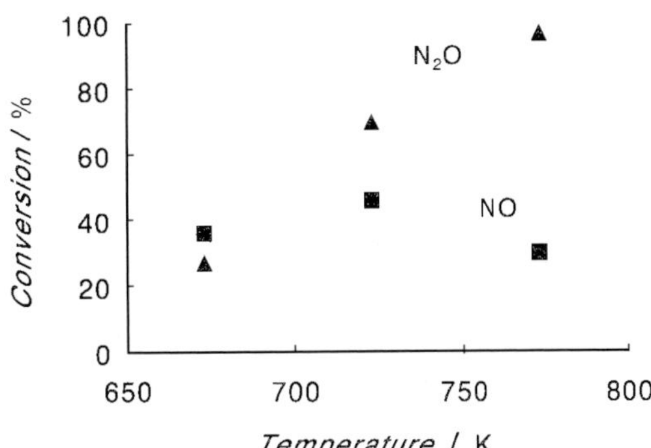

Figure 7. The simultaneous removal of NO and N_2O over Co-ZSM-5 (adapted from Ref. 77).

Yokoyama and Misono [78] concluded from their studies that Ce-exchanged ZSM-5 was more active than Cu-ZSM-5 at 573 K and at a low space velocity of 10 000 h^{-1}.

Li and Armor [79] studied NO reduction by CH_4 in the presence of excess O_2 over Ga-containing H-ZSM-5, H-mordenite, and γ-Al_2O_3. The "gallium" catalysts were prepared by a variety of methods, such as ion exchange of the ammonia form of the zeolite with gallium nitrate solutions and incipient wetness impregnation. The activity of all catalysts thus prepared were compared with the activity over Co-ZSM-5. The most active catalyst at temperatures ranging from 723 to 823 K was Ga (4.2 wt%) on H-ZSM-5 made by ion exchange. The activity was comparable to that of Co-ZSM-5. However, gallium catalysts are more selective than cobalt catalysts when water is absent in the feed. Water in the feed decreases the activity more, as compared to Co-ZSM-5 [79].

Methane was used as reductant over both palladium ion-exchanged H-ZSM-5 and palladium–cerium ion-exchanged H-ZSM-5 [80]. The conversion as a function of the temperature for Pd/H-ZSM-5 and Pd–Ce/H-ZSM-5 are roughly similar and show a maximum at 723 K. A conversion of about 70% for NO and 65% for CH_4 were observed at that temperature.

In Fig. 8 the NO conversion for the reduction with methane is shown as a function of the temperature for four metal-containing catalysts. The activity shows a maximum in the temperature range 723–773 K. The sequence of increasing activity is Pd > Ru > Pt > Cu.

Montreuil and Shelef [81] studied the SCR reaction over ZSM-5/Al_2O_3 supported on a cordierite monolithic substrate. Various oxygenated hydrocarbons, such as methanol, ethanol, propanol, acetaldehyde, acetone, methyl ethyl ketone, and 1,4-dioxane were used as reductant at 755 K [81]. The reaction extent with the various organic compounds was compared with that obtained when using propene as reductant. Propene was the superior reductant. Methanol as a reductant showed a very low NO conversion, whereas propanol showed the highest activity, albeit lower than that of propene. Upon adding oxygen to the feed the difference between propanol and propene vanishes.

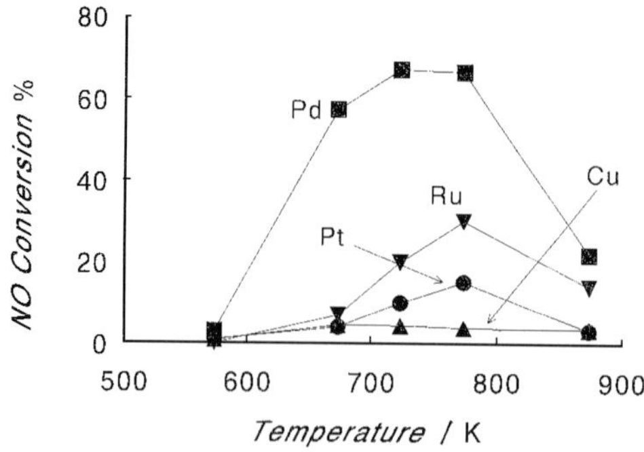

Figure 8. The reduction of NO with methane over H-ZSM-5 catalyst: $[NO]_i = 1000$ ppm, $[CH_4]_i = 2000$ ppm, $[O_2] = 2$ vol.%, balance He; catalyst weight 0.5 g; GSVH = 9000 h^{-1} (adapted from Ref. 80).

D Carbon-Supported Catalysts

Carbon supported catalysts, are also used as SCR catalysts. Treatment of activated carbon by sulfuric acid at reaction temperatures above 423 K enhances the activity for the SCR reaction at temperatures above 423 K. During this treatment surface oxides such as carboxyl and carbonyl groups are formed which serve as adsorption sites for NO, O_2, and NH_3. NO conversion increases significantly by increasing the gas-phase oxygen concentration.

Fe^{3+}-promoted active carbons were tested in the SCR reaction with ammonia as reductant [82]. Active carbon was oxidized in concentrated nitric acid solution, and, after washing and drying, impregnated with an aqueous solution of iron nitrate. The activity of the carbon-supported catalysts was dependent on the oxidative pretreatment of the support and on the drying and calcination steps during the preparation of the catalysts. It was found that the molar ratio of oxygen and carbon increases for higher oxidation temperatures.

Imai et al. [83] studied the reduction of NO by carbon present in activated carbon fibers. Particles of α-FeOOH were present on the fibers in order to absorb NO in a dimerized form. Iron hydroxide enhances NO decomposition and N_2 formation at 473 K. Intermediate steps were the disproportionation of NO into N_2O and NO_2. NO_2 was reduced by carbon to CO_2 and NO again.

E Promoted Oxide Catalysts

It has been shown by Bjorklund et al. [84] that vanadia on TiO_2/SiO_2 catalysts are promoted by iron and copper. Iron-promoted catalysts showed the highest activities compared to copper-promoted and nonpromoted catalysts. The activity increased from 92% NO conversion up to 97% after 200 h on stream. Introducing SO_2 into the gas resulted in an increase of activity for the iron-promoted catalyst and a rapid decrease of activity for the copper-promoted catalyst.

Weng and Lee [85] found that the number of active sites, and of Lewis and Brønsted acid sites as well, increase with increasing amounts of niobium oxide in vanadia/titania catalysts. The highest activity of the promoted catalysts was found at 573 K. At lower and higher temperatures ammonia is oxidized.

Chen and Yang [45] reported first results on sulfated titania as a SCR catalyst. Sulfated catalysts were obtained by means of two procedures. Oxidation of SO_2

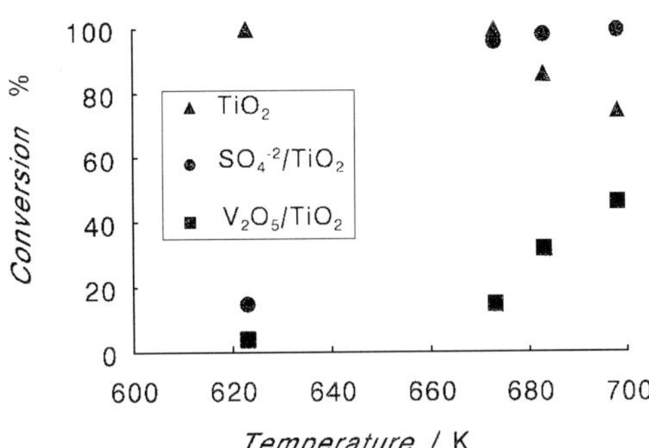

Figure 9. The conversion of NO over TiO_2, SO_4^{2-} on TiO_2, and V_2O_5 on TiO_2: $[NO]_i = [NH_3]_i = 1000$ ppm, $[O_2]_i = 2\%$, balance N_2; GHSV $= 15\,000\ h^{-1}$ (adapted from Ref. 45).

over the catalyst resulted in a sulfated catalyst. Sulfation was also obtained by passing a mixture of NO, NH_3, O_2, H_2O, and SO_2 over the catalyst. The resulting catalysts exhibited different behavior. Three materials were compared: TiO_2, SO_4^{2-} on TiO_2, and V_2O_5 on TiO_2. The results are shown in Fig. 9 [45]. The decrease of the NO conversion at higher temperatures is caused by the oxidation of ammonia.

Water has a negative effect on the SCR reaction because it competes with ammonia for chemisorption. In the presence and absence of water the rate constants at 673 K are about 17 and 27 $cm^3\ g^{-1}\ s^{-1}$, respectively. It was suggested that sulfation increases both Lewis and Brønsted acidity. The existence of Lewis sites was proved by adding water to the catalyst which results in an increase in Brønsted acidity. Brønsted sites were observed by means of infrared spectroscopy.

Hadjiivanov et al. [86] studied the compounds formed during adsorption of NO_2 on anatase by means of infrared spectroscopy. They detected a variety of compounds such as: (i) Ti–O_2NOH species; (ii) monodentate and bidentate nitrates coordinated with water; and (iii) nitrates and NO^+. The adsorption of NO_2 neutralizes the Lewis sites under formation of nitrates and generates Brønsted sites. Adsorption of NH_3 on NO_2–TiO_2 surfaces leads to the formation of NH_4^+.

F Chromia Catalysts

The reaction of NO and NH_3 in the presence of oxygen and the oxidation of ammonia were studied over chromia with respect to its morphology [87]. Amorphous chromia exhibits a very high activity in the SCR reaction at low temperatures (420 K). Still, the selectivity is lower than that over vanadia on titania [88]. The ratio of the conversions of NH_3 and of NO appeared to be greater than 1. Higher amounts of N_2O were formed over chromia compared with the amounts formed over vanadia on titania. A 90% conversion of NO was obtained at a temperature of 420 K. The apparent activation energy was found to range from 45 to 55 $kJ\ mol^{-1}$. All reactions were carried out in the absence of water. Amorphous chromia and crystalline α-chromia were compared for their behavior in the following gas mixtures: $NH_3 + NO + O_2$; $NH_3 + O_2$; $NH_3 + NO$ [87]. α-Chromia has a higher specific activity for the oxidation of ammonia into N_2 than the amorphous type, whereas the specific activity for the oxidation of ammonia into N_2O is equal for both types of catalysts for temperatures ranging from 393 to 473 K. The authors draw the conclusion that the oxidation sites for the oxidation of ammonia into N_2 are more abundant on α-chromia and, thus, that the oxidation resulting in N_2 is more structure-sensitive than the oxidation to form N_2O.

The reaction of NH_3 and NO produces mainly N_2O over both types of catalysts. The reaction rate of NH_3, NO, and O_2 over amorphous chromia appeared to be four times that over crystalline chromia. The amorphous chromia catalyst is very selective for N_2, whereas the crystalline chromia produces both N_2 and N_2O.

Schraml-Marth et al. [89] studied the two types of chromia by means of diffuse reflectance FTIR and Raman spectroscopy. Amorphous chromia has a higher density of labile oxygen sites than that of crystalline chromia in the temperature range 423–473 K and is accordingly more active.

G Miscellaneous Oxide Catalysts

Hosose et al. [90] carried out the reduction of NO using ethene as reductant over copper oxide on SiO_2/Al_2O_3 catalyst at temperatures ranging from 473 to 900 K. Conversion for ethene and NO was 100% and 15%, respectively, at temperatures above about 750 K.

The effect of water and SO_2 on the selective reduction of nitric oxide by propene over alumina supported Ag, In, Ga, Sn, and Zn were studied by Miyadera and Yoshida [91] (Fig. 10). These catalysts show high activities in the absence of both water and SO_2. In addition it was indicated that the activity in the absence of water is, in the order of decreasing activity, Ga > In > Ag > Zn > Sn. With water in the feed this sequence became Ag > In > Sn > Zn > Ga. The activity of the "gallium" catalyst decreased from 98% to 36%; SO_2 had little effect on the activity. From this study it was concluded that the Ag on Al_2O_3 catalyst was most promising.

Unsupported manganese oxides were used as catalysts in the SCR reaction at temperatures ranging from 385 to 575 K [92]. It was found that the activity per unit surface area was, in the order of decreasing activity, MnO_2 > Mn_5O_8 > Mn_2O_3 > Mn_3O_4 > MnO. However, Mn_2O_3 is preferred because it shows a very high selectivity with respect to the formation of N_2. The highest activity for most of the oxides was found at a temperature of about 475 K. Nitrous oxide occurs over crystalline manganese oxides.

Vanadium phosphorus oxide was found to be a catalyst for the reduction of NO with butane and in the absence of oxygen at temperatures ranging from 766 to

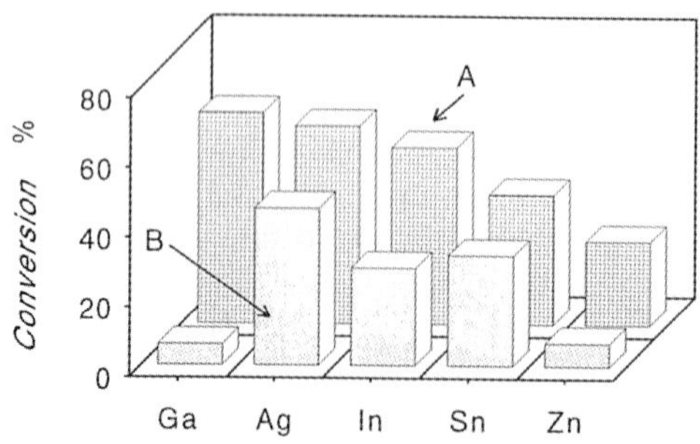

Figure 10. The conversion of NO into N_2 for five alumina-supported catalysts: (A) $[NO]_i$ = 1021 ppm, $[C_3H_6]_i$ = 667 ppm, $[O_2]_i$ = 10 vol.%; (B) A + $[H_2O]_i$ = 10 vol.%, $[SO_2]_i$ = 200 ppm, balance helium. Temperature 723 K, SV 6400 h^{-1} (adapted from Ref. 91).

816 K [93]. The NO conversion increases with increasing butane concentration. This catalyst contains vanadium in a stabilized 4^+ oxidation state. In this state vanadium is active in the NO decomposition.

It was observed that the oxides of lanthanum and copper supported on ZrO_2 show a very high SCR activity at 573 K [94]. The selectivity towards nitrogen was 80% and towards N_2O 20%. At 473 K the selectivity changes dramatically: for N_2 12% and for N_2O 88%. It was suggested that the oxides were in the form of La_2CuO_4. The high activity was ascribed to the high dispersion of copper on the support surface. It is assumed that the presence of lanthanum and the +2 valency of the copper ion are causes for this dispersion.

Perovskite-type mixed oxides such as $LaMO_3$ and $La_{1-x}Sr_xMO_3$ (M = Co, Fe, Mn, Cr, Ni; $0 < x < 0.4$) were applied to the reactions of NO and CO, CO and O_2, and CO and N_2O in a closed circulation system at 573 K [95]. The oxides were pretreated with O_2 at 573 K or 673 K for 1 h. The products from the reaction of CO and NO were N_2, N_2O, and CO_2, whereas from the reaction of N_2O and CO, N_2 and CO_2 were obtained. Increasing the value of x in $La_{1-x}Sr_xCoO_3$ resulted in a decrease of the reaction rate over $LaCoO_3$. For $x = 0.2$ the rate of the reaction was half that over $LaCoO_3$. It was suggested that for higher x values $La_{1-x}Sr_xCoO_3$ is more readily reduced by CO. The following overall reactions were proposed

$$2NO + CO \rightarrow N_2O + CO_2$$

$$N_2O + CO \rightarrow N_2 + CO_2$$

Lavados and Pomonis [96] studied the effect of substitution in the perovskites $La_{2-x}Sr_xNiO_{4-\lambda}$ on the reaction of CO and NO; x varied from 0 to 1.5 and λ from -0.15 to 0.505. Again, the activity decreased with increasing values of x.

It has been shown that small amounts of silica deposited on alumina enhance the catalytic activity for the SCR reaction using C_2H_4 as a reductant [97]. Silica was deposited on alumina by chemical vapor deposition by using alkoxy silanes such as $(CH_3)_xSi(OC_2H_5)_{4-x}$ ($x = 0, 2$) and $Si(OCH_3)_4$ at room temperature. Catalysts prepared from $(CH_3)_2Si(OC_2H_5)_2$ with a silicon loading of 2.8 wt% exhibit a conversion of about 55% at 800 K. Under these conditions about 40% of C_2H_4 was converted into CO and CO_2.

The possible role of isocyanate species in the reduction of NO by hydrocarbons over Cu/Cs on alumina catalysts has been reported [98]. Water inhibits the formation of isocyanate species. This effect is very small if ethyne or *n*-heptane are the reductants. The isocyanate species were detected by means of infrared spectroscopy. The isocyanate band appeared at 2234 cm^{-1}.

In Germany iron oxide/chromium oxide catalysts have been used since the 1960s. Flockenhaus [99] describes an application for SCR downstream in a natural-gas fired system; NO_x conversions of about 70% for a NH_3 : NO ratio of 0.8 were obtained.

Cant and Cole [100] investigated the photocatalytic reaction between nitric oxide and ammonia over TiO_2 wafers which were placed in a reaction chamber based on a standard UHV cross. These samples were ultraviolet illuminated while recording the infrared spectrum of the wafer. The reactions of NH_3 and NO and the reaction of NH_3 and O_2 were studied. Nitric oxide decomposes over illuminated TiO_2. The rate of this reaction, however, is lower than that of the reaction of NO and NH_3. The reaction of NH_3 and NO produces N_2 and N_2O in a fixed ratio of about 2. Nitrous oxide production appeared to be zero order in ammonia and showed a Langmuir-like dependence on oxidant pressure.

2.3.3 Mechanisms

Three types of vanadium-containing species are present at the surface of the vanadia on titania catalysts. Monomeric vanadyl, polymeric vanadates, and crystalline vanadia depending on the vanadia loading (Section 2.3.2). Moreover, Brønsted acid sites and Lewis sites are present at the surface of vanadia on titania catalysts. All species were needed to explain the mechanism of the SCR reaction. The active sites are related to previously mentioned vanadia/vanadium species. A variety of active sites were proposed such as: two adjacent V^{5+}=O groups or coordinated vanadyl centers [44, 53, 103, 104]; V^{5+}=O and surface hydroxyl groups [105]; Brønsted sites such as vanadium hydroxyl groups [45]; an oxygen vacancy and hydroxyl groups [106].

Several mechanisms or reaction schemes were proposed by various researchers based on their experimental results. In this section a summary of these mechanisms is given.

The elucidation of the reaction mechanism of the SCR reaction has been carried out using a variety of techniques. Transient studies with isotopically (oxygen-18 and nitrogen-15) labeled molecules have been performed [104, 107]. Spectroscopic studies of the working catalysts were performed by Went et al. [30] and Topsøe [108] using laser Raman spectroscopy and FTIR, respectively.

Two types of mechanisms are proposed in the literature: the Langmuir–Hinshelwood and the Eley–Rideal mechanisms. Takagi et al. [109] presented a study with V_2O_5 on γ-Al_2O_3. It was suggested that the reaction proceeds via the adsorbed species of NO_2 and NH_4^+ through a Langmuir–Hinshelwood mechanism. It has been shown by IR that ammonia was adsorbed as both NH_4^+ (1410 cm^{-1}) and physisorbed ammonia on alumina (1610, 1275 cm^{-1}). Using FTIR it was shown that NO adsorbed as NO_2.

However, several research groups have proposed the Eley–Rideal mechanism for describing the mechanism of the SCR reaction [45, 104, 105, 107]. Miyamoto et al. [107] proposed a dual-site Eley–Rideal mechanism which is a better description of the mechanism. In this mechanism ammonia is strongly adsorbed as NH_4^+. NO from the gas phase reacts with the adsorbed ammonia species to form N_2, H_2O, and V–OH. The latter is oxidized by oxygen from the gas phase or lattice oxygen to V^{5+}=O. The overall reaction equations are

$$NO + NH_3 + V{=}O \rightarrow N_2 + H_2O + V{-}OH$$

$$2V{-}OH + O \rightarrow 2V{=}O + H_2O$$

The mechanism of the selective reduction of nitric oxide with ammonia in the

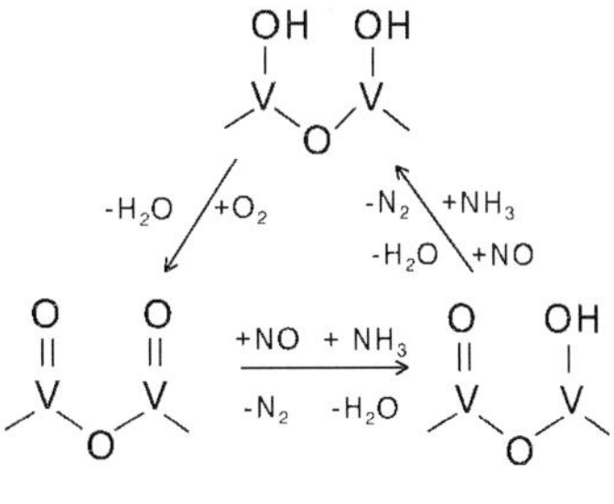

Figure 11. Summarized reaction mechanism of the selective reduction of NO with NH_3 in the presence of oxygen (adapted from Ref. 104).

presence of both labeled oxygen and ammonia over a series of catalysts consisting of unsupported V_2O_5, V_2O_5 on TiO_2, V_2O_5 on SiO_2/Al_2O_3, and V_2O_5 on Al_2O_3 has been reported by Janssen et al. [104]. The experimental results confirm that lattice oxygen participates in the reaction. Water is formed at two sites: the first part comes from the reaction of gaseous nitric oxide and adsorbed ammonia via an Eley–Rideal mechanism; the other part comes from a surface dehydration process. Two adjacent $V^{5+}{=}O$ groups were supposed to be the active site. From the work of Lietti and Forzatti [41] it appeared that the oxygen atoms of monomeric species are less reactive than those of polymeric metavanadate species.

NO oxidizes the prereduced catalyst to the same extent as oxygen, but at a much lower rate. During the reaction, most of the N_2 and N_2O is formed from one nitrogen atom from nitric oxide and from one nitrogen atom from the ammonia, irrespective of whether oxygen is present or absent.

No experimental evidence for isotopic scrambling of gaseous and lattice oxygen at temperatures ranging from 575 to 675 K was observed. Molecular oxygen oxidizes V–OH groups. Janssen et al. [104] have suggested that three combinations of the groups V=O and V–OH are present on the surface of the catalyst at the start of the reaction. These species are in equilibrium due to the mobility of hydrogen atoms of the OH groups. Figure 11 summarizes the reaction scheme as suggested by Janssen et al. [104].

Amide-like species NH_2 have also been shown to be present at the surface [30, 44, 104].

The formation of N_2O is also described in detail [55, 104, 105].

It has been suggested by Kantcheva et al. [105] that two molecules of NO adsorb on a doublet oxygen vacancy with simultaneous release of N_2 and N_2O into the gas phase. The oxygen vacancies are reoxidized more rapidly by O_2 than by NO. This explains the enhancing effect of oxygen on the rate of the overall reaction of NO, NH_3 and O_2.

During the SCR reaction V^{4+} species are present on the surface of the catalyst. Biffar et al. [110] showed that the concentration of V^{4+} in the catalyst is a function of the oxygen concentration in the gas phase (Fig. 12).

Bjorklund et al. [46] found that vanadia loadings up to 10–15 wt% did not influence the electrical conductance in TiO_2/SiO_2-supported vanadia catalysts. Above 15 wt% the electrical conductivity increases rapidly which may be ascribed to the presence of V^{4+}. Admitting NO or NH_3 or both to the catalyst, the conductivity

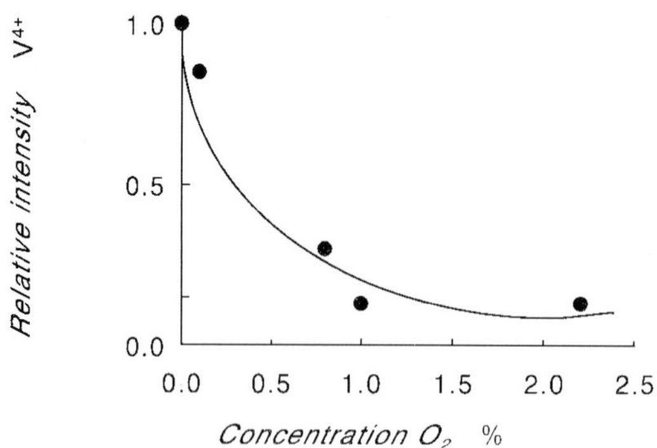

Figure 12. The relative ESR signal intensity of V^{4+} in a vanadia on titania catalyst as a function of the gas-phase oxygen concentration (adapted from Ref. 110).

increased by a factor of 3–4. Increasing the oxygen concentration from 0 to 1.5 vol. % resulted in a decrease of the conductivity, which is in agreement with the results of Biffar et al. [110].

Gasior et al. [111] stated that the amount of N_2 formed is not a measure of the concentration of V=O groups at the surface. In the absence of oxygen the following reactions occur:

$$V^{5+}\text{–OH} + NH_3 \rightarrow V^{5+}\text{–ONH}^4$$

$$2V^{5+}\text{–ONH}_4 + V^{5+}\text{–O–}V^{5+} + 2NO + 2e^-$$
$$\rightarrow 2N_2 + 3H_2O + 2V^{5+}\text{–OH} + V^{4+}\square V^{4+}$$

In the presence of oxygen the vacancy becomes oxidized according to

$$V^{4+}\square V^{4+} + \tfrac{1}{2}O_2 \rightarrow V^{5+}\text{–O–}V^{5+} + 2e^-$$

Odenbrand et al. [55] extended Gasior's model for the formation of N_2O. The active site was suggested to be an oxygen vacancy with one vanadyl group:

$$4V{=}O + 2V^{4+}\square V^{4+} + 4NH_3 = 4V\text{–}NH_2 + 4V^{5+}\text{–OH}$$

$$4V\text{–}NH_2 + 4NO = 4N_2O + 4H_2O + 2V^{4+}\square V^{4+}$$

Again, the oxygen vacancy is oxidized. Further oxidation of the amido group at higher temperatures leads to

$$4V\text{–}NH_2 + 3O_2 = 2N_2O + 4H_2O + 2V^{4+}\square V^{4+}$$

However, Bjorklund et al. [46] were not able to distinguish between the two mechanism proposed by Janssen et al. [104] and Gasior et al. [111] by using electrical conductance measurements. If ammonia adsorbs on vanadia at room temperature two bands at $1610\,cm^{-1}$ and at $1420\,cm^{-1}$ occurred in the infrared spectrum. These are assigned to coordinatively held ammonia and NH_4^+, respectively. Carrying out this experiment at 623 K only the band at $1610\,cm^{-1}$ was left indicating that ammonia is coordinatively bound to the surface as proposed by Janssen et al. [104].

Chen and Yang [45] concluded from their studies that the active sites for SCR reaction over sulfated titania are Brønsted sites and the mechanism is Eley–Rideal; 52% of the chemisorbed ammonia was active in the SCR reaction while 48% of the chemisorbed ammonia was inactive.

Ciambelli et al. [38] proposed a model for the vanadium oxide monolayer on titania. At loadings of vanadia below 6 wt%, a V^{4+} species prevails. At higher loadings the number of V^{5+} species increases, and thus the number of Brønsted sites grows.

Recently, Schneider et al. [112] characterized the active surface species of vanadia on titania by means of diffuse reflectance FTIR. Their results supported the mechanism proposed by Janssen et al. [104]. From the signal intensities at 1435 and 1660 cm^{-1} it was concluded that ammonia adsorbs on Brønsted sites [112]. The 1435 cm^{-1} band was assigned to the $\nu_4(F)$ bending mode of ammonia adsorbed on Brønsted sites. The corresponding $\nu_2(E)$ feature was observed as a shoulder at 1660 cm^{-1}.

Ammonia adsorption on Lewis sites is stronger than that on Brønsted sites [113]. In situ infrared spectroscopy has been used to monitor surface coverages by various species under reaction conditions. Temperature programmed desorption shows that no NO decomposition occurs in the temperature range 100–600 K. By means of in situ FTIR spectroscopy it was observed that the fractional surface coverages by ammonia on the Brønsted and Lewis acid sites were 0.26 and 0.39, respectively, at 573 K. No adsorption of NO was found. Moreover, it was stated that water does not block the sites for ammonia adsorption.

Odenbrand et al. [55] observed that the selectivity toward nitrogen increases upon addition of water, additionally, the formation of nitrous oxide is suppressed at temperatures below 775 K.

A kinetic model was proposed by Turco et al. [53] with a nitrosamide as an intermediate compound and vanadyl groups were supposed to be the active sites.

Went et al. [30] used TPD and laser Raman spectroscopy to determine the structure of the catalyst and of the adsorbed species. It was found that the specific activity of polymeric vanadate species was 10 times greater than that of monomeric vanadyls at 500 K. Monomeric species produce N_2 both in the presence and in the absence of oxygen, whereas polymeric species produce both N_2 and N_2O. The selectivity towards N_2O increases with increasing O_2 concentration in the feed.

Duffy et al. [106] carried out the reaction of ^{15}NO and $^{14}NH_3$ over V_2O_5 on TiO_2 and over α-Cr_2O_3. The major product over both catalysts is $^{14}N^{15}N$. A simple mechanism was proposed:

$$2V\square + V\text{–}OH + {}^{15}NO \rightarrow V{=}{}^{15}NH + 2V{=}O$$

The adsorbed NH group could be hydrolyzed into $^{15}NH_3$ and a V=O group. By-products were found: over vanadia on titania $^{14}N^{15}NO$, and over chromia $^{14}N_2$, $^{15}N_2$, and $^{15}N_2O$.

In situ FTIR has been used to study the reaction steps and intermediates in vanadia on titania catalysts [108]. The changes in concentration of surface sites and adsorbed species of the working catalyst were also obtained. A relationship between the concentration of the Brønsted acid sites and the NO_x conversion was found. Adsorption of ammonia results in two ammonia species. Bands at 3020, 2810, 1670, and 1420 cm^{-1} are assigned to adsorbed NH_4^+ species, whereas intensities at 3364, 3334, 3256, 3170 and 1600 cm^{-1} are responsible for coordinated NH_3. A mechanism was suggested in which, initially, ammonia is adsorbed on $V^{5+}{=}O$ forming an activated species. Then NO reacts with this species forming $V^{4+}{-}OH$, N_2, and H_2O. Subsequently the vanadia species is reoxidized by NO or O_2 in order to obtain $V^{5+}{=}O$.

The reaction of NO, NO_2, and NH_3 over a CuO/NiO on α-Al_2O_3 catalyst was studied by Blanco et al. [114]. Ammonia chemisorbs on the catalyst surface forming NH_2 species, whereas copper is reduced to Cu^+. Evidence for the presence of Cu^+ was obtained by means of XPS and FTIR. Then NO_2 oxidizes Cu^+ into Cu^{2+} to form NO. This adsorbed NO reacts with the NH_2 species through a Langmuir–Hinshelwood mechanism. NO present in the gas phase can react with the catalyst through an Eley–Rideal mechanism forming a reduced copper site. The Cu^+ sites are promptly reoxidized by O_2 or NO_2; O_2 plays a minor role in the reoxidation.

So far no exclusive reaction mechanism is found for the SCR reaction over vanadia on titania catalysts which can explain all experimental observations. The Eley–Rideal mechanism, however, is favored by many investigators. Ammonia is adsorbed on the surface on Brønsted or Lewis sites, as NH_4^+ ion or as an NH_2 species, respectively. Then NO from the gas phase reacts with the ammonia species forming nitrogen and water leaving oxygen vacancies behind. Thus, lattice oxygen is involved in the reaction mechanism. These vacancies are then occupied by oxygen from the gas phase. More information about the vanadium-containing species at the surface of vanadia on titania catalyst was obtained in the last decade which has helped to explain the reaction mechanism. Isolated vanadyl and polymeric vanadyl species were detected by means of spectroscopic techniques. Polymeric vanadyls are more reactive than isolated vanadyls. The composition of the products of the SCR reaction probably depends on the amount of surface species present at the surface. This issue has still to be clarified.

2.3.4 Deactivation of SCR Catalysts

The major concern in applying selective catalytic reduction is deactivation or poisoning of the catalyst. One cause of deactivation may be catalyst poisons present in the flue gases.

Clearly, a decrease in the activity of the catalyst for SCR leads to an increasing emission of NO_x and unreacted ammonia.

The main cause of deactivation are elements or compounds which chemically attack the catalytically active material or its support. Also, structural changes and pore blocking are important issues of deactivation. A variety of poison compounds containing elements such as halogens, alkali metals, alkaline earth metals, arsenic, lead, phosphorus, and sulfur are mentioned in the literature. As_2O_3 is the most severe poison in coal-fired power plant operation in Germany. In power plants equipped with wet-bottom boilers alkali metal oxides mostly remain in the molten ash, whereas As_2O_3 tends to escape into the flue gas and deposits on the catalyst.

In this section the following issues will be discussed:

(i) the types of catalyst for which deactivation has been observed;
(ii) the causes of deactivation;
(iii) what measures can be taken to prevent catalyst deactivation, or to enhance the lifetime of catalysts.

Two types of deactivation studies may be distinguished: those in which catalysts are doped with the catalyst poison and those in which the catalyst is deactivated by compounds present in flue gases.

In most studies the rate constant is used to compare the activity of a fresh catalyst and the catalyst after deactivation. The rate constant can be expressed as

$$k = -AV \ln(1 - \alpha) \tag{6}$$

assuming plug flow behavior in an integral reactor and a first order rate expression with respect to NO (AV is the area velocity; α is the conversion).

Most laboratory tests have been carried out using a fixed-bed reactor type. Gutberlet [115] compared 14 wet-bottom furnaces with complete ash recirculation with respect to deactivation of the V_2O_5/WO_3 on TiO_2 catalyst by As_2O_3. It was shown that the As_2O_3 content in the gas phase ranges from 0.001 to 10 mg m^{-3} As. moreover, the content of As in the gas phase was independent of the content of As in the coal. Schallert [116] described the results of the use of five pilot plants attached to coal fired power plants. One pilot plant was installed downstream from a flue gas desulfurization unit and four were used in a high-dust environment downstream from two dry-bottom and also two wet-bottom furnaces. The pilot plant consisted of two series in parallel of three honeycomb units each. Deactivation of the V_2O_5/WO_3 catalysts was observed for all types of furnaces used. The rate constant relative to that of the fresh catalyst after about 10 000 h of operation was 0.8 and 0.5 for dry-bottom and wet-bottom furnaces, respectively. The deactivation for a dry-bottom furnace was ascribed to the contamination of the surface of the catalyst with small fly-ash particles, whereas the deactivation observed by flue gas from a wet-bottom furnace could be ascribed to arsenic. No loss of catalytic activity was found with the reactor downstream from a flue gas desulfurization unit.

The deactivated catalyst was analyzed by energy dispersive X-ray analysis. At the entrance of the channels of the honeycomb catalyst enhanced amounts of SiO_2, Al_2O_3, Fe_2O_3, CaO, and K_2O were found over a length of 2 cm [130]. The composition resembles that of fly ash. Additionally, sulfur was found.

Vogel et al. [117] studied V_2O_5 on TiO_2-based catalysts with MoO_3 or WO_3 as the third compound. These catalysts were deactivated with As_2O_3 in the authors' laboratories and the SCR activities were compared with those of catalysts used in power plants equipped with wet-bottom or dry bottom furnaces. Figure 13 shows the results of this comparison. Chen and Yang [118] used smaller amounts of arsenious oxide in their deactivation experiments than Vogel et al. [117]. They observed a value of 0.47 for the ratio of $k:k_0$ for an arsenic level of about 0.12 ng g^{-1}.

From Fig. 13 it can be concluded that the deactivation of the catalyst which was used in the power plant cannot be ascribed exclusively to the presence of As_2O_3 in the flue gas. Moreover, it was established that the largest contribution to the decrease of $k : k_0$ occurred in the first 1000 h of operation. Most of the arsenic was

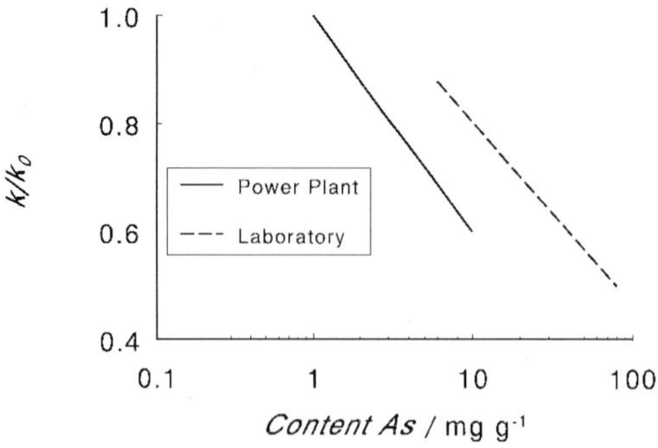

Figure 13. The relative first order rate constant as a function of the amount of arsenic in WO_3 or MoO_3 and V_2O_5 on TiO_2 (adapted from Ref. 117).

found in a 200 μm surface layer of the catalyst. In contrast to these observations Brand et al. [119] determined the concentration profile of arsenic and found a penetration depth of 50 μm after 1500 h.

Beeckman and Hegedus [50] poisoned a nine-channel monolith piece containing vanadia on titania in a reactor similar to that of Vogel et al. [117]. As_2O_5 vapor deposited on the catalyst surface forms an inactive monolayer. Several equations were given which describe the penetration of arsenic trioxide into the catalyst, the change in bulk gas-phase concentration along the reactor and the change in bulk gas-phase concentration of NO along the partially poisoned catalyst. Integrating these equations resulted in a relationship of the relative catalyst life and of the pore structure of the catalyst for flue gas containing 0.03 ppm As_2O_3.

Honnen et al. [121] proved by means of secondary ion mass spectrometry (SIMS) that arsenic was chemically bound to the surface of the catalyst and in the form of As_2O_3.

It is difficult to study arsenic-poisoned catalysts by XPS. Therefore, Lange et al. [122] explored the effect of arsenic on the support titania dominantly used in SCR catalysts. Arsenic-on-titania materials were prepared and studied by techniques such as XPS and diffuse reflectance infrared Fourier transform (DRIFT). Two preparation techniques were employed. Titania supports were impregnated with a solution of As_2O_3 in water, subsequently dried at 385 K and calcined at 673 K for 8 h. With a second method As_2O_5 was sublimed onto titania at 673 K in an autoclave for 8 h. It was suggested that the precursor As_2O_3 is oxidized into orthoarsenate(V) surface species which replaces the hydroxyl groups of the titania surface. Inaccurate results may be obtained from XPS because of the X-ray-induced reduction of As^{5+} into As^{3+}. Hilbrig et al. [123] published an article concerning the interaction of arsenious oxide with SCR catalysts. In that study the valence state of As, the chemical environment of As, and the interaction of As with the surface of the SCR catalyst, were determined by X-ray absorption near edge spectroscopy (XANES), extended X-ray absorption fine structure (EXAFS) and DRIFT. Again, the catalysts were deactivated in the laboratory by using the sublimation technique and by preparing physical mixtures of As_2O_3 and the catalyst. Commercially available catalysts and a catalyst from a SCR unit were used. The catalysts were WO_3 on TiO_2, MoO_3 on TiO_2, and physical mixtures.

It was shown by DRIFT and XAS that As_2O_3 was oxidized at the surface of all catalysts forming orthoarsenate(V) species. These species are anchored to the support and to MoO_3 and to WO_3 as well. There is evidence from XANES measurements that W–O–As bonds are present at the surface and that these types of bonds are the cause for the deactivation of SCR catalysts.

Hums [124] studied the effect of As_2O_3 on the phase composition of physical mixtures of V_2O_5 and MoO_3, and of V_2O_5, MoO_3, and TiO_2, by a transmission powder X-Ray diffraction. His main interest was to prove how As_2O_3 affects the catalyst and its phase composition. The mixtures of the oxides were prepared by heating the powders in quartz ampoules at 975 K for 70 h. The heated samples were then exposed to As_2O_3 in a quartz ampoules at 700 K again for 70 h. It was observed that As_2O_3 induces the phase transformation of $V_9Mo_6O_{40}$ into $V_6Mo_4O_{25}$, resulting in an enhanced concentration of V^{4+} ions. Because of the absence of V_2O_5 in the mixed oxide systems the authors concluded that the reducible metallic species present in MoO_3 and $V_9Mo_6O_{40}$ are the active sites in the SCR reaction. It is not known yet to what extent these oxide materials are active in the SCR reaction.

Rademacher et al. [125] used X-Ray photoelectron spectroscopy for the characterization of two types of catalysts which were exposed to flue gas of a slag tap furnace with 100% ash recirculation for 1800 h. These catalysts were MoO_3 /V_2O_5 on SiO_2 and WO_3/V_2O_5 on $SiO_2/P_2O_5/Al_2O_3$. Elements such as Ca, Si, S, As, P, and N were found. Both As^{3+} and As^{5+} were detected. A substantial shift of the XPS signals due to charging was observed. Two methods of correcting for charging have been used by Hopfengärtner et al. [126], using the binding energies of both C 1s and Au $4f_{7/2}$ as standards for energy calibration.

An investigation of the effect of potassium doping of submonolayer vanadia on titania catalysts has been carried out [40]. Potassium doping strongly reduces the conversion of nitric oxide. No reaction occurs at K loadings of 0.3 wt% and above (atomic ratio K : V = 0.47). Chen and Yang [118] and Chen et al. [120] prepared doped V_2O_5 on TiO_2 and WO_3/V_2O_5 on TiO_2 catalysts by incipient wetness impregnation using Li acetate (Ac), $NaNO_3$, KNO_3, RbAc, CsAc, $Ca(Ac)_2$, $Pb(Ac)_2$, P_2O_5, and As_2O_5, respectively. They determined intrinsic rate constants for the doped and undoped catalysts.

Figure 14 clearly shows the effect of the addition of potassium to the catalyst on the activity and of the effect of the presence of WO_3 in the catalyst thereon. The poison resistance to both alkali metal oxides and arsenious oxide is significantly increased by the presence of WO_3 in the WO_3/V_2O_5 on TiO_2 catalyst. The reason for this effect is that the Brønsted acid sites protect the catalyst for the attack of alkali. Moreover, WO_3 inhibits the oxidation of ammonia and the oxidation of SO_2 as well.

Chen and Yang [118] reported that the strength of the poison of alkali metals oxides follows the order, Cs > Rb > K > Na > Li for the SCR catalysts mentioned.

Buzanowski and Yang [49] reported a first-order rate constant of 1.09 cm^{-3} $g^{-1}s^{-1}$ for an atomic ratio K : V of 0.47. The experimental conditions were the same as those published by Chen and Yang [118], but without SO_2 in the flue gas. The extrapolated k value presented in the study of Chen and Yang [118] was about 7 cm^{-3} g^{-1} s^{-1}. Probably SO_2 has a promoting effect on the SCR reaction.

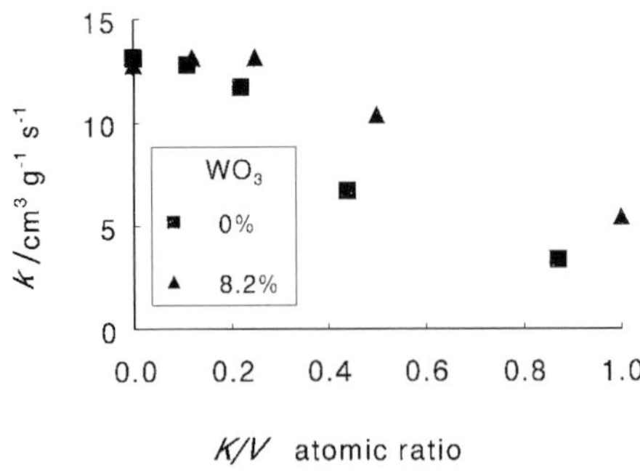

Figure 14. Intrinsic first-order rate constant over a potassium poison-doped V_2O_5 on TiO_2 catalyst and a WO_3/V_2O_5 on TiO_2 catalyst as a function of the atomic ratio K : V : $[NO]_i$ = $[NH_3]_i$ = $[SO_2]_i$ = 1000 ppm; $[O_2]_i$ = 2%; $[H_2O]_i$ = 8%; balance N_2; GHSV = 15 000 h^{-1}; T = 575 K (adapted from Ref. 118).

The presence of water and SO_2 in flue gas results in an increase of the rate constant. This was explained as a result of increased Brønsted acidity. Potassium is also a very effective promoter for the oxidation of ammonia and temperatures above 573 K.

Deactivation and promoting effects for HCl and chlorides have been studied [118]. Small amounts of NaCl serve as a promoter. At very high concentrations compounds such as VCl_4 and VCl_2 were observed in the off-gas. Three possibilities for the decline in activity of the SCR reaction were suggested:

(i) HCl consumes NH_3 forming NH_4Cl;
(ii) HCl and vanadia react forming VCl_2 and VCl_4;
(iii) NH_4Cl deposits at temperatures below 615 K.

It was conclusively, stated that dopants which enhance the Brønsted acidity are promoters, whereas those weakening it are poisons.

Waste incineration plants produce flue gases which may be hazardous for SCR catalysts. Tokarz et al. [127] detected a number of elements present in flue gas in the stack of a waste incineration plant. These elements were: Fe, Al, Mg, Ca, K, Na, and P in amounts ranging from 300 to 2000 ppm; Ba, Cr, Cu, Ni, Pb, Sr, Zn, and Zr in amounts ranging from 7 to 1200 ppm; traces of Be, Co, La, Li, Sc, Y, and Mn. In order to study the effect of these elements on activity, laboratory experiments were carried out in which fresh catalysts were impregnated with the nitrates or acetates of the elements by up to 3% by weight by incipient wetness. Catalyst samples with blends of these elements were also prepared. Mg, K, and Na showed the highest detrimental effect on the activity of the catalyst, which was ascribed to the sintering of the surface of the catalyst because of the reaction of these elements and V_2O_5.

Honnen et al. [121] found reversible deactivation of SCR catalysts downstream a flue gas desulphurization plant of a waste incinerator. The SO_2 concentration varies from 40 to 80 mg m^{-3} and results in the formation of ammonium sulfates at temperatures below 600 K. These sulfates could be removed easily by heating the catalyst to a temperature of about 660 K.

Dieckmann et al. [128] succeeded in restoring the activity of a catalyst which was reversibly deactivated by H_2SO_4 and NH_4HSO_4 by thermal treatment.

Recently, Gutberlet [129] summarized the main causes for deactivation of the SCR catalyst:

(i) poisoning by arsenic;
(ii) neutralization of acid sites by sodium and potassium;
(iii) formation of thin layers of $CaSO_4$;
(iv) poisoning by silicon.

Silicon containing layers of about 50 μm were found. Silicon layers are formed by the reaction of SiF_4 and water according to

$$SiF_4 + 2H_2O \rightarrow SiO_2 + 4HF$$

Silicon tetrafluoride is formed from the reaction of HF which is present in flue gas with fly ash or silicon-containing coatings.

2.3.5 SCR Catalyst Testing

Nowadays SCR is a well developed technology. Although combustion modification techniques are very effective for NO_x control, the applications of SCR have grown in the last decade. A variety of SCR catalysts has been tested, in particular V_2O_5 on titania catalysts [4]. Waste incinerators, diesel engines, and gas turbines are, or will be, equipped with NO_x-removal catalysts. Because of this broad field of application laboratory scale and bench-scale tests are required to establish the activity, selectivity, and stability as a function of time. One main problem of laboratory testing is the interpretation of the results for the prediction of the behavior of catalysts in large scale systems. Obtaining representative catalyst samples from large installations is also very difficult.

Several monolith geometries for SCR are shown in Fig. 15. Fig. 16 shows an example of a monolith.

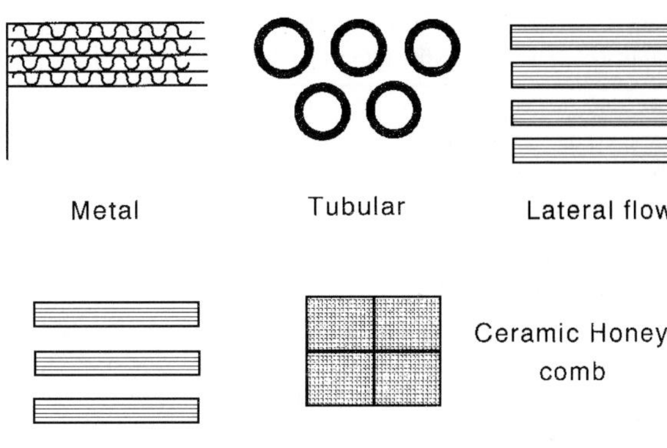

Figure 15. Different geometries of SCR catalysts (adapted from Ref. 4).

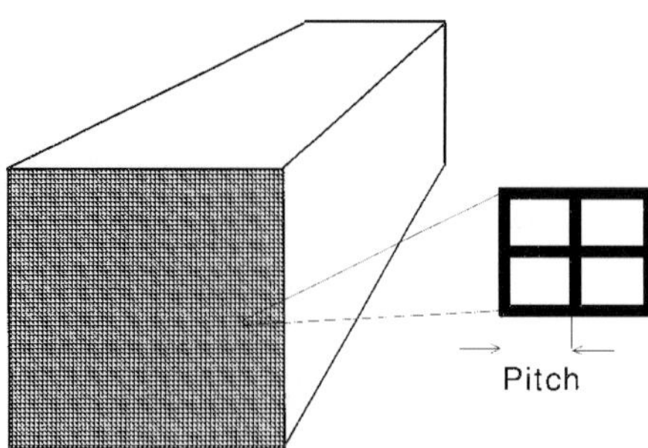

Figure 16. An SCR monolith, 70–100 cm by 15 cm^2 (see Tables 7 and 11).

The geometry and other parameters of honeycomb catalysts are given in Tables 7 (and 10 below). The behavior of the catalyst may be expressed in numbers related to activity, selectivity, and stability of the catalytic materials. Also, modeling the catalytic system is an important feature.

For laboratory tests only part of a honeycomb (4, 9 or 36 channels) or powdered material is required. For bench-scale and pilot-scale testing a whole catalyst element will be needed. These testing methods are discussed in this section.

All three types of reactors mentioned in Table 8 are used for the quality control during the preparation and the use of monolithic systems. Additionally, microreactors are used for the development of new catalysts, whereas bench-scale reactors are used for reactor design. Pilot-scale systems have been used for the determination of the activity during the lifetime of the monolith.

Additionally, in a microreactor the intrinsic kinetics and deactivation behavior of SCR catalysts is studied with flows up to 1.5 $m^3\,h^{-1}$. In both test facilities it is possible to vary all process parameters: temperature, the ammonia to nitric oxide feed ratio, the nitric oxide and sulfur dioxide concentrations, the space velocity, and the catalyst geometry. These techniques provide information for somewhat small areas and therefore should always be performed to complement bench- or laboratory-scale activity and selectivity measurements.

In principle, all laboratory equipment used for determining the activity of a catalyst has a gas-mixing system, an oven with the catalyst, and analyzing equipment.

Table 7. Several parameters of SCR catalysts [119].

Parameter	Number of cells			
	High dust		Low dust	
	20 × 20	22 × 22	35 × 35	40 × 40
Diameter (mm^2)	150^2	150^2	150^2	150^2
Length (mm)	1000	1000	800	800
Diameter of cell (mm)	6.0	6.25	3.45	3.0
Wall thickness (mm)	1.4	1.15	0.8	0.7
Surface area ($m^2\,m^{-3}$)	2.0	1.8	1.35	1.35
Porosity (%)	64	70	64	64

Table 8. Three types of SCR reactors.

Test equipment	Micro scale	Semimicro scale	Bench scale
Flow ($m^3\,h^{-1}$)	0.5–1.5	0.5–20	20–300
Diameter (mm)	10–50	70	200
Channels	3 × 3	6 × 6	20 × 20
Length (m)	0.3	0.5–2.3	1.1
Linear velocity ($m\,s^{-1}$)	0.2–0.7	0.6–4	0.4–5
Area velocity ($m^3\,m^{-2}\,h^{-1}$)	4	0.2	2

Powdered material or samples extracted from honeycombs or plate materials may be used.

The testing conditions are different from those in a catalyst element. In full-scale SCR plant linear velocities range from 4 to 10 $m\,s^{-1}$, whereas in laboratory equipment the linear velocity is between 0.8 and 2.3 $m\,s^{-1}$ [128].

The parameters, which should be determined are the concentration of NH_3, NO, O_2, SO_2, SO_3, and H_2O, the catalyst surface area, gas velocity, and the temperature. The analyzing equipment for NO is UV absorption, for NH_3 an ion selective electrode, and for SO_2 infrared spectroscopy. SO_3 is absorbed in isopropanol and then determined titrimetrically [128].

Equation (7) was used for calculating the rate constant assuming first order kinetics in NO

$$k_{NO} = -\frac{F}{S}\ln(1-\alpha) \tag{7}$$

Here, S is the specific surface area of the catalyst, α is the conversion and F is the gas flow. The term F/S is the area velocity. The rate constant depends on the type of catalyst material, the oxygen concentration, and the temperature. Additionally, it was found that the rate constant depends on the velocity in the channels, which implies that transport phenomena have an effect on the NO conversion.

Increasing the linear velocity in the channels of a honeycomb results in an increase of the rate constant (Fig. 17).

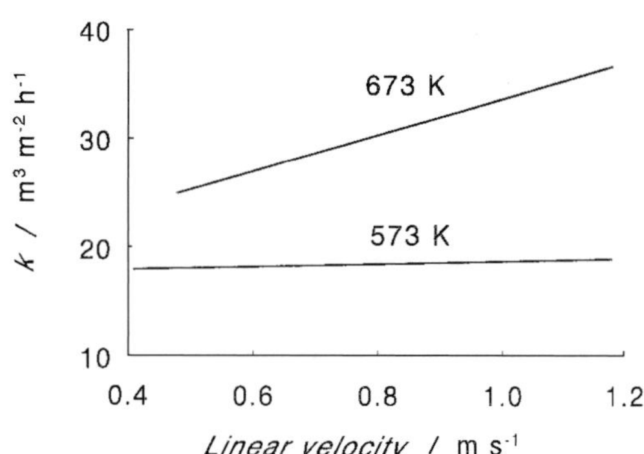

Figure 17. The rate constant as a function of the linear velocity in the channels of a honeycomb with a pitch of 7 mm (adapted from Ref. 128).

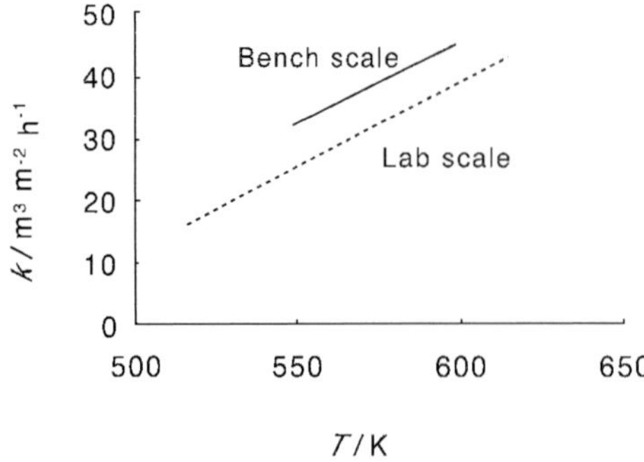

Figure 18. The rate constant as a function of temperature for triangular catalysts (adapted from Ref. 128).

Four channels were extracted from a catalyst element for laboratory testing [128]. The catalyst sample is placed in an oven and simulated flue gas is passed to the catalyst. For a laboratory test the amount of ammonia that has been absorbed by the catalyst is first determined from the absorption curve. Then the conversion of NO is determined as a function of the $NH_3 : NO$ molar ratio. In full-scale SCR plants this ratio is smaller than 0.8 to lower the ammonia slip.

The surface areas used in the catalyst test reactor are different for different catalyst geometries. Plate catalyst samples have a surface area of about $20\,cm^2$ and honeycomb types have a surface area of about $100\,cm^2$.

The rate constants determined by means of laboratory equipment may differ from those determined by bench-scale type equipment. This holds for catalysts with a triangular geometry and for catalysts with a pitch (channel width plus wall thickness) of 4 mm. Honeycombs with a pitch of 7 mm show good agreement between laboratory-scale and bench-scale measurements. The differences in rate constant are shown in Fig. 18.

For a fresh honeycomb the rate constant is independent of the location in the channel. However, for used honeycomb catalysts the rate constant at the entrance of the channel is lower than that at the exit. This effect is caused by the differences in the flow profile. At the entrance of the channel turbulence is observed which becomes laminar in the channel. Due to this turbulence at the entrance, elements such as Ca, S, and Si contaminate the surface of the catalyst (see Section 1.2.4 and Fig. 24).

To check the behavior of SCR elements they have to be tested at regular intervals. For this purpose a test programme was developed which includes catalyst activity and selectivity tests in a bench-scale system with surface analysis and characterization of catalysts using highly sensitive surface techniques. By combining these experimental techniques valuable information on ageing and fouling processes on the catalysts is obtained.

From this catalyst performance test the lifetime of SCR catalysts may be predicted, thus providing a replacement strategy for SCR elements. Honeycomb elements or plate-type elements are placed in the oven of a bench-scale setup. With this method the integral activity of an element is obtained.

Dieckman et al. [128] have developed a test to measure the activity of a catalyst

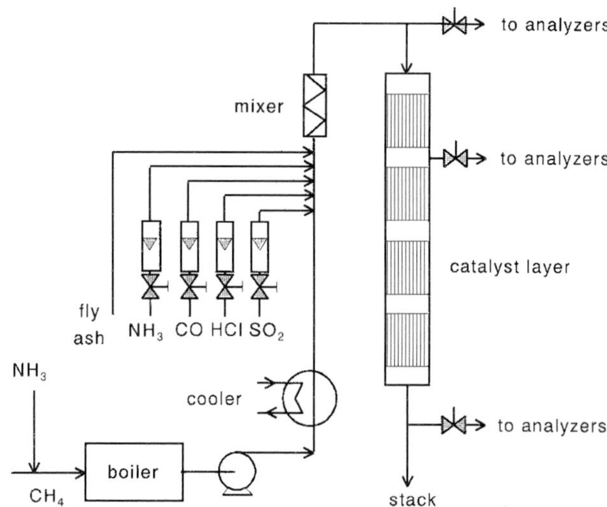

Figure 19. Bench-scale setup (adapted from Ref. 130).

very quickly with real flue gas. They installed a bypass in the air preheater in which a honeycomb could be installed.

Meijer and Janssen [131] described tests to evaluate issues such as activity, selectivity, fouling lifetime of the catalyst, the surface composition and concentration profiles in depth of honeycomb catalysts.

Activity and selectivity performance tests with catalysts of different geometry are performed in bench- and laboratory-scale flow systems.

The bench-scale setup is shown in Fig. 19. Flue gas is generated by a natural gas boiler, in which ammonia is burnt for NO_x production. Flue gas amounts of 100 to $200\,m^3\,h^{-1}$ are passed to complete honeycomb elements. Optionally, NH_3, SO_2, CO, HCl, HF, and fly ash can be added to the flue gas, simulating real flue gas.

After passing a heat exchanger, ammonia, sulfur dioxide, carbon monoxide or other components can be added. Downstream the mixer flue gas is introduced into the reactor in which the SCR element is situated. Sampling for gas analysis is performed at different points in the system. In the behch-scale setup, up to a maximum of six of these elements can be placed in series. In the bench-scale the amount of fly ash can be varied. The linear gas velocity is similar to that in commercial SCR honeycombs. In this setup the concentration of NO, NO_2, N_2O, O_2, SO_2, and SO_3 in both the reactant and product gas is measured continuously or semicontinuously with the techniques mentioned before. Furthermore, the pressure drop over the element is monitored. Most of the NO_x is converted by the first two catalyst elements. At high ammonia : NO_x ratios ammonia slip occurs. With the experience built up with different honeycomb elements in this bench-scale setup a model was developed for prediction of the NO_x removal efficiency in high dust SCR reactors [132]. In this model all important process variables are considered. Additionally, an effort is made to include the effect of catalyst poisoning, thus providing a reliable prediction of the lifetime behavior.

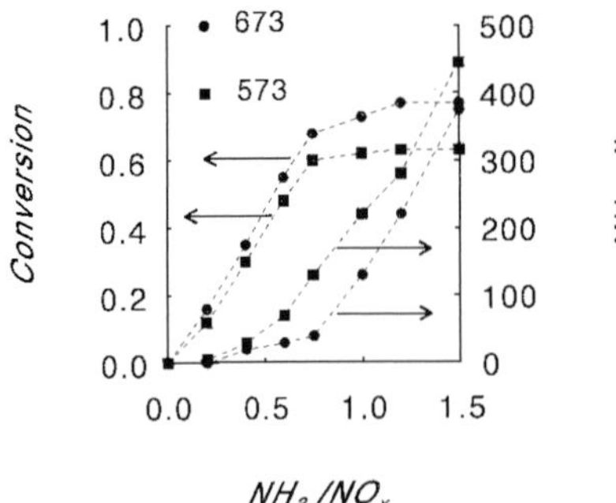

Figure 20. The NO_x conversion as a function of the ratio $NH_3 : NO_x$ and as a function of the concentration of unreacted ammonia at 573 and 673 K respectively after 29 000 h on stream in a coal-fired power plant (adapted from Ref. 131).

Figure 20 shows the results of an activity test of a honeycomb element and the analysis of unreacted ammonia.

At $NH_3 : NO_x$ ratios smaller than 1, NO_x conversion increases linearly with increasing ratio. The reaction rate depends on the concentration of ammonia. For ratios higher than 1, the reaction rate is dependent on the concentration of NO_x. For these two ratios two rate equations may be defined (eqs 8 and 9).

$$-\frac{d[NO]}{d\tau} = -\frac{d[NH_3]}{d\tau} = k_{NH_3}[NH_3] \quad (NH_3 : NO_x < 1) \tag{8}$$

$$-\frac{d[NO]}{d\tau} = k_{NO}[NO] \quad (NH_3 : NO_x > 1) \tag{9}$$

Newly developed catalysts that show promising results from tests in the laboratory have to be tested in a real flue gas environment.

Location of the SCR reactor upstream of the electrostatic precipitators asks for catalysts that show low attrition behavior. Flue gas contains about 10–30 $g\,cm^{-3}$ dust. However, downstream the FGD the amount of dust is 30–50 $mg\,m^{-3}$.

Brand [119] developed a testing facility attached to the flue gas channel. Flue gas is isokinetically pumped through the catalyst. Also, Hein and Gajewski [133] have described pilot plants which are attached to a power plant. The flue gas flow is about 500 $m^3\,h^{-1}$. One important aspect is that the results from the laboratory test can be compared with those obtained from pilot plant tests.

In a full-scale SCR unit two or three layers of honeycomb elements are used. In these layers several elements are placed for ex situ testing at regular time intervals [129]. Testing these samples in bench-scale equipment produces overall rate constants which depend on the location in the SCR reactor. Moreover, information is obtained when catalyst layers have to be exchanged.

2.3.6 Modeling

Monolith material has been mainly used for emission control and for cleanup of gases which contain large amounts of dust. The main features of monoliths are: the large external surface area, the low pressure drop, uniform flow, the low axial dis-

Table 9. Materials for monolithic supports [134].

Type	Composition
α- and γ-Alumina	Al_2O_3
Cordierite	$2MgO \cdot 2Al_2O_3 \cdot 5SiO_2$
Mullite	$3Al_2O_3 \cdot SiO_2$
Cordierite–mullite	$2MgO \cdot 2Al_2O_3 \cdot 5SiO_2–3Al_2O_3 \cdot SiO_2$
Magnesium aluminate	$MgO \cdot Al_2O_3$
Mullite–aluminum titanate	$3Al_2O_3 \cdot SiO_2–Al_2O_3 \cdot TiO_2$
Silica	SiO_2
Silicon carbide	SiC
Silicon nitride	Si_3N_4
Titania	TiO_2
Zeolites	$Al_2O_3 \cdot SiO_2$
Zirconia/magnesium aluminate	$ZrO_2/MgO \cdot Al_2O_3$
Metals	Fe, Cr, Al, Y

persion, and the low radial heat flow [134]. Monoliths have been successfully applied in automobile exhaust applications and in SCR.

Square-channelled monolith or plate-type vanadia on titania catalysts are commonly used in SCR units of power plants. A typical monolith block (Fig. 16) has the geometrical parameters given in Tables 7 and 10.

Now, both low-surface and high-surface area and metallic monoliths are commercially available. A variety of materials used as monolithic supports are summarized in Table 9. Structural parameters of typical honeycomb catalysts used for SCR of NO are shown in Table 10.

Irandoust and Andersson [134] described the modeling of three major flow patterns: single-phase flow, cross flow, and two-phase flow. Work on monoliths has been performed with single-phase flow and the main issues for this type of flow are pressure drop, entrance effects, axial dispersion, and mass and heat transfer. The catalyst effectiveness is generally mass-transfer limited.

The entrance length z/d for a fully developed velocity profile, for a concentration profile, or for a corresponding thermal entry length are given as 0.05 *Re*, 0.05 *Re Sc*, 0.05 *Re Pr*, respectively. The symbols are summarized in Table 11.

The Reynolds number for metal monolith is

$$Re = \frac{4v\rho}{\alpha\mu} \tag{10}$$

Table 10. Structural parameters of typical honeycomb catalysts used for SCR of NO (see Fig. 16).

Parameter	Dimensions (cm)
Length	70–100
Square	15×15
Channel diameter	0.3–1.0
Wall thickness	0.05–0.15

Table 11. Symbols used.

Symbol	Definition	Dimensions
A	cross-sectional area of the cell	m^2
a	surface area per volume of porous material	$m^2\ m^{-3}$
D_h	hydraulic diameter	m
D_e	effective diffusivity of NO	$m^2\ s^{-1}$
d	channel diameter	m
f	friction factor	–
J_{NO}	molar flux of NO	$mol\ m^{-2}\ s^{-1}$
k	first-order rate constant	$m\ s^{-1}$
K_{NH_3}	adsorption constant	$m^3\ mol^{-1}$
k_m	film mass transfer coefficient	$m\ s^{-1}$
L	total length of monolith	m
$[NO]_t$	total concentration of NO	$mol\ m^{-3}$
ΔP	pressure	Pa
Pr	Prandl number	–
Re	Reynolds number	–
Sc	Schmidt number	–
SV	space velocity	s^{-1}
X	conversion	–
x	distance measured from the surface into the wall	m
y_{NO}	mol fraction of NO	–
v	linear fluid velocity	$m\ s^{-1}$
z	axial coordinate	m
α	geometric surface area per volume	m^{-1}
ε	void fraction	–
μ	gas viscosity	$kg\ m^{-1}\ s^{-1}$
ϕ	Thiele modulus	–
ρ	gas density	$kg\ m^{-3}$
σ	perimeter, length of the cell	$m\ cell^{-1}$

For gases the entrance effect may be neglected when the diameter of the channels is smaller than 3 mm.

Metallic monoliths have, in contrast to ceramic monoliths, a high heat conductivity. Metallic monoliths have a lower pressure drop than ceramic monoliths due to a small wall thickness.

The pressure drop is

$$\Delta P = \frac{32\mu L v}{d^2} \tag{11}$$

It is assumed that gases within a channel of the monolith are well mixed in the direction perpendicular to the wall and plug flow axially. Buzanowski and Yang [49] have presented an analytical solution for the overall NO conversion as an explicit function of space velocity and other parameters for a monolithic catalyst under isothermal conditions. Three monoliths were prepared, one of which was poisoned with K_2O.

The flux equation for NO in the porous wall is

$$J_{NO} = -D_e[NO]_t \frac{dy_{NO}}{dx} \tag{12}$$

Here J_{NO} is the molar flux (mol m^{-2} s^{-1}), D_e is the effective diffusivity of NO (m^2 s^{-1}), y_{NO} is the mol fraction of NO, $[NO]_t$ is the total concentration (mol m^{-3}) and x is the distance measured from the surface into the wall (m).

The mass balance equation for NO in the wall is

$$\frac{dJ_{NO}}{dx} = -ka[NO] \tag{13}$$

where k is the first order rate constant and a is the surface area per volume of porous material (m^2 m^{-3}).

Combining these two equations yields a differential equation which has to be solved including adequate boundary conditions. The overall conversion of NO as a function of the film mass-transfer coefficient, the effective diffusivity, the linear velocity, and the first-order rate constant, is obtained with structural parameters of the honeycomb.

The conversion of NO can be expressed as

$$X = 1 - e^{-[(\sigma L/uA)(1/B)]} \tag{14}$$

The term B includes k_m, D_e and φ, the film mass transfer coefficient (m s^{-1}), the effective diffusivity (m^2 s^{-1}), and the Thiele modulus, respectively (see Table 11). Film diffusion limitation was small. The results of the calculations agreed with the experimental results.

Beeckman and Hegedus [50] determined the intrinsic kinetics over two commercial vanadia on titania catalysts. A mathematical model was proposed to compute NO and SO_2 conversions and the model was validated by experimental values. Slab-shaped cutouts of the monolith and powdered monolith material were used in a differential reactor. The cutouts contained nine channels with a length of 15 cm and with a channel opening and wall thickness of 0.60 and 0.13 cm, respectively. The SCR reaction over a 0.8 wt% V_2O_5 on titania catalyst was first-order in NO and zero-order in NH_3.

A Langmuir–Hinshelwood-type rate expression was used:

$$R_{NO} = k_0 e^{-E/RT}[NO]\frac{K_{NH_3}[NH_3]}{1 + K_{NH_3}[NH_3]} \tag{15}$$

From model calculations it appeared that the pore structure is of great importance for the activity. To improve the activity of the monolith catalyst, silica was used as base material for the monolith. The pore structure was managed by the technology used. Silica was coated with titania upon which vanadia was precipitated. The activity of this new catalyst type improved according to the calculated 50%.

2.4 Industrial Experience and Development of SCR Catalysts

2.4.1 Power Plants

Selective catalytic reduction was introduced in Japan by Nakajima and co-workers [135] in 1973. About 100 combustion facilities with a total capacity of 34 GW are equipped with SCR; the same number holds for Germany [130].

Most of the industrial experience of SCR applications has been gathered in Japan and Germany. SCR has been used for post-combustion control in power plants, nitric acid plants, gas turbines, and waste incinerators.

SCR catalysts are Pt, $CuSO_4$, Fe_2O_3, Cr_2O_3, MnO_x, NiO, Co_2O_3, WO_3, V_2O_5, MoO_3, and zeolites [110]. Zeolites are the best choice for application downstream from a gas turbine. However, with respect to application in power plants the most suitable catalysts are those based on WO_3, V_2O_5, and MoO_3. Pt is easily poisoned by SO_2 in rhe flue gas. Iron oxide-based catalysts catalyze the oxidation of SO_2 into SO_3 and iron sulfates are formed. At higher temperatures Cr_2O_3 oxidizes ammonia into nitric oxide. MnO_x-, NiO-, and Co_2O_3-containing catalysts are easily poisoned by sulfuric acid, whereas zeolites are deactivated by water.

The main problems met in the application of SCR technology are extensively described in the review by Bosch and Janssen [4] and the contribution by Gutberlet and Schallert [136].

Monolithic honeycomb have found major application in the selective catalytic reduction process (Fig. 16). The pressure drop is about 250–1000 Pa. The SCR reactor asks for a flat velocity profile of the flue gas. Therefore, a layer of dummy honeycombs is installed for flow-straightening purposes.

The plate-type catalyst has many advantages over honeycomb systems: less pressure loss; higher erosion resistance; less sensitivity towards fouling by dust; higher mechanical and thermal stabilities [133].

Mixing of ammonia with flue gas should be carried out with care because a very low slip is allowed of 5 ppm ammonia at an NO_x removal efficiency of 80%. SCR systems in the United States are designed for 80–90% NO_x removal with an ammonia slip of 10–20 ppm [137].

The type of catalyst to be used in a power plant depends on the location in the flue gas cleanup system of the plant and the sulfur content of the fuel. The minimum temperature is 473 K and the maximum temperature about 673 K. At higher temperatures ammonia is oxidized into undesireable byproducts and the catalyst may sinter and thus become deactivated.

Most vanadia-type catalysts are located downstream from the economizer and before the air preheater where the temperature of the flue gas ranges from 573 to about 700 K. Load variations of the power plant require catalysts to operate over a broad temperature range.

Figure 21 shows three possibilities for the positioning of the SCR unit in the flue

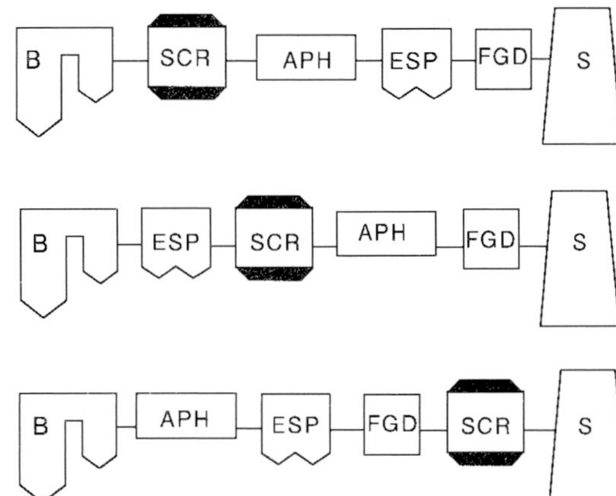

Figure 21. Three locations for SCR in a power plant. B, boiler; APH, air preheater; ESP, electrostatic precipitator; FGD, flue-gas desulfurization; S, stack.

gas treatment system of a coal-fired utility boiler. These are now discussed, together with their advantages and disadvantages.

In the first option, flue gas which leaves the boiler is cleaned by means of SCR; this is the "high-dust system". In the "low-dust system" the SCR unit is mounted downstream of the electrostatic precipitator (ESP) and before the flue gas desulfurization unit (FGD). A third location may be found between the FGD and the stack. A fourth option, which is not shown, is placing the catalysts in the air preheater. Advantages and disadvantages for both the high- and the low-dust systems are given in Table 12. In the high-dust application, soot blowers are needed to remove the accumulated fly ash on the catalyst. For low-dust applications high temperature electric precipitators are required, otherwise the flue gas from the precipitator has to be reheated. Figure 22 shows the installed capacity of SCR for applications of both high-dust and downstream flue gas desulfurization units (FGD) as a function of time. The concentration of NH_3 downstream from the SCR reactor in a high-dust application should be as low as possible because of fouling of both the air preheater and fly ash.

The main causes for ammonia slip are:

(i) heavy fouling of the catalyst;
(ii) deactivation of the catalyst;

Table 12. Comparison of the behavior of the SCR catalysts in high- and low-dust systems [138].

Feature	High dust	Low dust
Erosion by dust	high	low
Dust deposited in reactor	less	more
Ammonium bisulphate deposit in APH	less	more
Contamination of fly ash by ammonium compounds	possible	none
Costs of the ESP	less	more

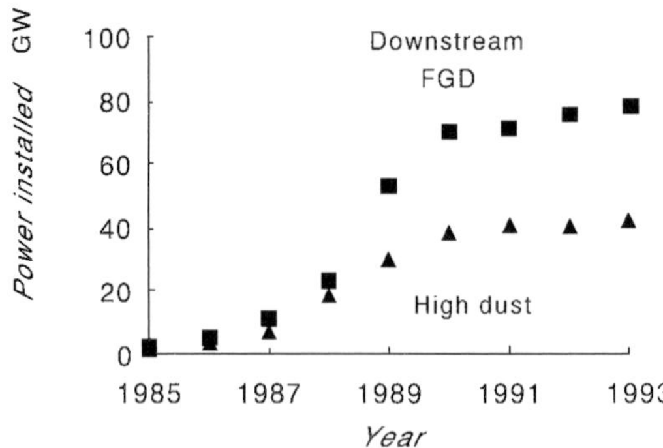

Figure 22. The installed capacity in Germany of SCR as a function of time (adapted from Ref. 139).

(iii) inhomogeneous distribution of ammonia in flue gas;
(iv) unequal ammonia distribution when two flue-gas streams in parallel have been used [129].

Ammonia causes not only problems of slip but also of transportation and storage. A solution to these problems is the transportation and storage of aqueous ammonia.

Recently, Gutberlet and Schallert [136] described the following important issues in SCR catalysis:

(i) the chemical and physico-chemical properties of SCR catalysts;
(ii) the chemical engineering design of the catalyst volume;
(iii) the deterioration of the catalytic activity with time;
(iv) the ammonia slip because of NH_3 maldistribution and side reactions;
(v) the effect of the SCR plant on downstream equipment.

There is a positive effect of the SCR unit on downstream equipment such as flue gas desulfurization, which produces gypsum. In the absence of a SCR unit, water from a FGD unit contains hydroxyl amine disulfonic acid which has been formed from NO_2, water, and SO_2. This compound acts probably as an inhibitor for the oxidation of calcium sulfite into calcium sulfate [129].

Another advantage of SCR on FGD is that mercury remains on the SCR catalyst as Hg^{2+}. Gypsum is thus not contaminated with mercury. The Hg^{2+} may be easily removed by washing the catalyst with water. When placing the SCR unit downstream a FGD unit, compounds such as N–S compounds and Hg_2Cl_2 may pollute the water of the FGD [140].

The main disadvantages of a high dust application are:

(i) the catalyst may be plugged by dust, fly ash deposits will occur due to inhomogeneous dust and velocity distributions;
(ii) the oxidation of SO_2 into SO_3 is catalyzed;
(iii) deposits of ammonium bisulfate on the catalyst occur at temperatures below 573 K;

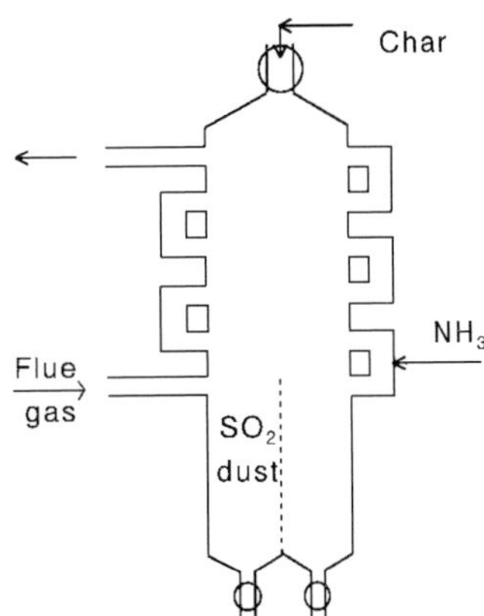

Figure 23. Simultaneous removal of NO_x, SO_2, and dust (adapted from Ref. 141).

(iv) deposits of ammonium bisulfate occur in cooler parts of the air preheater;
(v) alkaline deterioration of the catalyst, i.e. poisoning by potassium;
(vi) deposition of calcium sulfate; deactivation by arsenic in case of wet bottom boilers with fly ash recirculation; $HgCl_2$ is formed on the catalyst.

Another important effect of SCR on downstream equipment was described by Gutberlet and Schallert [136]. Cooling of the flue gas at the exit of the SCR reactor leads the formation and deposition of aerosols. A position of these aerosols pass the spray tower scrubbers of the flue gas desulfurization unit and acid particles are emitted.

For retrofitting of existing power plants or waste incinerators it can be more economical to install the SCR reactor downstream from the desulfurization unit. This is called the low-dust application of SCR. One advantage is that required catalyst volumes are lower and, due to the low SO_2 concentration, a very active catalyst may be used. Contamination of fly ash (NH_4HSO_4) and waste water of the FGD are avoided. Additionally, flue gases of several boilers may be cleaned by one unit.

The flue gas has to be reheated to the optimum temperature of the catalyst. Expensive regenerative heat exchangers are needed. Today, however, low-temperature catalysts are available [56].

Chars (coke) are also used at low temperatures (363–393 K). Kassebohm and Wolfering [141] developed a reactor system for the simultaneous removal of NO_x and SO_2. It is a moving-bed reactor, the catalyst is char (Fig. 23) In the application of this reactor type for flue gas cleanup of a waste incinerator, the char removes both SO_2 and HCl. Moreover, very low levels of mercury and polychlorinated hydrocarbons were found at the exit of the reactor.

Meijer and Janssen [130, 131] studied SCR monolithic catalysts by means of depth profiling (SEM/EDS). Depth profiles of W, Ti, Si, and S were obtained near the entrance and at the exit of a honeycomb element after 29 000 h of operation in a coal-fired power plant. The profiles, shown in Fig. 24, shows the behavior of the

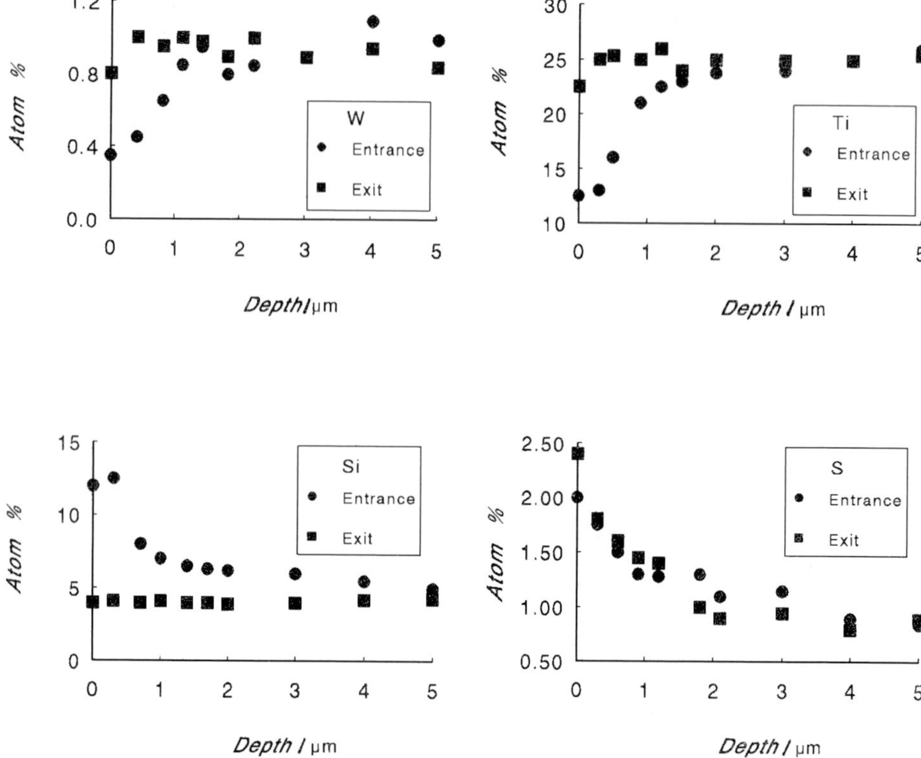

Figure 24. Depth profiles of W, Ti, Si, and S were obtained near the entrance and at the exit of a honeycomb element after 29 000 h on stream in a coal-fired power plant. Three honeycomb elements were removed from the SCR reactor of a power plant (adapted from Ref. 131).

four elements at the exit and entrance of the honeycomb to be different. The concentration of W and Ti is lower at the entrance of the channel of a honeycomb. The reason for this is that compounds containing sulfur and silicon accumulate at the entrance. The amount of sulfur found at the entrance is comparable with that at the exit.

A variety of elements were detected by XPS. These were Mg, Ba, Ca, N, S, and As. Silicon and boron were determined by SIMS. Sulfur was accumulated both at the entrance of the honeycomb and at the exit. However, the increasing effect of sulfur was more pronounced at the entrance. It was suggested that the source of boron was the gas phase and not fly ash particles.

For the simultaneous removal of SO_x and NO_x from waste gas at low temperatures (383–443 K) activated coke is a promising option [101]. SO_2 reacts with water and oxygen to form sulfuric acid which is trapped on a column with coke. The coke catalyst accumulates the acid and is regenerated regularly. The desulfurized gas is then passed to a reactor which contains a coke catalyst activated with sulfuric acid and NO_x is removed with ammonia in the presence or absence of oxygen [101].

It was shown that the rate of N_2 formation was increased with increasing ammonia concentration and that the presence of water (20 vol. %) reduces this rate. This suggests that the adsorption of water and ammonia are competitive. The SO_x present in waste gas forms sulfuric acid and deactivates the catalyst. The main reason for deactivation of the catalyst was ascribed to a decreasing BET surface area by the formation of sulfuric acid and by the formation of $(NH_4)_2SO_4$.

Kasaoka et al. [102] prepared vanadia catalysts supported with titania, activated carbon, and a mixture of carbon and titania, as supports for the simultaneous removal of SO_2 and NO_x at temperatures ranging from 400 to 425 K. The vanadia on titania catalyst was most appropriate. SO_2 from flue gas is oxidized to SO_3 and forms sulfuric acid. Ammonia reacts with sulfuric acid forming $(NH_4)_2SO_4$ and NH_4HSO_4. The catalysts were regenerated with water after treating the catalysts with gaseous ammonia to neutralize the acid sites on the catalyst.

In Denmark the new SNOX process has been developed by Agerholm et al. [153]. Flue gases with a temperature of about 653 K are cleaned from NO_x by means of selective catalytic reduction. The gas is reheated up to 673 K and introduced to an SO_2 converter which is located downstream from the SCR reactor. SO_2 is oxidized into SO_3 which is converted to sulfuric acid. The ammonia slip from the SCR reactor is oxidized simultaneously. Advantages of the SNOX process are:

(i) no chemicals have to be used for the desulfurization;
(ii) no solid or liquid wastes are produced;
(iii) high ratios of $NH_3:NO_x$ are allowed because unreacted ammonia is oxidized over the "SO_2" catalyst.

Ohlms [164] described the DESONOX process which was developed by Degussa (Germany). The concept of the process is the same as the process described by Agerholm et al. [153].

2.4.2 Nitric Acid Plants

For the production of nitric acid, ammonia passes over a platinum catalyst and is oxidized to nitric oxide. The NO is further oxidized into NO_2 which is absorbed in water to form HNO_3. The removal of NO_x in the exhaust from nitric acid plants differs from that of power plants because of the different $NO:NO_2$ ratio. For nitric acid plants and power plants this ratio is 1 and 19, respectively.

In Germany, catalysts for nitric acid plants were developed in the sixties [142]. Catalysts such as iron oxide/chromium oxide and supported vanadia were used. Off-gas from nitric acid plants contains about 2000 ppm NO_x and 3 vol. % O_2. The gas flows to be cleaned are about 100 000 $m^3\,h^{-1}$. A reduction of the amount of NO_x from 10 000 to 400 $mg\,m^{-3}$ NO_2 is achievable at space velocities ranging from 3000 to 10 000 h^{-1}.

Andersen et al. [143] published results on the removal of NO_x over supported platinum catalysts. Other effective catalysts included palladium, ruthenium, cobalt, and nickel. Maximum conversion of NO_x was acquired at a temperature of about

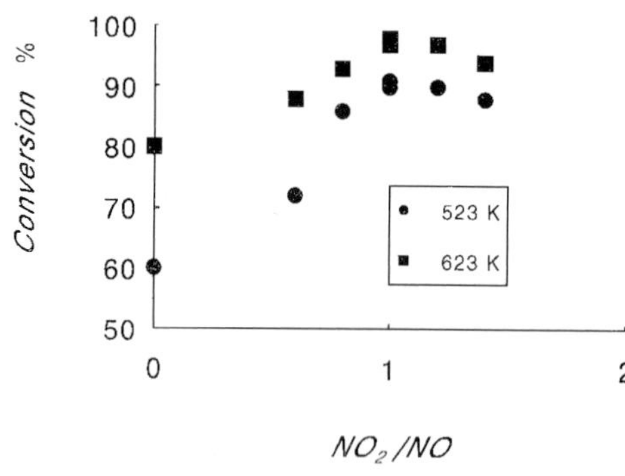

Figure 25. The conversion of NO_x as a function of the molar ratio NO_2 : NO (adapted from Ref. 144).

470 K. At higher temperatures the conversion decreases due to the oxidation of ammonia.

Avila et al. [144] studied the behavior of a V_2O_5/WO_3 on TiO_2 catalyst for the reduction of NO_x from stack gases of a nitric acid plant. They found that the activity becomes independent of the amount of oxygen present in the feed at NO_2 : NO ratios above 0.5 and at oxygen concentrations above 1% by volume. The relationship for NO_x conversion as a function of the molar ratio NO_2 : NO goes through a maximum for NO_2 : NO ratios close to unity, at temperatures of 523 and 623 K, and at a gas hourly space velocity of 40 000 h^{-1} (Fig. 25). The rate determining step is the oxidation of NO into NO_2. A mechanism suggested by these authors was the formation of ammonium complexes which react with absorbed NO and NO_2 [144].

Blanco et al. [114] describe a CuO/NiO on γ-alumina catalyst for conversion of equimolar mixtures of NO and NO_2 with ammonia and oxygen.

2.4.3 Gas Turbine Application

Figure 26 shows a gas turbine setup. In the conventional system air enters the combustor where it is mixed with fuel. The mixture burns at temperatures of about 2075 K. Gas from the combustor is cooled for passing to the turbine inlet.

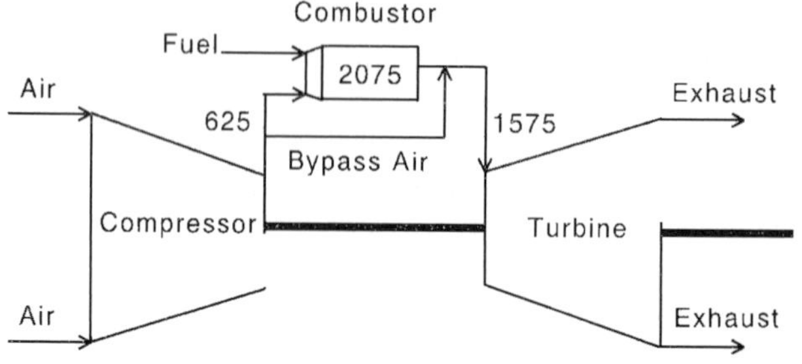

Figure 26. A conventional gas turbine combustion system; temperatures in Kelvin (adapted from Ref. 145).

Table 13. NO_x emissions (in ppm) of two types of gas turbines, equipped with various control techniques, as a function of the control technology [145].

Control technology	Natural gas	Distillate oil
Uncontrolled	100–430	150–680
Wet injection	25–42	42–65
Low-NO_x burner	25–42	65
SCR	5–9	15–68

The emission of NO_x by gas turbines in the United States are estimated to be 1 million t per year [146]. Conventional combustors fired with natural gas show NO_x emissions ranging from 100 to 430 ppm, whereas distillate oil-fired gas turbines have emissions ranging from 150 to 680 ppm. In Table 13 the NO_x emission data are summarized.

Water or steam injection into the flame lowers the temperature and, hence, the NO_x emissions are decreased. The reduction varies for each turbine and ranges from 60% to 94%. With low-NO_x combustors emissions of 25 ppm may be achieved.

Two options are available for the removal of NO_x. These are SCR downstream from the gas turbine or catalytic combustion (Section 3) in the gas turbine itself.

The exhaust temperature from a gas turbine ranges from 753 to 800 K. In cogeneration/combined-cycle operation the gas is cooled down to the operating temperatures of SCR (500–673 K). The SCR system downstream from a gas turbine may result in an increase of pressure resulting in a decreasing power output of the gas turbine. If carbon monoxide has to be removed, the oxidation catalyst should be installed before the SCR reactor. Problems arise if the flue gas contains SO_2.

In California (US), a 385 MW cogeneration facility comprising four gas turbine combined cycle trains is in operation [137]. Separate fuels such as natural gas, butane, and refinery fuel gas may be used. The refinery gas contains 800 ppm (by volume) sulfur. Sulfur was removed from the fuel before it was burnt in the gas turbine. The gas cleanup system includes a CO oxidation catalyst and downstream from this catalyst a SCR system.

Catalytic combustion for gas turbines is an important tool for lowering NO_x emissions from gas turbines. Multistage catalytic reactors for gas turbines have shown ultralow emissions, namely 0.5 ppm NO_x, 0.8 ppm CO, and 1.7 ppm unburned hydrocarbons from natural gas fuel [145].

In Fig. 27, a configuration of a gas turbine with a catalytic combustor is shown. In the system equipped with a catalytic combustor, fuel and air react over the surface of the catalyst-forming radicals. The combustion goes to completion outside the monolith. The catalysts used in this concept are monolithic supported palladium catalysts with ZrO_2 and cordierite as the support [147]. The most active catalyst for natural gas combustion is PdO [147]. Other types are alumina-supported BaO, La_2O_3, CaO, SrO, CeO_2, or MgO [148]. Below temperatures of 1073 K, decomposition of PdO into metallic palladium does not occur. The high efficiency can be ascribed to its ability to dissociate oxygen molecules in ions by forming a surface oxide [149].

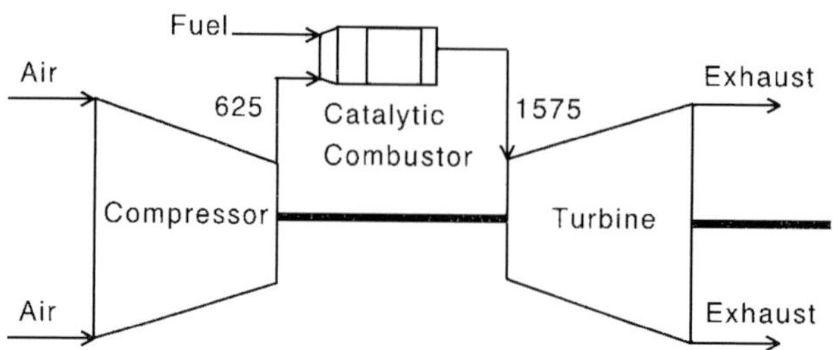

Figure 27. A catalytic gas turbine combustion system; temperatures in Kelvin (adapted from Ref. 145).

The catalyst should be resistant against thermal shocks and thus hot spots have to be avoided. In addition, the pressure drop in the monolith should be as low as possible.

2.4.4 Waste Incineration

Flue gas from waste incinerators is composed of NO_x, CO, HCl, HF, and SO_2; Metals such as Cd, Hg, Zn, and Pb, and organic compounds such as dioxines and furanes are also observed. The flue gas cleanup system downstream a waste incinerator is comparable to that of a power plant. Again, a number of locations of SCR may be distinguished leading to flue gases with different compositions. These are sulfur, alkaline metals, HCl, HF, dioxines and furanes containing gases with no dust, flue gas without sulfur and dust, and clean flue gas.

Hagenmaier and Mittelbach [150] used a vanadia on titania catalyst at 523 K. The activity decreased by 30% after 6000 h on stream. The catalyst was contaminated with Hg, Pb, Cd, As, and sulfates. Additionally, it was found that dioxines and furanes were oxidized over the catalyst (Section 1.2.5).

2.4.5 New Ideas

Lefers and Lodder [151] reported a new reactor type for SCR of NO_x. In Figure 28, a scheme is shown including this reactor. The shaft-like reactor is divided into compartments by horizontal grids. A catalyst with particle sizes ranging from 0.5 to 5 mm is introduced into the reactor with a gas stream. The length of the reactor is about 3 m and the throughput was 200 $m^3\,h^{-1}$. The flue gas was produced by a small boiler. The amount of catalyst used was about 8 kg [152]. The flow is sufficient to keep the particles in a floating condition. The floating bed has little tendency to form gas bubbles, or to form channels in the bed due to stabilization by the grids. An advantage of this reactor type over that of a monolith reactor type is that the catalyst may be easily removed.

Hardison et al. [154] describe a process which uses a fluidized catalyst bed. The process is designed for operating temperatures between 560 K and 672 K. Experi-

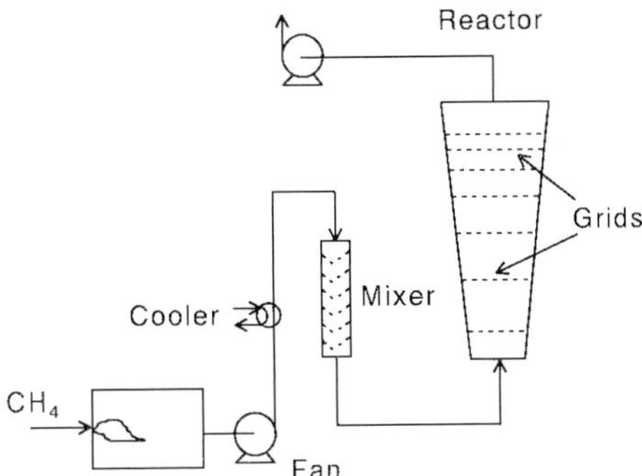

Figure 28. The floating-bed reactor system (adapted from Ref. 152).

ence was gathered since 1975 in treating flue gas from a lead-melting furnace and emissions from several plants processing chlorinated organic materials such as vinyl chloride monomer and phosgene.

The schematic flow diagram of the system employed for hydrocarbon fume abatement is shown in Fig. 29. Operating characteristics of the system are GHSV 5000–10 000, pressure drop 2500 Pa, mol ratio $NH_3 : NO_x$ 0.8–0.9, NH_3 and NO_x 99% and 85%, respectively.

Jüttner et al. [155] developed an electrochemical method for the simutaneous removal of SO_2 and NO_x. The process is shown schematically in Fig. 30. Flue gas is passed to an absorption column containing dithionite ions ($S_2O_4^{2-}$). Dithionites are powerful reducing agents. The dithionite is oxidized to sulphite and NO is reduced to compounds such as NH_4^+, NH_2HSO_3, and NH_2OH, while SO_2, is not absorbed. Also, small amounts of N_2O are formed. The solution is fed to the cathodic compartment of the electrochemical cell.

The rest of the flue gas, including SO_2, is passed to the anode compartment which contains PbO. SO_2 is oxidized into SO_4^{2-}, which forms sulfuric acid.

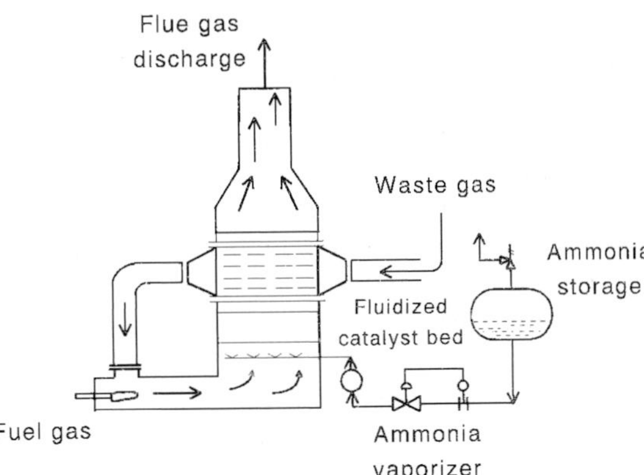

Figure 29. Schematic flow diagram of a fluidized bed SCR downstream from a waste gas incinerator (adapted from Ref. 154).

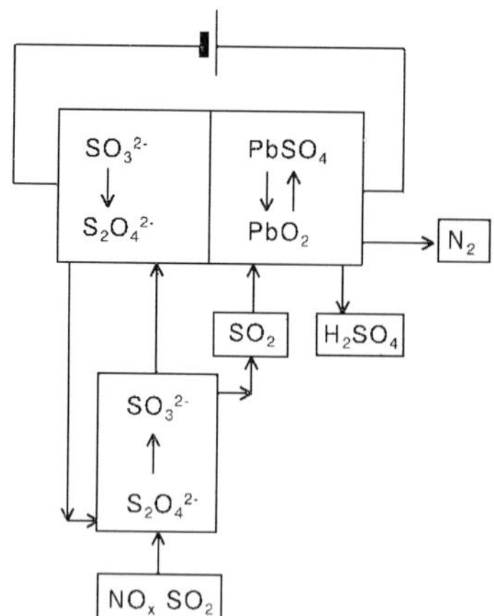

Figure 30. The PbO–dithionite process for the simultaneous removal of SO_2 and NO_x (adapted from Ref. 155).

2.5 Volatile Organic Compounds

A diversity of volatile organic compounds are emitted from a variety of stationary sources such as solvent evaporation, organic chemical manufacturing and miscellaneous processes, e.g. catalyst regeneration and gasoline marketing.

A variety of catalysts are known for gas cleanup: the noble metals Pt and Pd, oxides of such elements as V, W, Cr, Mn, Cu, and Fe, and zeolites.

Catalysts for the oxidation of volatile organic compounds (VOC) are generally supported platinum or palladium catalysts. Copper oxide, vanadium oxide and chromium oxide are suitable for the oxidation of halogenated compounds.

The reviews by Spivey [3] and by Jennings et al. [156] are excellent sources for further details on catalytic incineration of volatile organics emissions. Spivey [3] describes two types of techniques for removal of VOC from off-gases, namely one without preheater and one with a direct flame preheater. From an economically point of view it is more beneficial to carry out the catalytic oxidation at lower temperatures. In a catalytic incinerator, sometimes called an afterburner, VOCs are oxidized into carbon dioxide and water. The efficiency is about 70–90%. The incinerator has a preheat burner, a mixing chamber, a catalyst bed, and a heat recovery equipment. Temperatures of about 590 K are sufficient for the destruction of VOCs. Various catalyst geometries have been used: metal ribbons, spherical pellets, ceramic rods, ceramic honeycombs, and metal honeycombs. Precious metals such as platinum and palladium are often used in catalytic incinerators.

If the off-gas contains compounds such as CH_2Cl_2, CH_3CCl_3, H_3PO_4, HCN, or H_2S the catalyst may be poisoned. Most commercial systems are of the afterburner/catalyst configuration.

Schmidt [157] presented a variety of applications of catalytic off-gas cleanup. For instance the exhaust of Diesel engines contains aldehydes, CO, NO_x, soot, and polycyclic aromatic hydrocarbons. Three-way catalysts are not suitable for purifi-

Table 14. Some applications of catalytic afterburning [157].

Type	Problem	Example
Food processing	odor control	smoke houses, purification of food
Medicine	removal of organic contaminants	
Polymer processing	hazardous organics	synthesis of form-aldehyde, vinyl chloride
Coating operations	organic solvents	lacquered ware

cation because of the high oxygen content of the off-gases. Some applications are shown in Table 14. Flue gas from waste incinerators contains compounds such as HCl, SO_2, SO_3, Hg, Pb, Cd, As, and chlorinated hydrocarbons.

Hagenmaier [158] studied the catalytic effect of metals, oxides and carbonates on the decomposition of chloroaromatics such as polychlorinated dibenzo-p-dioxins and polychlorinated dibenzofuranes. Copper catalyses dechlorination/hydrogenation of these compounds at temperatures of about 553 K. After heating for 15 minutes the amount of most polychlorinated compounds was below 0.1 ng.

Hagenmaier and Mittelbach [150] also studied the influence of dedusted flue gas from a waste incinerator on an SCR catalyst. The relative activity of the titania-supported catalyst decreased after 2000 h on stream at 513 K and remained constant during 6000 h at 533 K. The activity drop for 513 K is higher than that for 533 K. Dioxins are converted over SCR catalysts at temperatures above 573 K. High conversions were found for low ammonia concentrations. These low concentrations occur at the exit of the SCR catalyst.

Tichenor and Palazzolo [159] tested a range of design and operating parameters on a variety of compounds and mixtures. The temperature range was 523–723 K. Some conclusions of the study are that VOC destruction efficiency increases with increasing temperature and that individual VOCs have different efficiencies.

Kosusko et al. [160, 161] reviewed studies for the destruction of groundwater air-stripping emissions. The catalyst consisted of a precious metal deposited on a ceramic honeycomb. Typical organic compounds in air stripping overhead are pentane, cyclohexane, trichloroethylene, benzene, cumene, etc. Catalyst deactivation occurs when H_2S is present in the ground water or when aerosols are stripped. Figure 31 shows an example of a system used for the purification of stripped air. Air is treated countercurrently with a contaminated water stream. Downstream from the demister the contaminated gas is heated by an external source such as the combustion of natural gas. Part of the contaminants will be oxidized.

In a Claus plant plant H_2S is converted into sulfur; however, the conversion is not complete (94–98%). About 1% H_2S and 0.5% COS remain in the off-gas due to the thermodynamics of the Claus equilibrium reaction. Van Nisselrooy and Lagas [162] developed a catalytic process, called Superclaus, which is based on bulk sulfur removal in a conventional Claus section, followed by selective catalytic oxidation of the remaining H_2S to elemental sulfur. Iron oxides and chromium oxides supported

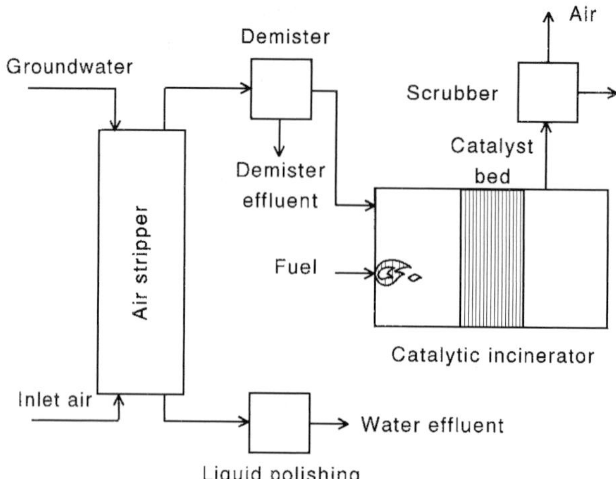

Figure 31. Schematic diagram of the catalytic destruction of air-stripping emissions (adapted from Ref. 161).

on α-Al_2O_3 are active and selective catalysts for the reaction

$$H_2S + \tfrac{1}{2}O_2 \rightarrow 1/nS_n + H_2O$$

Conversions of up to 99 vol. % were achieved.

Kettner and Lübcke [163] compared a catalytic incinerator with a thermal incinerator for off-gas of a Claus plant (Table 15). The sulfur compounds were oxidized to SO_2. The comparison is shown in Table 15.

2.6 Future Trends

In the preceding sections an overwiev is given of the application of catalysts in environmental technology, focusing on stationary sources. This last section overviews new developments in environmental catalysis and the challenges for fundamental research. One new application of catalysis is in thermal biomass conversion and

Table 15. The catalytic and thermal incineration of off-gas from a Claus plant [163].

Feature	Catalytic	Thermal
Gas flow ($m^3\,h^{-1}$)	80 000	80 000
Inlet temperature (K)	400	400
Reactor temperature (K)	600	1 070
Residence time (s)	2	0.5
Pressure drop (Pa)	2 000	800
Catalyst volume (m^3)	45	
H_2S outlet ($mg\,m^{-3}$)	<10	<10
CO conversion (%)	10	45
NO_x emission ($kg\,h^{-1}$)	1	4
Energy consumption (kWh)	7 000	38 000

other types of wastes conversion. The energy in biomass and domestic waste may be obtained by combustion. However, the upgrading of biomass into more valuable fuels is also a very promising possibility [9]. Upgrading may be carried out by means of thermal techniques such as pyrolysis, combustion, gasification, and liquefaction.

Pyrolysis and liquefaction may be used for the production of fuels for both electricity generation and chemical synthesis. However, these fields of catalyst application are still in their infancy.

Catalysis will play an important role in meeting environmental requirements for gasification. The control of nitrogen and sulfur compounds is crucial to avoid emissions of NO_x and SO_x.

The main goals for the application of environmental catalysis are minimizing both gas cleaning costs and emission requirements.

Preshaped catalysts, monolithic structures, are found to have many large-scale applications downstream from gas turbines, heaters, and in afterburners. Catalytic membranes are under development.

Spent catalysts are considered hazardous waste within the United States [137]. Most manufacturers take back the spent catalysts. However, if the export of hazardous waste is prohibited by legislation then new problems for the user arise.

The catalytic approach for protecting and improving our environment has had its successes. This will encourage the scientific community to proceed with its efforts to understand catalysis and its contribution to a cleaner environment.

References

1. J. N. Armor, *Appl. Catal.* **1992**, *1*, 221–256.
2. J. F. Armor, *Chem. Mat.* **1994**, *6*, 730–738.
3. J. J. Spivey, *Ind. Eng. Chem. Res.* **1987**, *26*, 2165–2180.
4. H. Bosch, F. Janssen, *Catal. Today* **1987**, *2*, 369–532.
5. G. C. Bond, S. Flamerz Tahir, *Appl. Catal.* **1991**, *1*, 1–31.
6. F. Janssen, *Catal. Today* **1993**, *16*, 155–287.
7. T. Inui, A. Miyamoto, K. Aika, *Catal. Today* **1991**, *10*, 1–118.
8. P. Ruiz, F. Thyrion, B. Delmon, *Catal. Today* **1993**, *17*, 1–383.
9. A. V. Bridgewater, *Appl. Catal.* **1994**, *116*, 5–47.
10. A. Amirnazmi, J. E. Benson, M. Boudart, *J. Catal.* **1973**, *30*, 55–65.
11. Y. J. Li, J. N. Armor, *Appl. Catal.* **1992**, *1*, L21–L29.
12. Y. Li, W. K. Hall, *J. Catal.* **1991**, *129*, 202–215.
13. J. Valyon, W. K. Hall, J. Catal. **1993**, *143*, 520–532.
14. M. Iwamoto, S. Yokoo, K. Sakai, S. Kagawa, *J. Chem. Soc., Faraday Trans.* **1981**, *77*, 1629–1638; M. Iwamoto, H. Yahiro, Y. Mine, S. Kagawa, *Chem. Lett.* **1989**, 213–216, M. Iwamoto, H. Yahiro, K. Tanda, N. Mizuno, Y. Mine. S. Kagawa, *J. Phys. Chem.* **1991**, *95*, 3727–3730, M. Iwamoto, H. Yahiro, N. Mizuno, W. Zhang, Y. Mine, H. Furukawa, S. Kagawa, *J. Phys. Chem.* **1992**, *96*, 9360–9366.
15. F. Tabata, *J. Mat. Sci. Lett.* **1988**, *7*, 147–148.
16. J. R. Hardee, J. W. Hightower, *J. Catal.* **1984**, *86*, 137–146.
17. R. T. Yang, N. Chen, *Ind. Eng. Chem. Res.* **1994**, *33*, 825–831.
18. F. Janssen, *Thesis*, University Twente, the Netherlands **1987**.
19. G. C. Bond, S. Flamerz Tahir, L. van Wijk, *Catal. Today* **1987**, *1*, 229–243.
20. H. Bosch, F. Janssen, F. van den Kerkhof, J. Oldenziel, J. van Ommen, J. Ross, *Appl. Catal.* **1986**, *25*, 239–248.

21. J. W. Geus, in *Studies in Surface Science and Catalysis III – Scientific Bases for the Preparation of Heterogeneous Catalysts*, **1993**, (Eds: G. Poncelet, P. Grange, P. A. Jacobs), Elsevier, Amsterdam, p. 1–33.
22. E. Vogt, A. van Dillen, J. Geus, F. Janssen, in *Proceedings 9th Intern. Congress on Catalysis*, (Eds: M. J. Philips, M. Teman) Calgary, Canada, **1988**, vol. 4, p. 1976.
23. E. Vogt, A. Boot, A. van Dillen, J. Geus, F. Janssen, F. van den Kerkhof, *J. Catal.* **1988**, *114*, 313–320.
24. A. Baiker, P. Dollenmeier, M. Glinski, A. Reller, *Appl. Catal.* **1987**, *35*, 365–380.
25. A. Baiker, B. Handy, J. Nlokl, M. Schraml-Marth, A. Wokaun, *Catal. Lett.* **1992**, *14*, 89–99.
26. G. C. Bond, S. Flamerz, *Appl. Catal.* **1989**, *46*, 89–102.
27. B. E. Handy, A. Baiker, M. Schraml-Marth, A. Wokaun, *J. Catal.* **1992**, *133*, 1–20.
28. J. Nickl, D. Dutoit, A. Baiker, U. Scharf, A. Wokaun, *Ber. Bunsen Ges. Phys. Chem.* **1993**, *97*, 217–228.
29. R. Mariscal, J. M. Palacios, M. Galan-Fereres, J. L. G. Fierro, *Appl. Catal.* **1994**, *116*, 205–219.
30. G. T. Went, L. Leu, A. T. Bell, *J. Catal.* **1992**, *134*, 492–505.
31. H. Barten, F. Janssen, F. v. d. Kerkhof, R. Leferink, E. T. C. Vogt, A. J. van Dillen, J. W. Geus, in *Preparation of catalysts IV*, (Eds: B. Delmon, P. Grange, P. A. Jacobs, G. Poncelet), Elesevier, Amsterdam, **1987**, p. 103.
32. B. E. Brent, A. Baiker, M. Schraml-Marth, A. Wokaun, *J. Catal.* **1992**, *134*, 1–20.
33. P. Ciambelli, G. Bagnasco, L. Lisi, M. Turco, G. Chiarello, M. Musci, M. Notaro, D. Robba, P. Ghetti, *Appl. Catal.* **1992**, *1*, 61–77.
34. F. Janssen, *J. Thermal Anal.* **1991**, *37*, 1281–1288.
35. G. T. Went, L. Leu, A. T. Bell, *J. Catal.* **1992**, *134*, 479–491.
36. U. S. Ozkan, Y. Cai, M. W. Kumthekar, L. Zhang, *J. Catal.* **1993**, *142*, 182–197.
37. G. Ramis, G. Busca, C. Cristianl, L. Lietti, P. Forzatti, F. Bregani, *Langmuir* **1992**, *8*, 1744–1749.
38. P. Ciambelli, G. Bagnasco, L. Lisi, M. Turco, G. Chiarello, M. Musci, M. Notaro, D. Robba, P. Ghetti, *Appl. Catal.* **1992**, *1*, 61–77.
39. J. P. Solar, P. Basu, M. P. Shatlock, *Catal. Today* **1992**, *14*, 211–224.
40. L. Lietti, P. Forzatti, G. Ramis, G. Busca, F. Bregani, *Appl. Catal.* **1993**, *3*, 13–35.
41. L. Lietti, P. Forzatti, *J. Catal.* **1994**, *147*, 241–249.
42. A. Baiker, A. Wokaun, *Naturwissenschaften* **1989**, *76*, 178–170.
43. J. Nickl, C. Schild, A. Baiker, M. Hund, A. Wokaun, *Fres. J. Anal. Chem.* **1993**, *346*, 79–83.
44. G. Ramis, G. Busca, F. Bregani, *Catal. Lett.* **1993**, *18*, 299–303.
45. J. P. Chen, R. T. Yang, *J. Catal.* **1993**, *139*, 277–288.
46. R. B. Bjorklund, C. U. I. Odenbrand, J. G. M. Brandin, H. L. H. Andersson, B. Liedberg, *J. Catal.* **1989**, *119*, 187–200.
47. U. S. Ozkan, Y. Cai, M. W. Kumthekar, *Appl. Catal.* **1993**, *96*, 365–381.
48. U. Baltensperger, M. Ammann, U. K. Bochert, B. Eichler, H. W. Gaggeler, *J. Phys. Chem.* **1993**, *97*, 12325–12330.
49. M. A. Buzanowski, R. T. Yang, *Ind. Eng. Chem. Res.* **1990**, *29*, 2074–2078.
50. J. W. Beeckman, L. Hegedus, *Ind. Eng. Chem. Res.* **1991**, *30*, 969–978.
51. M. Kotter, H.-G. Lintz, T. Turek, D. L. Trimm, *Appl. Catal.* **1989**, *52*, 225–235.
52. H.-G. Lintz, T. Turek, *Appl. Catal.* **1992**, *85*, 13–25.
53. M. Turco, L. Lisi, R. Pirone, *Appl. Catal.* **1994**, *3*, 133–149.
54. N. Y. Topsøe, T. Slabiak, B. S Clausen, T. Z. Srnak, J. A. Dumesic, *J. Catal.* **1992**, *134*, 742–746.
55. C. U. Odenbrand, P. L. T. Gabrielsson, J. G. M. Brandin, L. A. H. Andersson, *Appl. Catal.* **1991**, *78*, 109–123.
56. F. Goudriaan, C. M. A. M. Mesters, R. Samson, *Proceedings of the 1989 Joint EPA-EPRI Symposium on Stationary Combustion NO_x control*, **1989**, San Francisco, March 6–9.
57. C. U. Odenbrand, L. A. H. Andersson, J. G. M. Brandin, S. T. Lundin, *Appl. Catal.* **1986**, *27*, 363–377.
58. O. B. Lapina, A. V. Nosov, V. M. Mastlkhm, K. A. Dubkov, V. V. Mokrinski, *J. Mol. Catal.* **1994**, *87*, 57–66.

59. S. Szakacs, G. J. Altena, T. Fransen, J. G. van Ommen, J. R. H. Ross, *Catal. Today* **1993**, *16*, 237–245.
60. J. S. Hepburn, H. G. Stenger, C. E. Lyman, *J. Catal.* **1991**, *128*, 48–62.
61. J. Kaspar, C. de Leitenburg, P. Fornasiero, A. Trovarelli, M. Graziani, *J. Catal.* **1994**, *146*, 136–143.
62. T. Katona, L. Guczi, G. A. Somorjai, *J. Catal.* **1992**, *135*, 434–444.
63. M. F. H. van Tol, M. A. Quinlan, F. Luck, G. A. Somorjai, B. E. Nieuwenhuys, *J. Catal.* **1991**, *129*, 186–194.
64. S. J. Lombardo, M. Slinko, T. Fink, T. Löher, H. H. Madden, F. Esch, R. Imbihl, *Surface Science* **1992**, *269/270*, 481–487.
65. N. M. H. Janssen, B. E. Nieuwenhuys, M. Ikai, K. Tanaka, A. R. Cholach, *Surf. Sci.* **1994**, *319*, L29–L33.
66. G. Zhang, T. Yamaguchi, H. Kawakami, T. Suzuki, *Appl. Catal.* **1992**, *1*, L15–L20.
67. G. Centi, S. Perathoner, Y. Shioya, M. Anpo, *Res. Chem. Intermed.* **1992**, *17*, 125–135.
68. G. Moretti, *Catal. Lett.* **1994**, *23*, 135–140.
69. M. Shelef, C. N. Montreuil, H. W. Jen, *Catal. Lett* **1994**, *26*, 277–284.
70. R. Gopalakrishnan, P. R. Stafford, J. E. Davidson, W. C. Hecker, C. H. Bartholomew, *Appl. Catal.* **1993**, *2*, 165–182.
71. F. Radtke, R. A. Koeppel, A. Baiker, *Appl. Catal.* **1994**, *107*, L125–L132.
72. J. O. Petunchi, G. Sill, W. K. Hall, *Appl. Catal.* **1993**, *2*, 303–321.
73. J. O. Petunchi, W. K. Hall, *Appl. Catal.* **1993**, *2*, L17–L26.
74. T. Komatsu, N. Nunokawa, I. S. Moon, T. Takahara, S. Namba, T. Yashima, *J. Catal.* **1994**, *148*, 427–437.
75. R. Burch, S. Scire, *Catal. Lett.* **1994**, *27*, 177–186.
76. Y. J. Li, J. N. Armor, *Appl. Catal.* **1992**, *1*, L31–L40.
77. Y. J. Li, J. N. Armor, *Appl. Catal.* **1993**, *3*, 55–60.
78. C. Yokoyama, M. Misono, *Chem. Lett.* **1992**, 1669–1672.
79. Y. J. Li, J. N. Armor, *J. Catal.* **1994**, *145*, 1–9.
80. Y. Nishizaka, M. Misono, *Chem. Lett.* **1993**, 1295–1298.
81. C. N. Montreuil, M. Shelef, *Appl. Catal.* **1992**, *1*, L1–L8.
82. T. Grzybek, H. Papp, *Appl. Catal.* **1992**, *1*, 271–283.
83. J. Imai, T. Suzuki, K. Kaneko, *Catal. Lett.* **1993**, *20*, 133–139.
84. R. B. Bjorklund, S. Järäs, A. Ackelid, C. U. I. Odenbrand, H. L. H. Andersson, J. G. M. Brandin, *J. Catal.* **1991**, *128*, 574–580.
85. R. Y. Weng, J. F. Lee, *Appl. Catal.* **1993**, *105*, 41–51.
86. K. Hadjiivanov, V. Bushev, M. Kantcheva, D. Klissurski, *Langmuir* **1994**, *10*, 464–471.
87. H. E. Curry-Hyde, A. Baiker, *Appl. Catal.* **1992**, *90*, 183–197.
88. H. E. Curry-Hyde, A. Baiker, *Ind. Eng. Chem. Res.* **1990**, *29*, 1985–1989.
89. M. Schraml-Marth, A. Wokaun, H. E. Curry-Hyde, A. Baiker, *J. Catal.* **1992**, *133*, 415–431.
90. H. Hosose, H Yahiro, N. Mizuno, M. Iwamoto, *Chem. Lett.* **1991**, 1589–1860.
91. T. Miyadera, K. Yoshida, *Chem. Lett.* **1993**, *2*, 1483–1486.
92. F. Kapteijn, L. Singoredjo, A. Andreini, *Appl. Catal.* **1994**, *3*, 173–189.
93. M. C. Kung, *Catal. Lett.* **1993**, *18*, 111–117.
94. N. Mizuno, M. Yamato, M. Tanaka, M. Misono, *J. Catal.* **1991**, *132*, 560–563.
95. N. Mizuno, M. Tanaka, M. Misono, *J. Chem. Soc., Faraday Trans.* **1992**, *88*, 91–95.
96. A. K. Lavados, P. J. Pomonis, *Appl. Catal.* **1992**, *1*, 101–116.
97. N. Okazaki, T. Kohno, R. Inoue, Y. Imizu, A. Trada, *Chem. Lett.* **1993**, *7*, *1195–1198.*
98. Y. Ukisu, S. Sato, A. Abe, Y. Yoshida, *Appl. Catal.* **1993**, 147–152.
99. C. Flockenhaus, *Gas Wärme Intern.* **1987**, *36*, 210–214.
100. N. W. Cant, J. R. Cole, *J. Catal.* **1992**, *134*, 317–330.
101. K. Kusakabe, M. Kashima, S. Morooka, Y. Kato, *Fuel* **1988**, *67*, 714–718.
102. S. Kasaoka, E. Sasaoka, H. Iwasaki, *Bull. Chem. Soc. Jpn.* **1989**, *62*, 1226–1232.
103. M. Inomata, A. Miyamoto, Y. Murakami, *J. Catal.* **1980**, *62*, 140–148.
104. F. Janssen, F. van den Kerkhof, H. Bosch, J. R. H. Ross, *J. Phys. Chem.* **1987**, *91*, 5921–5927; *J. Phys. Chem.* **1987**, *91*, 6633–6638.

105. M. Kantcheva, V. Bushev, D. Klissurski, *J. Catal.* **1994**, *145*, 96–106.
106. B. L. Duffy, H. E. Curry-Hyde, N. W. Cant, P. F. Nelson, *J. Phys. Chem.* **1993**, *97*, 1729–1732.
107. A. Miyamoto, K. Kobayashi, M. Inomata, Y. Murakami, *J. Phys. Chem.* **1982**, *86*, 2945–2950.
108. N. Y. Topsøe, *Sci.* **1994**, *265*, 1217–1219.
109. M. Takagi, T. Kawai, M. Soma, T. Onishi, K. Tamaru, *J. Phys. Chem.* **1976**, *2*, 430–431.
110. W. Biffar, R. Drews, K. Hess, R. Lehnert, W. D. Moß, O. Scheidsteger, *Brennstoffen, Wärme, Kraft* **1986**, *38*, 211–216.
111. M. Gasior, J. Haber, T. Machej, T. J. Czeppe, *J. Mol. Catal.* **1988**, *43*, 359–369.
112. H. Schneider, S. Tschudin, M. Schneider, A. Wokaun, A. Baiker, *J. Catal.* **1994**, *147*, 5–14.
113. T. Z. Srnak, J. A. Dumesic, B. S. Clausen, E. Törnqvist, N.-Y. Topsøe, *J. Catal.* **1992**, *135*, 246–262.
114. J. Blanco, P. Avila, J. L. G. Fierro, *Appl. Catal.* **1993**, *96*, 331–343.
115. H. Gutberlet, *VGB Kraftwerkstechnik* **1988**, *3*, 287–293.
116. B. Schallert, *VGB Kraftwerkstechnik* **1988**, *4*, 432–440.
117. D. Vogel, F. Richter, J. Sprehe, W. Gajewski, H. Hofmann, *Chem.-Ing.-Tech.* **1988**, *60*, 714–715.
118. J. P. Chen, R. T. Yang, *Appl. Catal.* **1992**, *80*, 135–148.
119. R. Brand, B. Engler, W. Honnen, P. Kleine-Möllhoff, E. Koberstein, H. Ohmer, *Chemische Industrie* **1989**, *1*, 12–16.
120. J. P. Chen, M. A. Buzanowski, R. T. Yang, J. E. Cichanowicz, *J. Air Waste Manage. Assoc.* **1990**, *40*, 1403–1409.
121. W. Honnen, R. Brand, B. Engler, P. Kleine-Möllhoff, E. Koberstein, *Dechema-Monographien Band 118*, VCH Verlagsgesellschaft, **1989**, 3–16.
122. F. Lange, H. Schmelz, H. Knözinger, *J. Electr. Spec. Rel. Phen.* **1991**, *57*, 307–315.
123. F. Hilbrig, H. E. Göbel, H. Knözinger, H. Schmelz, B. Lengeler, *J. Catal.* **1991**, *129*, 168–176.
124. E. Hums, *Res. Chem. Intermed.* **1993**, *19*, 419–441; *Ind. Eng. Chem. Res.* **1991**, *30*, 1814–1818.
125. I. Rademacher, D. Borgman, G. Hopfengärter, G. Wedler, E. Hums, G. W. Spitznagel, *Surf. Interf. Anal.* **1993**, *20*, 43–52.
126. G. Hopfengärtner, D. Borgmann, I. Rademacher, G. Wedler, E. Hums, G. W. Spitznagel, *J. Elec. Spec. Rel. Phenom.* **1993**, *63*, 91–116.
127. M. Tokarz, S. Järås, B. Persson, in *Catalyst Deactivation 1991* (Eds: C. H. Bartholomew, J. B. Butt) Elsevier, Amsterdam, The Netherlands, **1991**, p. 523.
128. A. Dieckmann, H. Gutberlet, B. Schallert, *VGB Kraftwerkstechnik* **1987**, *12*, 1204–1213.
129. H. Gutberlet, *VGB Kraftwerkstechnik* **1994**, *1*, 54–59.
130. F. Janssen, R. Meijer, *Catalysis Today* **1993**, *16*, 157–185.
131. R. Meijer, F. J. J. G. Janssen, *VGB Kraftwerkstechnik* **1993**, *10*, 914–917.
132. J. B. Lefers, P. Lodder, G. D. Enoch, *Chem. Eng. Technol.* **1991**, *14*, 192–200.
133. D. Hein, W. Gajewski, *VGB Kraftwerkstechnik* **1987**, *2*, 118–122.
134. S. Irandoust, B. Andersson, *Catal. Rev.-Sci. Eng.* **1988**, *30*, 341–392.
135. F. Nakajima, M. Takeuchi, S. Matsuda, S. Uno, T. Mori, Y. Watanbe, N. lmanari, Japanese Patents **1973**, 1010563, 1034771, 1115421, 1213543.
136. H. Gutberlet, B. Schallert, *Catal. Today* **1993**, *16*, 207–236.
137. D. Cobb, L. Glatch, J. Ruud, S. Snyder, *Environm. Progr.* **1991**, *10*, 49–59.
138. J. Ando, *Review of the Japanese NO_x abatement technology for stationary sources*, in: NO_x Symposion, (Eds: O. Rentz, F. Iszla, M. Weibel) Karlsruhe, **1985**, Paper A1.
139. H. Krüger, *VGB Kraftwerkstechnik* **1991**, *71*, 371–395.
140. H. Gutberlet, H.-J. Dieckmann, B. Schallert, *VGB Kraftwerkstechnik* **1991**, *6*, 584–591.
141. B. Kassebohm, G. Wolfering, *VGB Kraftwerkstechnik* **1991**, *71*, 404–408.
142. E. Richter, in *Fortschritte bei der thermischen, katalytischen und sorptiven Abgasreinigung*, VDI-Verlag GmbH, Düsseldorf, **1989**, p. 121–156.
143. H. C. Andersen, W. J. Green, D. R. Steele, *Ind. Eng. Chem.* **1961**, *53*, 199–204.
144. P. Avila, C. Barthelemy, A. Bahamonde, J. Blanco, *Atmospher. Environm.* **1993**, *3*, 443–447.
145. I. Stambler, *Gas Turbine World* **1993**, 32–44.
146. R. B. Snyder, K. R. Durkee, W. J. Neuffer, in: *85th Annual Meeting & Exhibition* **1992**, Kansas City, Missouri, June 21–26, p. 2–15.

147. R. A. Dalla Betta, K. Tsurumi, T. Shoji, US Patent 5 258 349 **1993**.
148. L. D. Pfefferle, W. C. Pfefferle, *Catal. Rev.-Sci. Eng.* **1987**, *29*, 219–267.
149. K. Garbowski, C. Feumi-Jantou, N. Mouaddib, M. Primet, *Appl. Catal.* **1994**, *109*, 277–291.
150. H. Hagenmaier, G. Mittelbach, *VBG Kraftwerktechnik* **1990**, *70*, 491–493.
151. J. B. Lefers, P. Lodder, US Patent 5 158 754 **1992**.
152. L. H. J. Vredenbregt, P. Lodder, G. D. Enoch, F. J. J. G. Janssen, in: *Non-CO_2 Greenhouse Gases* (Eds: J. van Ham et al.) Kluwer, The Netherlands, **1994**, p. 369–375.
153. B. Agerholm, K. Jelsbak, B. Sander, *VBG Kraftwerktechnik* **1990**, *70*, 337–241.
154. L. C. Hardison, G. J. Nagl, G. E. Addison, *Environm. Progress* **1991**, *10*, 314–318.
155. K. Jüttner, G. Kreyssa, K. H. Kleifges, R. Rottmann, *Chem.-Ing.-Tech* **1994**, *66*, 82–85.
156. M. S. Jennings, N. E. Krohn, R. S. Berry, M. A. Palazzolo, R. M. Parks, K. K. Fidler, *Catalytic Incineration for Control of Volatile Organic Compounds Emissions*, **1985**, Noyes Publications, New Jersey, USA.
157. T. Schmidt, *Techn. Mitt.* **1989**, *82*, 360–369.
158. H. Hagenmaier, in *Fortschritte bei der thermischen, katalytischen und sorptiven Abgasreinigung*, **1989**, VDI-Verlag, GmbH Düsseldorf, p. 239–254.
159. B. A. Tichenor, M. A. Palazzolo, *Environm. Progress* **1987**, *6*, 172–176.
160. M. Kosusko, C. M. Nunez, *J. Air Waste Manage. Assoc.* **1990**, *40*, 254–259.
161. M. Kosusko, M. E. Mullins, K. Ramanathan, T. N. Rogers, *Environm. Progress* **1988**, *7*, 136–142.
162. P. F. M. T. van Nisselrooy, J. A. Lagas, *Catal. Today*, **1993**, *16*, 263–271.
163. R. Kettner, T. Lübcke in: *Fortschritte bei der thermischen, katalytischen und sorptiven Abgasreinigung*, **1989**, VDI-Verlag, GmbH, Düsseldorf, p. 255–274.
164. N. Ohlms, *Catalysis Today* **1993**, *16*, 247–261.

3 Catalytic Combustion

R. L. Garten, R. A. Dalla Betta and J. C. Schlatter, Catalytica Combustion Systems, 430 Ferguson Drive, Bldg 3, Mountain View, CA 94043-5272/USA

3.1 Introduction

Catalytic combustion is a process in which a combustible compound and oxygen react on the surface of a catalyst, leading to complete oxidation of the compound. This process takes place without a flame and at much lower temperatures than those associated with conventional flame combustion [1, 2]. Due partly to the lower operating temperature, catalytic combustion produces lower emissions of nitrogen oxides (NO_x) than conventional combustion. Catalytic combustion is now widely used to remove pollutants from exhaust gases, and there is growing interest in applications in power generation, particularly in gas turbine combustors.

Catalytic combustion applications can be classified as either primary or secondary pollution control, that is, emissions prevention or emissions clean-up. The most common example of catalytic combustion for emissions clean-up is the catalytic converter in the exhaust system of automobiles. Catalytic combustion is also increasingly used for the removal of volatile organic compounds (VOCs) from industrial exhaust streams. The use of catalytic combustion in exhaust gas clean-up is discussed in other sections of this Handbook; this section deals only with primary control applications.

Applications of catalytic combustion in primary pollution control mainly involve the control of NO_x emissions in natural gas-fueled gas turbines and boilers. NO_x emissions are atmospheric pollutants that cause serious pulmonary problems and are also precursors to the formation of ozone, another lung irritant. NO_x emissions are therefore strictly regulated in many countries. Restrictions are especially severe in areas that have serious air quality problems, including California, Texas, the northeastern area of the United States, Japan and parts of Europe.

There is increasing interest in the use of gas turbines for power generation while at the same time more stringent regulations on NO_x emissions are being implemented in many areas. This situation provides an opportunity for catalytic combustion and has stimulated much research on catalytic combustors in recent years. Because of their potential commercial importance this section will largely focus on gas turbine applications of catalytic combustion.

3.1.1 Gas Turbine Applications

The worldwide demand for power is projected to grow by 2–2.5% per year over the next decade, with a projected requirement of 650–750 gigawatts of new power generating capacity. Natural gas-fueled gas turbines are expected to provide a significant portion of the growing world demand for electric power generation because natural gas burns more cleanly than other fossil fuels and produces less carbon dioxide, a greenhouse gas, per unit of power produced. Gas turbines also have shorter construction lead times and lower installation costs per kilowatt than alternative methods of power generation, and their modularity allows incremental additions to existing capacity.

Increasingly stringent environmental regulations on NO_x emissions must be met for existing and new gas turbines in many areas, however. In some urban areas in the United States and Japan new gas turbines are required to achieve NO_x levels of less than 5 parts per million by volume (ppm) in order to obtain an installation permit. It is likely that similar restrictions will be extended to many other areas in the near future.

A brief survey of the methods available to reduce NO_x emissions in gas turbines indicates the advantages of catalytic combustion when these emissions must be reduced to the single digit levels required in many areas.

In conventional gas turbine combustors (diffusion flame combustors) the fuel : air ratio varies widely. The flame stabilizes in a region of high fuel : air ratio, forming localized areas (hot spots) in which the temperature can reach 1700–1800 °C. These hot spots produce NO_x levels in the 100–200 ppm range. Water or steam injection can reduce flame temperatures and thereby lower NO_x emissions to about 25–42 ppm, but this significantly increases operating costs due to the requirement for highly purified water.

NO_x emissions can also be reduced by using a dilute fuel/air mixture in so-called lean premixed combustion. Air and fuel are premixed at a low fuel : air ratio before entering the combustor, resulting in a lower flame temperature and hence lower NO_x emissions. Lean premix combustors (sometimes called dry low NO_x combustors) based on this concept have generally achieved NO_x levels of 25 ppm in commercial operation. To achieve lower NO_x levels, however, this technology must overcome some significant hurdles. As the fuel: air mixture is increasingly diluted, the flame temperature approaches the flammability limit and the flame becomes unstable. The flame instability produces noise and vibration that can reduce the combustor life, increase maintenance costs and adversely impact the operational reliability of the turbine [3].

Further reductions in NO_x levels can be achieved by removing NO_x from the turbine exhaust using selective catalytic reduction (SCR) with ammonia. SCR is an effective method for NO_x control that can reduce NO_x levels to about 10 ppm, usually in combination with water or steam injection or lean premix combustors. However, SCR is an expensive technology and the storage and handling of ammonia, a toxic chemical, also pose problems. Possible future restrictions on ammonia emissions may also limit the application of this technology.

Although a combination of lean premix combustion and SCR technology can achieve NO_x levels below about 10 ppm, the resulting cost is very high and is likely to be prohibitive, especially for small turbines. Such costs can range from \$20 000 to \$50 000 per megawatt per year, depending on turbine size [4]. In contrast, catalytic combustion has the potential to reduce NO_x emissions to ultralow levels (less than 1 ppm) at much lower cost, while avoiding the problems of lean premix combustors. The reduction of NO_x emissions to extremely low levels is based on the kinetics of NO_x formation, as described in the next section.

3.1.2 Basis for NO_x Prevention

In the combustion of fuels that do not contain nitrogen compounds, NO_x compounds (primarily NO) are formed by two main mechanisms, the thermal mechanism proposed by Zeldovitch and the prompt mechanism [5]. In the thermal mechanism NO is formed by the oxidation of molecular nitrogen through the following reactions:

$$O + N_2 \rightarrow NO + N$$

$$N + O_2 \rightarrow NO + O$$

$$N + OH \rightarrow NO + H$$

The prompt mechanism is initiated by hydrocarbon radicals, predominantly through the reaction

$$CH + N_2 \rightarrow HCN + N$$

The HCN and N are converted rapidly to NO by reaction with oxygen and hydrogen atoms in the flame.

The prompt mechanism predominates at low temperatures under fuel-rich conditions, whereas the thermal mechanism becomes important at temperatures above 1500 °C. Due to the onset of the thermal mechanism the formation of NO_x in the combustion of fuel/air mixtures increases rapidly with temperature above 1500 °C and also increases with residence time in the combustor (Fig. 1). The thermal and prompt mechanisms are discussed in detail in the review by Miller and Bowman [5].

In catalytic combustion of a fuel/air mixture the fuel reacts on the surface of the catalyst by a heterogeneous mechanism. The catalyst can stabilize the combustion of ultra-lean fuel/air mixtures with adiabatic combustion temperatures below 1500 °C. Thus, the gas temperature will remain below 1500 °C and very little thermal NO_x will be formed, as can be seen from Fig. 1. However, the observed reduction in NO_x in catalytic combustors is much greater than that expected from the lower combustion temperature. The reaction on the catalytic surface apparently produces no NO_x directly, although some NO_x may be produced by homogeneous reactions in the gas phase initiated by the catalyst.

The effect of catalytic combustion on NO_x formation is shown very clearly in hybrid systems in which part of the fuel is burned by catalytic combustion and the

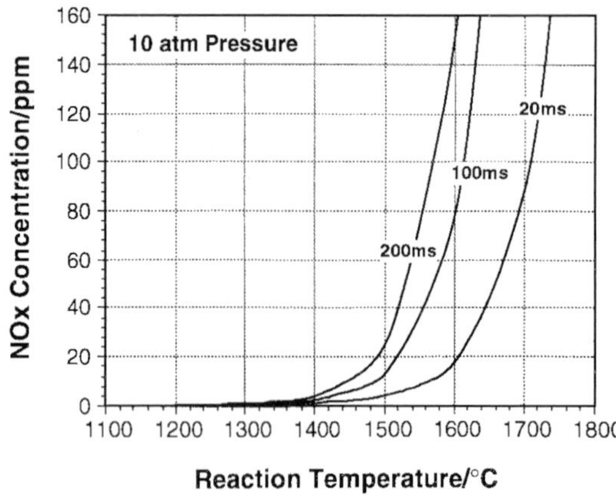

Figure 1. NO_x concentration as a function of temperature calculated with kinetic model for homogeneous CH_4 combustion similar to that in Ref. 5.

remainder is burned downstream of the catalyst by homogeneous combustion. In these systems the reduction in NO_x is directly proportional to the fraction of the fuel that is burned in the catalytic reaction [6]. Presumably, reducing the amount of fuel available for the homogeneous reaction reduces proportionately the pool of radicals, thus reducing the formation of both thermal and prompt NO_x.

3.2 Application Requirements

This section discusses certain key features of the operating conditions for gas turbines and how these features determine the requirements for a catalytic combustor.

3.2.1 Operating Conditions

In a gas turbine the air is compressed in a compressor section prior to addition of the fuel. For modern industrial gas turbines operating at pressures of 12–16 bar the compressive heating results in an air temperature of 350–410 °C at the catalyst inlet. Some combustion catalysts currently under development can operate at inlet temperatures in the range 380–390 °C [7]. Catalysts with very high activity are required in order to achieve ignition temperatures in this range.

For gas turbines operating at pressures below about 15 bar the gas temperature at the catalyst inlet will be below 380 °C and so the catalyst will not have sufficient activity to initiate catalytic combustion. In these cases a portion of the fuel can be burned in a conventional burner upstream of the catalyst in order to raise the temperature at the catalyst inlet. This process would generate some NO_x. An important goal in the development of combustion catalysts for gas turbines is therefore to lower the ignition temperature still further to minimize the use of a preburner in the current generation of gas turbines. The next generation of gas turbines will operate at pressure ratios of about 20, so the catalyst inlet temperature will increase significantly and preburners should no longer be necessary, except at start-up.

The temperature of the gas entering the power turbine has increased significantly in recent years in order to improve turbine efficiency. Older gas turbines operate at turbine inlet temperatures up to 1100 °C, but more recent turbines typically have turbine inlet temperatures of about 1300 °C. Further increases to about 1400–1450 °C are planned in the next generation of gas turbines. The catalytic combustor must therefore be designed to produce a gas temperature of 1300 °C or above at the combustor outlet. This presents a significant design challenge, because most catalytic materials will not tolerate such high temperatures.

Another important feature of gas turbine operation is the very high velocity of the fuel/air mixture. For example, in a full scale catalytic combustor designed by General Electric [8], the air flow through the 51 cm diameter unit was 23 kg s^{-1}. This corresponds to a gas hourly space velocity (GHSV) of 300 000 h^{-1} at 12 bar pressure and 450 °C. The catalyst must obviously be able to operate effectively at these very high gas velocities.

3.2.2 Catalyst Characteristics

The operating conditions for the gas turbine determine many of the desirable characteristics of the catalyst used in the catalytic combustor. The catalyst must be sufficiently active to operate at the lowest possible inlet temperature, preferably below 400 °C, in order to minimize the use of the preburner. Unless the catalyst temperature is limited by some means, the catalyst must also be able to withstand the combustor outlet temperature of 1300 °C. The catalyst must also operate efficiently at the very high gas velocities in the combustor.

The catalyst lifetime should be at least 1 year (8760 h) because it would be undesirable to change the catalyst more frequently. Turbines are usually inspected every year, and a catalyst could be changed at this point without additional disruption of operations, provided the change is relatively straightforward. Ideally, the catalyst lifetime should be at least 3 years, as gas turbines undergo a major overhaul and detailed inspection every 3 years.

Poisoning of the catalyst by contaminants in the fuel/air mixture is one of the major factors affecting catalyst lifetime. Sulfur compounds in the natural gas fuel are possible catalyst poisons. Dust or other contaminants in the air ingested by the turbine could be deposited on the catalyst and mask the active sites or could react with and deactivate the catalyst. In coastal areas salt from sea air is a potential catalyst poison. The catalyst should be sufficiently resistant to these airborne contaminants so that performance is maintained for at least one year.

The catalyst should also be resistant to thermal shock, that is, a sudden increase or decrease in temperature. Rapid temperature changes occur during start-up or shut-down of the turbine. The most serious thermal shocks occur upon sudden loss of the turbine load. If the turbine load is lost (by opening a circuit breaker, for example) the fuel must be shut off immediately to prevent overspeeding and destruction of the turbine. The air continues to flow, however, so the temperature of the catalyst drops very rapidly. Under these conditions the catalyst temperature can fall 1000 °C in 100 ms, which poses severe problems for ceramic materials. Most ceramic components evaluated for use in combustion catalysts have been found to shatter following a thermal shock of this magnitude.

The catalytic combustor should be capable of meeting the NO_x emission levels required in the most strictly regulated areas. As mentioned above, certain areas of Japan and the United States now require NO_x emissions to be limited to 5 ppm, so this should be the emissions target for any programme to develop catalytic combustors for gas turbines.

3.3 Features of Catalytic Combustion

3.3.1 Surface Temperatures

At low temperatures the oxidation reactions on the catalyst are kinetically controlled and the catalyst activity is an important parameter. As the temperature increases, the buildup of heat on the catalyst surface due to the exothermic surface reactions produces ignition and the catalyst surface temperature jumps rapidly to the adiabatic flame temperature of the fuel/air mixture (Fig. 2) [2]. At the adiabatic flame temperature, oxidation reactions on the catalyst are very rapid, and the overall steady state reaction rate is determined by the rate of mass transfer of fuel to the catalytic surface. The bulk gas temperature rises along the reactor because of heat transfer from the hot catalyst substrate and eventually approaches the catalyst surface temperature.

The steady state temperature of the catalyst surface under mass-transport-limited conditions can exceed the adiabatic flame temperature if the rate of mass transport of fuel to the surface is faster than the rate of heat transport from the surface. The ratio of mass diffusivity to heat diffusivity in a gas is known as the Lewis number. Reactor models [9] show that for gases with a Lewis number close to unity, such as carbon monoxide and methane, the catalyst surface temperature jumps to the adiabatic flame temperature of the fuel/air mixture on ignition. However, for gases with a Lewis number significantly larger than unity the rate of mass transport to the surface is much faster than the rate of heat transport from the surface, and so the wall temperature can exceed the adiabatic gas temperature. The extreme case is

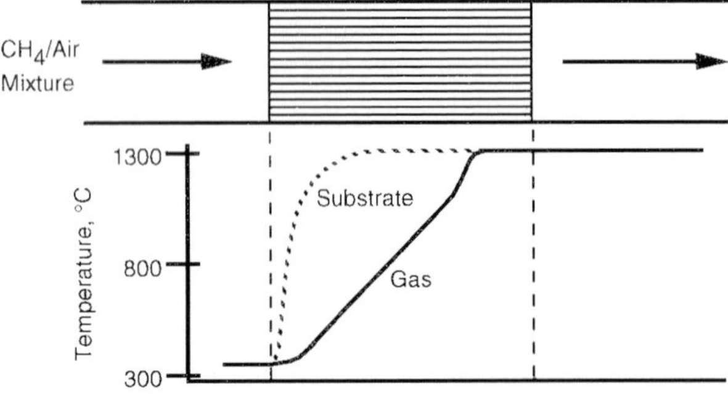

Figure 2. Schematic temperature profiles for catalyst (substrate) and bulk gas in a traditional catalytic combustor.

the combustion of hydrogen, for which the Lewis number is about four. In this case the catalyst surface temperature can exceed the adiabatic gas temperature by more than 100 °C [2].

As the catalyst surface temperature is equal to the adiabatic flame temperature after ignition, it is independent of the overall conversion in the combustion reaction. It follows that the catalyst surface temperature cannot be reduced simply by limiting the conversion (by using a short reactor or a monolith with large cells, for example). Therefore, unless some other means of limiting the catalyst surface temperature is used, the catalytic materials must be able to withstand the adiabatic flame temperature of the fuel/air mixture during the combustion reaction. For the present generation of gas turbines this temperature will be equal to the required turbine inlet temperature of 1300 °C, which presents severe problems for existing combustion catalysts.

3.3.2 Role of Gas Phase Reactions

In addition to the surface reactions the catalyst can promote homogeneous reactions in the gas phase, if the temperature is high enough. The homogeneous reactions could be initiated either by heat release from the catalytic surface or by free radicals formed on the surface that escape into the gas phase. The gas phase reactions can significantly increase the overall reaction rate in the combustor and must be taken into account in the modeling and design of catalytic combustors that operate at high temperatures.

The temperature at which gas phase reactions must be taken into account will depend on the nature of the fuel and the fuel : air ratio. A recent simulation of the catalytic combustion of methyl chloride using a model that included a detailed gas phase reaction mechanism found that the gas phase reactions start to make a significant contribution to the overall reaction rate at wall temperatures above 700 °C [10]. In the case of methane, the primary constituent of natural gas, there is evidence that gas phase reactions start to play a role at wall temperatures above 1000 °C [11].

Thus if the catalyst wall temperature is held below 1000 °C to avoid sintering or vaporization of the catalytic materials, the reaction rate for methane combustion will not be boosted by gas phase reactions and the task of attaining complete combustion is carried out only by the catalyst. More catalyst (i.e. a longer reactor) will be required to attain complete combustion when the reaction is purely catalytic than when the catalytic reaction is assisted by gas phase reactions.

3.4 Design Approaches

3.4.1 Fully Catalytic

The traditional approach to catalytic combustion is to burn all or most of the fuel/air mixture in the catalyst in order to minimize NO_x production. This means that the catalyst will attain the adiabatic flame temperature of the fuel/air mixture, as discussed in Section 3.3.1. The adiabatic flame temperature cannot be less than the required turbine inlet temperature. For older, low-efficiency gas turbines this temperature would be about 1000 °C. However, in modern, high-efficiency gas turbines the catalyst must be able to withstand temperatures of at least 1300 °C and even higher temperatures in the next generation of gas turbines.

The major challenge in fully catalytic combustion is to develop a catalyst that will tolerate temperatures of 1300 °C and above. At these temperatures vaporization and sintering of the catalyst components present severe problems, as discussed in Section 3.5. So far no catalyst has been found that combines high activity with long-term stability at 1300 °C, and the prospects for developing an acceptable catalyst for this operating temperature must be considered to be remote. The problems become even more severe with the higher inlet temperatures for the next generation of gas turbines.

The problems encountered in developing catalysts for fully catalytic combustion have led to the development of various design approaches in which the catalyst temperature stays below the combustor outlet temperature. These approaches are described in the following sections.

3.4.2 Staged Fuel

Toshiba Corp. has developed a hybrid catalytic combustor in which only part of the fuel is reacted in the catalyst and the remainder is mixed with the catalyst exit gases and burned in a homogeneous reaction [12]. The very lean fuel/air mixture in the catalytic combustion stage keeps the catalyst temperature below 1000 °C. The homogeneous combustion of the fuel in the second stage brings the combustor outlet temperature up to 1300 °C. Small-scale reactor tests of this system at atmospheric pressure gave maximum catalyst temperatures in the range 800–900 °C and NO_x emissions below 3 ppm.

In this hybrid system it is essential to obtain good mixing of the second-stage fuel with the catalyst exit gases before the mixture ignites. If mixing is imperfect, some combustion will take place at high fuel : air ratios, producing local high temperatures and high NO_x levels. At atmospheric pressure, the ignition delay time is long enough for the fuel and exit gases to become well mixed before ignition. However, at higher pressures the ignition delay time is much shorter and the fuel ignites before it becomes well mixed with the exit gases, resulting in high NO_x levels. The principal challenge in this approach is therefore to obtain better mixing of the second-stage fuel with the catalyst exit gases within the very short ignition delay time at high pressures.

3.4.3 Staged Air

Another type of staged combustion approach involves introducing the air rather than the fuel in two stages. In the first stage a fuel-rich fuel/air mixture is reacted over a catalyst. The fuel will only partially react because there is insufficient air for complete combustion, so the temperature rise will be limited. The products of this stage will be predominantly carbon monoxide and hydrogen if the fuel is methane or natural gas. The products of the fuel-rich combustion are then rapidly mixed with additional air to produce a fuel-lean mixture. This mixture is burned in a homogeneous reaction to complete the combustion.

Tests of the staged air approach at the NASA Lewis Research Center gave NO_x levels at the exit of the catalytic reactor in the range 2–14 ppm [13]. This is presumably prompt NO_x that is formed in homogeneous reactions under fuel-rich conditions immediately downstream of the catalytic reactor. Some NO_x could be removed by reaction with the carbon monoxide and hydrogen produced by partial oxidation of the fuel. As with the staged fuel approach, rapid and complete mixing of the gases in the second stage is essential in order to avoid additional NO_x formation.

3.4.4 Catalytic plus Homogeneous Combustion

A new approach to catalytic combustion developed by Catalytica and Tanaka Kikinzoku Kogyo K. K. combines catalytic and homogeneous combustion in a multistage process [14]. In this approach, shown schematically in Fig. 3, the full fuel/air mixture required to obtain the desired combustor outlet temperature is reacted over a catalyst. However, the temperature rise over the catalyst is limited by a self-regulating chemical process described below. The catalyst temperature at the inlet stage therefore remains low and the catalyst can maintain very high activity over long periods of time. Because of the high catalyst activity at the inlet stage, ignition temperatures are low enough to allow operation at, or close to, the compressor discharge temperature, which minimizes the use of a preburner (Section 3.2.1). The outlet stage brings the partially combusted gases to the temperature required to attain homogeneous combustion. Because the outlet stage operates at

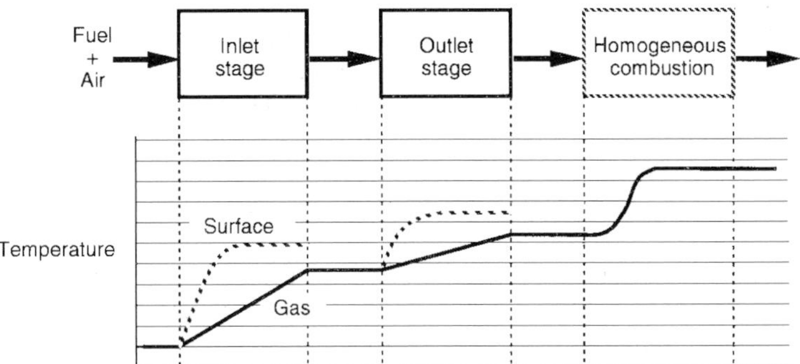

Figure 3. Schematic temperature profiles for Catalytica catalytic combustion system in which the wall temperature is limited and complete combustion occurs after the catalyst.

a higher catalyst temperature, the stable catalyst in this stage will have a lower activity than the inlet stage catalyst. However, as the gas temperature in this stage is higher, the lower activity is adequate. In the final stage, homogeneous gas phase reactions complete the combustion of the fuel and bring the gases to the required combustor outlet temperature.

The temperature rise in the inlet stage is limited by taking advantage of the unique properties of palladium combustion catalysts. Under combustion conditions palladium can be either in the form of the oxide or the metal. Palladium oxide is a highly active combustion catalyst, whereas palladium metal is much less active. Palladium oxide is formed under oxidizing conditions at temperatures higher than 200 °C, but decomposes to the metal at temperatures between 780 °C and 920 °C, depending on the pressure. So when the catalyst temperature reaches about 800 °C the catalytic activity will suddenly fall off due to the formation of the less active palladium metal, preventing any further rise in temperature. The catalyst essentially acts as a kind of chemical thermostat that controls its own temperature.

In test runs simulating actual gas turbine conditions this approach produced less than 1 ppm NO_x with a combustor outlet temperature of 1400 °C [14]. Even at a combustor outlet temperature of 1500 °C NO_x levels were only 2.2 ppm, indicating that this approach will also be applicable to the next generation of gas turbines with higher inlet temperatures. In addition, the low catalyst temperature allows the use of metal substrates, which are resistant to thermal shock, and produces a durable, high-activity catalyst system.

The heat transfer characteristics of the catalyst walls are of major importance in this approach because the catalyst temperature is so low that gas phase reactions do not occur and the only way in which the gas can attain the required catalyst outlet temperature is by heat transfer from the walls.

3.5 Catalytic Materials

The catalyst system in a catalytic combustor usually consists of three components: the substrate, which acts as a support for the other two components; the washcoat, which provides a high-surface-area material on which to distribute the active component; and the active component itself, referred to here as the catalyst. This section discusses the materials requirements for each of these components and the extent to which these requirements have been met in existing catalyst systems.

3.5.1 Substrates

The substrate in a combustion catalyst system is usually a ceramic or metal honeycomb monolith. This type of monolith is generally made up of an array of parallel channels, with the catalytic material deposited on the channel walls. Other configurations, such as crosswise corrugated (herringbone) structures [15] are also possible, and may present advantages in combustion applications. Monolith reactors provide a substrate with high geometric surface area and low pressure drop (high throughput). Cybulski and Moulijn [16] have published a comprehensive review of monolith catalytic reactors and their applications, including catalytic combustion.

Ceramic substrates made of cordierite ($2MgO \cdot 5SiO_2 \cdot 2Al_2O_3$) or mullite ($3Al_2O_3 \cdot 2SiO_2$), for example, are capable of withstanding the high combustor outlet temperatures (>1300 °C) in the current generation of gas turbines. Advanced ceramics such as silicon carbide, silicon nitride and aluminum titanate are stable up to 1600–1800 °C. Details of the properties of ceramic materials that could be used to fabricate substrates in catalytic combustors can be found in the reviews of Zwinkels et al. [17] and Cybulski and Moulijn [16].

The principal disadvantage of ceramic monoliths in gas turbine applications is their relatively low resistance to thermal shock. Most ceramics will fracture during the very rapid temperature drop that occurs after a turbine trip, when the turbine load is suddenly lost (see Section 3.2.2).

New types of ceramic composites with high thermal shock resistance have recently been developed that show some promise for gas turbine applications. These composites consist of a ceramic matrix reinforced by ceramic fibers or platelets inside the matrix. The fibers pull out of the matrix during fracture to resist crack propagation. Such composites can be readily fabricated using a new process developed by Lanxide Corporation [18]. The process uses directed oxidation reactions of molten metals to grow a ceramic matrix around a reinforcing material.

Monoliths made of metal foils can also be used as substrates in combustion catalysts [19, 20]. The metal is generally an iron- or nickel-based steel containing small amounts of aluminum. The aluminum diffuses to the surface on heating and oxidizes to form an adherent alumina layer. This alumina layer gives the alloy high oxidation resistance and is essentially self-healing as it arises from diffusion from the bulk material. It also provides good adhesion for the alumina washcoat.

Metal monoliths have a number of advantages compared to ceramics. They are more robust than ceramic monoliths and have excellent resistance to thermal shock, provided that the metal is not oxidized at high temperatures. The foil sheet that is used to manufacture the walls of the metal monolith is much thinner than the walls in a ceramic monolith (typically 0.05 mm in the metal vesus 0.25 mm in the ceramic). The thinner walls give a larger surface to volume ratio which produces a smaller pressure drop for a given mass transfer limited performance.

The disadvantage of metallic monoliths is that their maximum operating temperatures are substantially below the 1300 °C combustor outlet temperature required for current gas turbines. They can therefore only be used in combustor designs that limit the catalyst wall temperature in some way, such as the hybrid combustors described above in Section 3.4.

3.5.2 Washcoat

The substrate is coated with a high surface area material, known as a washcoat, which acts as a support for the active component. The support must be able to maintain its surface area under the operating conditions for the catalyst. However, most oxide supports sinter rapidly above 1200 °C and transform into low-surface-

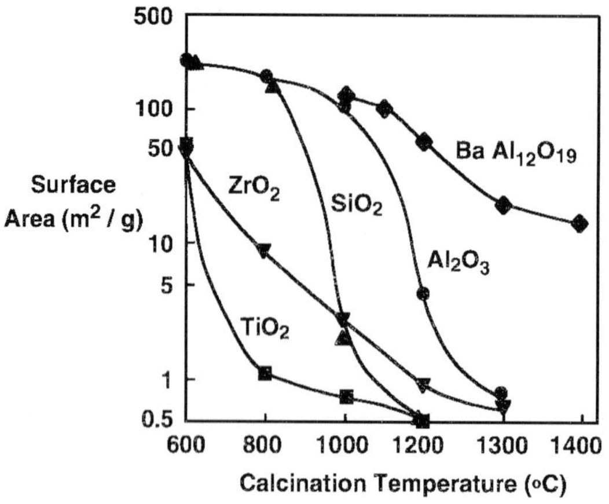

Figure 4. Variation of surface area with temperature for different oxide supports.

area materials (Fig. 4). In particular, neither silica nor alumina, the traditional support materials, have the high temperature stability required for catalytic combustion. Alumina undergoes a phase change above 1000 °C from the high-surface-area γ- phase to the low-surface-area α- phase. Other oxides, such as titania and zirconia, also undergo phase changes to low-surface-area phases at relatively low temperatures.

Attempts to prepare washcoats with high temperature stability have centered on stabilizing alumina by the addition of other metal oxides. Alkaline earth oxides significantly increase the stability of alumina. As shown in Fig. 4 barium hexa-aluminate, $BaO \cdot 6Al_2O_3$ retains a surface area of 15–20 $m^2\,g^{-1}$ at temperatures as high as 1300 °C [21]. The layered structure of barium hexaaluminate (similar to β-alumina) apparently retards the diffusion of aluminum ions. Alumina can also be stabilized by forming mixed oxides with rare earths, such as lanthanum. Some of these mixed oxides have layered β-alumina structures, similar to barium hexa-aluminate, and show similar high-temperature stability [22]. In other mixed oxides with lanthanum the stabilization was attributed to formation of a surface lanthanum aluminate, which prevented sintering by retarding the diffusion of surface oxygen ions [23]. The inhibition of ionic transport by the additive appears to be the key to the formation of stabilized aluminas.

The washcoat should also have low volatility at the catalyst operating temperature. Many oxides have significant vapor pressures above 1200 °C, however, leading to loss of washcoat over time. Computer models give the percentage loss over time as a function of washcoat vapor pressure [24]. These models show that for a catalyst operating at 1300 °C, 10 bar pressure and typical gas velocities present in a gas turbine combustor, 50% of a silica washcoat would be lost in 5 h, whereas only 0.1% of an alumina washcoat would be lost in 1000 h. Although pure silica is very volatile, silica in mixed oxides could have significantly lower volatility.

The washcoat should be inert towards the substrate and the active component of the catalyst at the operating temperature. It should nevertheless have good adhesion with the substrate and excellent cohesion in order to withstand the high gas velocities and the vibration that is associated with gas turbines.

3.5.3 Catalyst

The catalyst should have high activity for the oxidation of methane and other hydrocarbons in natural gas in order to obtain ignition at temperatures close to the compressor outlet temperature. At the same time the catalyst should maintain high surface area at the operating temperature.

The noble metals, particularly platinum and palladium, are the most active catalysts for methane combustion. Table 1 gives the activity per unit area for methane oxidation at 400 °C (a typical ignition temperature) for platinum and palladium, together with the activities of several oxide catalysts. The activity per unit area is converted to the activity on a weight basis by multiplying by the surface area per gram at 1000 °C. The per gram specific rate is much higher for these supported metals than for the oxide catalysts in Table 1. The per gram specific rate is the important rate for comparison as the combustor will contain a given weight (or, more accurately, a given volume) of catalyst.

Perovskites such as $LaCoO_3$ and Sr-doped $LaCoO_3$ have been investigated extensively as combustion catalysts [17] but, as can be seen from Table 1, these oxides sinter rapidly at 1000 °C giving low-surface-area materials with a low activity on a per gram basis.

The alkaline earth hexaaluminate supports mentioned in Section 3.5.2 have been made catalytically active by the addition of transition metal ions [25]. Manganese substitution produced the highest catalytic activity. The manganese-substituted aluminates maintain high surface areas at 1000 °C (see Table 1), but their activity per unit area is so low that the activity on a per gram basis remains much lower than that of the noble metals. The same applies to the Cu/La-substituted alumina included in Table 1. Note that surface areas at 1300 °C will be significantly lower

Table 1. Comparison of methane oxidation rates for several supported metal and oxide catalysts; rates measured at 400 °C in 2% CH_4 in air at 1 bar pressure.

Material	Areal rate ($\times 10^{-7}$ mol m^{-2} s^{-2})	Surface area at 1000 °C (m^2 g^{-1})	Specific rate ($\times 10^{-7}$ mol g^{-1}s^{-1})
Pd/Al_2O_3	140	2[a]	300
Pt/Al_2O_3	50	1[a]	50
Co_3O_4	4.4	3	13
$LaCoO_3$	0.36	3	1
$La_{0.5}Sr_{0.5}CoO_3$	1.2	6	7
$Cu/La{-}Al_2O_3$	0.09	80	7
$Sr_{0.8}La_{0.2}MnAl_{11}O_{19}$	0.045	70	3

[a] Surface area of the active metal only.

than those at 1000 °C; the surface area of the manganese hexaaluminate was $18\,m^2\,g^{-1}$ at 1300 °C and fell to $5\,m^2\,g^{-1}$ during long-term testing over 8000 h [26].

The catalyst should have a very low vapor pressure at the operating temperature in order to avoid loss of catalytic material. The vapor pressure of palladium at 1300 °C is about an order of magnitude lower than that of platinum. Computer modeling [24] shows that at 1300 °C 50% of platinum will be lost over 30 min and 50% of palladium over 30 h. Clearly, neither of these metals would be usable at this temperature. Loss of oxygen is a problem with oxides such as Co_3O_4 at high temperatures and can also be a problem with mixed oxides such as the perovskites, due to the presence of simple oxides as impurities.

It is also important to avoid the formation of compounds between the catalytic material and the support. In the case of oxides, this would be a major problem. For example, cobalt or copper oxides on alumina would form aluminate spinels, such as $CoAl_2O_4$, with significantly lower catalytic activity. The catalyst should also the resistant to poisoning by contaminants in the natural gas or the air, as mentioned in Section 3.2.2.

3.6 Reactor Modeling

Monolith reactors have been modeled extensively for combustion applications, with most of the models relating to the combustion of pollutants in exhaust gases. Details of the various models for combustion in monolith reactors can be found in the review by Cybulski and Moulijn [16]. The models have been important tools in the design of catalysts for the control of pollutants, especially automobile exhaust catalysts. Many of the insights obtained through modeling the combustion of pollutants are transferable to fuel-combustion catalysts.

The growing interest in gas turbine applications of catalytic combustion has promoted the development of models specifically aimed at these applications. For example, an informative model for catalytic combustors for gas turbine applications has been developed by Groppi et al. [27]. This model is a lumped-parameter model, simulating steady state conditions in a combustor monolith channel using coupled material and energy balances. The model was used to examine possible design improvements that could be obtained by variation of the monolith geometry. According to the model, ignition occurs more rapidly for larger channel diameters and for square rather than round channel shapes. This is due to reduced gas-phase cooling of the monolith walls for these geometries, resulting in a more rapid increase in wall temperature as the catalytic reaction releases heat. The model also indicated that a series of short monoliths in a segmented configuration can lead to a higher bulk gas temperature and improved combustor performance compared to a single long monolith. The channel openings and shapes for each segment can be chosen to give optimum performance, i.e. more rapid ignition and faster heating of the bulk gas (graded cell configuration).

In addition to the steady state models it is also necessary to develop models that simulate the transient response of the combustor to the frequent changes in operating conditions. A transient catalytic combustor models developed by Ahn et al. [28]

showed that metal monoliths have faster response times to transient changes in operating conditions than ceramic monoliths, because the higher surface to volume ratio of the metal monoliths gives them lower thermal inertia. This is another indication of the superiority of metal monoliths in catalytic combustion applications.

A model has also been developed to simulate the interaction between the channels in a monolith combustor [29]. Interactions involving heat transfer between the channels could occur in the case of a fuel maldistribution in the monolith inlet, with a localized zone of high fuel concentration. Again, metal monoliths transfer energy between the channels more rapidly than ceramic monoliths, leading to higher overall reaction rates in the case of a fuel maldistribution.

3.7 Status and Outlook

After several decades of research, catalytic combustion to eliminate emissions from gas turbines is nearing practical application. New hybrid systems in which combustion is initiated over a temperature-limiting catalyst and completed downstream in a homogeneous process hold promise for overcoming many of the problems encountered in earlier systems in which combustion occurred entirely in the catalyst. These new systems have been successfully demonstrated under turbine operating conditions at full scale in combustor test stands. The next and most important demonstration will be in an actual turbine environment, and it seems very likely that this will occur within the next few years. Indeed, the future of catalytic combustion for pollution prevention in gas turbines appears to be very bright over the next decade and beyond.

References

1. D. L. Trimm, *Appl. Catal.* **1983**, *7*, 249–282.
2. L. D. Pfefferle, W. C. Pfefferle, *Catal. Rev.-Sci. Eng.* **1987**, *29*, 219–267.
3. R. Swanekemp, *Power* **1995**, *June*, 15.
4. *Alternative Control Techniques Document – NO_x Emissions from Stationary Gas Turbines*, United States Environmental Protection Agency, EPA-453/R-93-007, **1993**.
5. J. A. Miller, C. T. Bowman, *Prog. Energy Combust. Sci.* **1989**, *15*, 287–338.
6. T. Griffen, W. Weisenstein, A. Schlegel, S. Buser, P. Benz, H. Bockhorn, F. Mauss in *Proceedings of the 2nd International Workshop on Catalytic Combustion*, Tokyo, Japan, **1994** (Ed.: H. Arai), p. 138–141.
7. R. A. Dalla Betta, J. C. Schlatter, M. Chow, D. K. Yee, T. Shoji, *Proceedings of the 2nd International Workshop on Catalytic Combustion*, Tokyo, Japan, **1994** (Ed.: H. Arai), p. 154–157.
8. K. W. Beebe, M. B. Cutrone, R. N. Matthews, R. A. Dalla Betta, J. C. Schlatter, Y. Furuse, T. Tsuchiya, *International Gas Turbine and Aeroengine Congress*, Houston, Texas, **1995**, Paper ASME-95-GT-137.
9. R. L. Heck, J. Wei, J. R. Katzer, *AIChE J.* **1976**, *22*, 477–484.
10. L. D. Pfefferle, *Proceedings of the 2nd International Workshop on Catalytic Combustion*, Tokyo, Japan, **1994** (Ed.: H. Arai), p. 78–87.
11. Catalytica Inc., unpublished results.
12. T. Furuya, K. Sasaki, Y. Hanakata, K. Mitsuyasu, M. Yamada, T. Tsuchiya, Y. Furuse, *Proceedings of the 2nd Intenational Workshop on Catalytic Combustion*, Tokyo, Japan, **1994** (Ed.: H. Arai), p. 162–165.

13. J. Rollbuhler, *27th Joint Propulsion Conference*, Sacramento, California, **1991**, paper AIAA-91-2463.
14. R. A. Dalla Betta, J. C. Schlatter, S. G. Nickolas, M. K. Razdan, D. A. Smith, *Int. Gas Turbine and Aeroengine Congress*, Houston, Texas, **1995**, paper ASME-95-GT-65.
15. G. Gaiser, *Heat Transfer 1994, Proceedings of the 10th International Heat Transfer Conference*, **1994** (Ed. G. F. Hewitt), p. 261–266.
16. A. Cybulksi, J. Moulijn, *Catal. Rev.-Sci. Eng.* **1994**, *36*, 179–270.
17. M. Zwinkels, S. G. Jaras, P. G. Menon, *Catal. Rev.-Sci. Eng.* **1993**, *35*, 319–358.
18. G. H. Schiroky, A. W. Urquhart, B. W. Sorensen, *International Gas Turbine and Aeroengine Congress*, Toronto, Canada, **1989**, paper ASME 89-GT-316.
19. C. J. Pereira, K. W. Plumlee, *Catal. Today* **1992**, *13*, 23–32.
20. A. S. Pratt, J. A. Cairns, *Plat. Met. Rev.* **1977**, *21*, 74–83.
21. M. Machida, K. Eguchi, H. Arai, *Bull. Chem. Soc. Jpn.* **1988**, *61*, 3659–3665.
22. A. Kato, H. Yamashita, S. Matsuda, *Stud. Surf. Sci. Catal.* **1988**, *44*, 25–32.
23. H. Schaper, E. B. M. Doesburg, L. L. van Reijen, *Appl. Catal.* **1983**, *7*, 211–220.
24. Catalytica Inc., unpublished results.
25. M. Machida, K. Eguchi, H. Arai, *J. Catal.* **1987**, *120*, 377–386.
26. H. Arai, K. Eguchi, M. Machida, T. Shiomitsu in *Catalytic Science and Technology* (Eds: S. Yoshida, N. Takazawa, T. Ono), Kodansha, Tokyo, **1990**, Vol. 1, p. 195–199.
27. G. Groppi, E. Tronconi, P. Forzatti, *Catal. Today* **1993**, *17*, 237–250.
28. T. Ahn, W. V. Pinczewski, D. L. Trimm, *Chem. Eng. Sci.* **1986**, *41*, 55–64.
29. S. T. Kolaczkowski, D. L. Worth, *Proceedings of the 2nd International Workshop on Catalytic Combustion*, Tokyo, Japan, **1994** (Ed.: H. Arai), p. 100–103.

4 Catalytic Routes to Hydro(chloro)fluorocarbons

Z. Ainbinder, L. E. Manzer and M. J. Nappa, Dupont, Central Research und Development, Experimental Station, P.O. Box 80262, Wilmington, DE 19880-0262/USA

4.1 Background

The first commercial catalytic system, $SbCl_5$, for the production of chlorofluorocarbons (CFCs) was based on the pioneering work of Swartz [1] during the 1890s. DuPont developed and commercialized CFCs during the 1930s. Since then, these marvelously stable, nontoxic and nonflammable compounds have found their way into many aspects of modern lifestyle [2]. Refrigeration, air conditioning, energy conserving foams, cleaning of electronic circuit boards, and firefighting are just a few of the applications of CFCs. However, during the 1980s, the incredible stability of CFCs was scientifically linked to the depletion of the earth's ozone layer by the NASA Ozone Trends Panel. This created a major challenge for industrial catalytic scientists and engineers to identify, develop, and commercialize an entirely new family of products that were environmentally safer, yet still satisfied the needs of society.

The CFC replacements need to be nontoxic, nonflammable, and have significantly lower, or zero ozone depletion potentials. Many organic or aqueous-based systems, that do not contain chlorine or fluorine, have been developed for some applications, whereas others will still use hydrochlorofluorocarbons (HCFCs) and hydrofluorocarbons (HFCs). Unlike hydrocarbon catalysis, the presence of hydrogen, chlorine and fluorine in the same molecule creates a very large number of isomeric possibilities. As a result of many years of careful study, the early list of >800 potential candidates has been narrowed down to less that a dozen viable molecules and their blends and azeotropes [3]. Many of these are now being commercially manufactured as society approaches the January 1996 phaseout date of the Montreal Protocol.

This section reviews some of the literature reported for the synthesis of CFC alternatives. It is not possible to be comprehensive in the space available, so the references are selected to be informative and to outline the extensive and exciting types of heterogeneous catalysis available. A more comprehensive review has been published elsewhere [4].

4.2 Catalytic Transformations

Carbon–fluorine bonds can be prepared by a variety of transformations through the use of both homogeneous and heterogeneous catalysis. This review is meant to cover only heterogeneous catalysts, yet that does not diminish the value of homogeneous systems.

4.2.1 Addition of HF

HCFC-123, HCFC-124, and HFC-125 have been found to be useful as refrigerants, either individually or as azeotropic mixtures with other components. These compounds can be prepared by contacting hydrogen fluoride and tetrachloroethylene (PCE) (eq 1) with selected catalysts.

$$\underset{}{HF} + \underset{\text{PCE}}{CCl_2{=}CCl_2} \longrightarrow \underset{123}{CHCl_2CF_3} + \underset{124}{CHClFCF_3} + \underset{125}{CHF_2CF_3} \tag{1}$$

Vapor phase catalysts include $Cr_2O_3/MgO/Al_2O_3$ [5], Cr^{+3}/AlF_3 [6], and Zn/Al_2O_3 [7]. The production of HFC-125 can be minimized if a catalyst comprising a selected metal (e.g. Cr, Mn, Ni or Co) supported on a high fluorine content (greater than 90% AlF_3) alumina is used [8].

A convenient method of preparing HFC-152a, useful in refrigerant blends, is by the reaction of vinyl chloride (VCl) with HF (eq 2). An 86% conversion of vinyl choride with a 99% selectivity to HFC-152a has been reported for the vapor phase reaction of vinyl chloride with HF using an alumina plus transition metal catalyst [9].

$$HF + \underset{\text{VCl}}{CH_2{=}CHCl} \longrightarrow \underset{\text{152a}}{CH_3CHF_2} \tag{2}$$

Table 1. CFC Substitutes Under Development or in Production.

Market	Current CFC	CFC alternative
Refrigerants	CFC-12 (CF_2Cl_2)	HFC-134a (CF_3CFH_2) HCFC-22 (CHF_2Cl) HFC-32 (CH_2F_2) HFC-125 (CF_3CF_2H) HCFC-124 (CF_3CHFCl) HFC-152a (CH_3CHF_2)
Blowing agents	CFC-11 ($CFCl_3$)	HCFC-141b (CH_3CFCl_2) HCFC-123 (CF_3CHCl_2) HCFC-22 (CHF_2Cl)
Cleaning agents	CFC-113 ($CF_2ClCFCl_2$)	blends/azeotropes new compounds hydrocarbons

HCFC-141b, HCFC-142b and HFC-143a can all be prepared by the reaction of HF with vinylidene chloride (VCl_2; eq 3). In the vapor phase using an HF-activated Al_2O_3 catalyst, the product selectivity can be dramatically altered using bismuth as a dopant. An HF-activated Al_2O_3 catalyst affords a >99% selectivity to HCFC-141b [10], whereas the same catalyst doped with bismuth affords a >99% selectivity to HFC-143a [11]; both conversions are essentially quantitative.

$$\underset{}{HF} + \underset{VCl_2}{CH_2{=}CCl_2} \longrightarrow \underset{141b}{CH_3CCl_2F} + \underset{142b}{CH_3CClF_2} + \underset{143a}{CH_3CF_3} \tag{3}$$

HFC-227ea, which has been proposed as a replacement for Halon fire extinguishants [12], can be prepared by the addition of HF to hexafluoropropene (HFP; eq 4) over either an activated carbon catalyst [13] or a CrO_2F_2 catalyst [14].

$$HF + \underset{HFP}{CF_2{=}CFCF_3} \longrightarrow \underset{227ea}{CF_3CHFCF_3} \tag{4}$$

4.2.2 Halogen Exchange

Halogen exchange is the operation whereby a C–Cl bond is converted to a C–F bond using a catalyst and HF as the fluorine source and HCl as the chlorine product (not shown in the equations below). When an alkene is the starting material, the initial step is HF addition to the double bond. In many cases this is followed by halogen exchange to make more highly fluorinated analogs. In the previous section, HF was added to perchloroethylene to make HCFC-121, which was then converted directly to HCFC-123, HCFC-124, and HFC-125 by halogen exchange. It is also possible to start with HCFC-121 or HCFC-122 (eq 5).

$$\underset{121}{CHCl_2CCl_2F} \longrightarrow \underset{122}{CHCl_2CClF_2} \longrightarrow \underset{123}{CHCl_2CF_3} \tag{5}$$

This has the advantage of less catalyst deactivation since there is a lower alkene concentration in the reactor. A combination of fluorinated chromium and magnesium oxide shows high selectivity for HCFC-123 (64%) and HCFC-124 (15%) even after 6 months of using HCFC-122 as the feedstock [15]. Other catalysts, such as $CrCl_3$ [16], $CoCl_2/CeCl_3$[17], $NiCl_2$ [18], and $CoCl_2/MgCl_2$ [19], all supported on γ-Al_2O_3 give a high selectivity for HCFC-123 (83–86%).

Although it is possible to start with perchloroethylene, HCFC-121, or HCFC-122 and convert these to HCFC-123 in the liquid phase using a TaX_5 [20] or $SbCl_5$ [21] catalyst, it is only in the vapor phase over a heterogeneous catalyst that the conversion to the more highly fluorinated HCFC-124 or HFC-125 (eq 1) occurs to any appreciable extent. The rate of fluorination decreases as the fluorine substitution increases in the series HCFC-121 to HFC-125. The rate-determining step in the formation of HFC-125 is the fluorination of HCFC-124, and Coulson has shown that the fluorination of HCFC-123 to HFC-125 using Co/γ-Al_2O_3 occurs sequentially on the catalyst surface [22]. Other metals on γ-Al_2O_3 have been claimed to fluorinate HCFC-123 to HFC-125, such as Zn and Cr [23] as well as a coextrudate of Cr_2O_3 and Al_2O_3 [24]. Chromium oxide on a carbon support [25], and a catalyst made from Cr_2O_3 and MgO or $Mg(OH)_2$ [26] show high selectivity for HFC-125 in the fluorination of HCFC-123. Chromium oxide low in alkali metals is known to be highly active for converting HCFC-123 to HCFC-124 and has long catalyst life [27].

The conversion of $CHCl_3$ to HCFC-22 (eq 6) has been practiced for many years since HCFC-22 is a precursor to tetrafluoroethylene, a valued fluoromonomer. During the conversion a small amount of HFC-23 is formed.

$$\underset{20}{CHCl_3} \longrightarrow \underset{21}{CHCl_2F} \longrightarrow \underset{22}{CHClF_2} \longrightarrow \underset{23}{CHF_3} \tag{6}$$

Milks reported the conversion of $CHCl_3$ to predominantly HFC-23 using Ni- and Co-activated Al_2O_3 (73% and 84%, respectively) [28]. In related work, Cu supported on Al_2O_3, gave 90% HFC-23 [29]. Ruh and Davis subsequently reported that hydrated CrF_3 activated by reaction with oxygen gives a high selectivity for HFC-23 (91%) [30]. Similarly, a process for producing fluorinated hydrocarbons using an improved chromium oxide catalyst made by treating a chromium hydroxide paste with water or steam before being dried and calcined, gives predominantly HFC-23 (37%) [31]. While carbon alone can catalyze the partial fluorination of $CHCl_3$, $FeCl_3$/C is highly active and selective for HFC-23 (87%) [32].

HCFC-22 can also be converted to HFC-23 by disproportionation (section 4.2.3) which has the advantage of lower temperature and no HF is used. However, HCFC-21, which is a product of the reaction will have to be recycled back to HCFC-22. For example, Guo and Cai reported the formation of HFC-23 in high yield (>99%) using AlF_3, Bi/La, and Co-activated AlF_3 [33].

HFC-134a has been found to be especially useful as a substitute for CFC-12 in automobile air conditioners. A fairly direct preparation method is shown in eqs (7) and (8); TCE = trichloroethylene. The reactions shown are typically done separately in several reaction zones. HCFC-133a can be prepared using catalysts such as Cr/Mg [34], Cr_2O_3/AlF_3 [35], $AlCl_xF_y$ [36], Zn/fluorinated alumina [37], and Zn/Cr [38, 39].

$$\underset{TCE}{CHCl{=}CCl_2} + HF \longrightarrow \underset{133a}{CH_2ClCF_3} + 2\,HCl \tag{7}$$

$$\underset{133a}{CH_2ClCF_3} + HF \longrightarrow \underset{134a}{CH_2FCF_3} + HCl \tag{8}$$

This conversion is equilibrium limited [40], and a large excess of HF is required to get a high conversion of HCFC-133a to HFC-134a. Using a Cr_2O_3 catalyst and increasing the HF : 133a ratio from 3.3 to 4, increases the HFC-134a selectivity from 16% to 18.2%, while decreasing the quantity of alkenes formed [41]. It is believed that alkenes at low HF ratios lead to carbon deposits and deactivation of the catalyst by blinding active sites. The co-feeding of chlorine to a chromium oxide-catalyzed reaction has been used to extend catalyst life, presumably by removing this carbon coating [42]. In lieu of chlorine, oxygen can be added. The oxygen reacts with HCl in a Deacon-like reaction to liberate Cl_2 and H_2O, and the Cl_2 functions as described above [43]. Removal of the HCl has the additional benefit of driving the equilibrium reaction to form more HFC-134a. A commercial process using a high HF ratio, needed to drive the equilibrium, will result in high energy costs, and the use of multiple distillation columns and complex recycle loops can effectively circumvent this [44].

Deactivation occurs via coking as described above and potentially by catalyst over-fluorination. In lieu of co-feeding oxygen as described previously, air can be used in a separate step to reactivate used catalyst [45]. To further remove coke or to reduce the fluorine content, a treatment with water vapor is also useful [46]. Many of the catalysts capable of fluorinating HCFC-133a are chromium based, such as the Cr_2O_3 examples described above. Many metals in combination with Cr are also effective such as Al, Co, Mn, Mg [47], Zn [48], as well as others. Alumina or aluminum fluoride catalysts can also be used [49]. The conversion of TCE to HFC-134a using a single reactor zone for both reaction stages, which utilizes recycle of the HFC-133a reaction product using catalysts based on trivalent chromium has been disclosed [50].

As in the synthesis of HFC-134a from HCFC-133a, the conversion of CH_2Cl_2 to HFC-32 (CH_2F_2) (eq 9) using HF is an equilibrium-limited reaction, and NaF has been used with Cr_2O_3 to remove HCl and drive the equilibrium to form more HFC-32 [51].

$$\underset{30}{CH_2Cl_2} \longrightarrow \underset{32}{CH_2F_2} \tag{9}$$

Many different catalysts have been reported for this transformation, ranging from $CrCl_3/C$, $CrCl_3/Al_2O_3$, $NiCl_2/C$ (all with aqueous HF [52]), Co/Al_2O_3 [53], Bi_2O_3-$CrCl_3/AlF_3$, Bi_2O_3-$CoCl_3/AlF_3$, Bi_2O_3-$MnCl_2/AlF_3$ [54], AlF_3, CrF_3/AlF_3, $FeCl_3/C$, and CrF_3/C [55]. In these latter cases, the most important factor was the HF : CH_2Cl_2 ratio, not the catalyst or the temperature, suggesting that the reaction is equilibrium, not kinetically, limited with very little temperature dependence on the equilibrium.

4.2.3 Isomerizations

CFC-113 (CCl_2FCClF_2) had been a large-volume fluorochemical used in solvent applications. The successful isomerization of CFC-113 to a useful intermediate, CFC-113a (CF_3CCl_3), has enabled some producers to utilize existing facilities to rapidly develop HCFC and HFC alternatives. A common catalyst for the isomerization of CFC-113 was described by Miller et al. [56], but the heterogeneous reaction between CFC-113 and $AlCl_3$, which undoubtedly made some CFC-113a, was originally reported by Henne and Newman [57]. Henne described a reaction in which $AlCl_3$ reacts with CFC-113 to give CFC-112a, by extracting fluorine from the organic reactant (eq 10).

$$\underset{113}{2\,CClF_2CCl_2F} + AlCl_3 \longrightarrow \underset{112a}{2\,CClF_2CCl_3} + AlClF_2 \tag{10}$$

An active catalyst is formed in this step which can also disproportionate the CFC-112a to CFC-111, CFC-113a, hexachloroethane, and CFC-114a (eq 11).

$$\underset{112a}{CClF_2CCl_3} \longrightarrow \underset{111}{CCl_2FCCl_3} + \underset{113a}{CF_3CCl_3} + \underset{110}{CCl_3CCl_3} + \underset{114a}{CF_3CFCl_2} \tag{11}$$

The activation and isomerization reactions were looked at in much detail by Okuhara [58] and Paleta et al. [59]. Okuhara made some pertinent observations such as (a) the isomerization did not occur with partially deteriorated aluminum chloride, (b) the rate of isomerization is sensitive to stirring efficiency, and (c) the rate of isomerization is retarded by highly chlorinated organics, e.g. CCl_4. Paleta et al. reported that the behavior of the catalyst in both activations and isomerizations depends on the kind of commercial aluminum chloride used. This has recently been supported by claims that catalyst life and efficacy are related to the surface area of the $AlCl_3$ catalyst precursor, which varies from batch to batch of commercial material [60].

The isomerization of CFC-114 to CFC-114a (eq 12) is more difficult than the isomerization of CFC-113 since disproportionation of CFC-114a to CFC-113a and CFC-115 competes with isomerization, leading to yield losses.

$$\underset{114}{CClF_2CClF_2} \longrightarrow \underset{114a}{CF_3CCl_2F} \tag{12}$$

AlX_3, activated with trace amounts of Cr and Mn powders, isomerizes CFC-114 in the liquid phase under pressure to yield 73% $C_2Cl_2F_4$ (114a : 114 = 9.6) [61]. A large fraction (27%) of the product was CFC-113a, coming from the disproportionation of CFC-114a. In the vapor phase, an alumina catalyst activated with Cr and Mg was used to disproportionate CFC-114 to CFC-113a and CFC-115, while simultaneously feeding Cl_2 and HF to fluorinate the CFC-113a to CFC-114a [62]. The net result is the conversion of CFC-114 to CFC-114a and CFC-115.

Aluminum halide catalysts are not effective isomerization catalysts for hydrogenated chlorofluorocarbons because of HCl elimination to form alkenes which are catalyst poisons. HF elimination from hydrofluorocarbons is not as facile, thus AlF_3 [63] and $AlCl_xF_yO_z$ $(x + y + 2z = 3)$ [64] are effective catalysts for the isomerization of 134 to 134a (eq 13).

$$\underset{134}{CHF_2CHF_2} \longrightarrow \underset{134a}{CF_3CH_2F} \tag{13}$$

Prefluorinated Cr_2O_3, an effective fluorination catalyst, also isomerizes 134 to 134a in the vapor phase [65].

While effective at isomerizing 134 to 134a, Cr_2O_3 is too active for the selective isomerization of 124a to 124 (eq 14) and many byproducts are formed.

$$\underset{124a}{CF_2ClCHF_2} \longrightarrow \underset{124}{CF_3CHClF} \tag{14}$$

However, γ-Al_2O_3 fluorided with CCl_2F_2 selectively isomerizes 124a to 124 in a heterogeneous liquid phase reaction [66]. In the vapor phase, γ-Al_2O_3 fluorided with CCl_4 will interconvert 124a and 124 depending on the temperature of reaction [67].

Just as in the isomerization of 124a, aluminum oxyhalides are effective in the isomerization of 123a (eq 15).

$$\underset{124a}{CF_2ClCHF_2} \longrightarrow \underset{124}{CF_3CHClF} \qquad (15)$$

$AlF_{0.2}O_{1.6}$, made by calcining $AlF_3 \cdot H_2O$ containing TFE fibrils, isomerizes HCFC-123a at 40 °C with 95% conversion after 1 h and 88% conversion after 3 h [68]. Cr_2O_3 or $CrCl_3/Al_2O_3$ isomerizes HCFC-123a to HCFC-123 with higher amounts of disproportionation products, HCFC-133a, and CFC-113a than Al_2O_3, but with less alkene formation due to HX elimination [69].

4.2.4 Disproportionations/Conproportionations

Disproportionation or dismutation is the reaction between two of the same molecules resulting in an exchange of atoms and the formation of at least two new molecules. Conproportionation is the reverse of this, where at least two different molecules react to form two molecules, which may be of the same kind. In conproportionations, it is common to view one of the reactants as a substrate and the other as a fluorinating agent. Chromium catalysts are proficient at disproportionation, and this feature has been exploited in the synthesis of CFC-114a, HCFC-123, HCFC-124, and HFC-125 (eqs 16 and 17). An MgO or $Mg(OH_2)/Cr_2O_3$ catalyst [70] catalyzes the conproportionation reaction between CFC-113a and HCFC-133a to make CFC-114a, with byproduct formation of trichloroethylene.

$$\underset{113a}{CCl_3CF_3} + \underset{133a}{CH_2ClCF_3} \longrightarrow \underset{114a}{CCl_2FCF_3} + \underset{TCE}{CHCl{=}CCl_2} \qquad (16)$$

$$\underset{123}{CHCl_2CF_3} + \underset{133a}{CH_2ClCF_3} \longrightarrow \underset{124}{CHClFCF_3} + \underset{TCE}{CHCl{=}CCl_2} \qquad (17)$$

HCFC-123 reacts with HCFC-133a over the Mg/Cr catalyst to give HCFC-124, also with TCE formation. At different feed ratios and over a Cr_2O_3 catalyst, CFC-113a and HCFC-133a react to form HCFC-123 as the primary product [71] (eq 18).

$$\underset{113a}{CF_3CCl_3} + \underset{133a}{CF_3CH_2Cl} \longrightarrow \underset{123}{CF_3CHCl_2} \qquad (18)$$

Whereas γ-Al_2O_3 and $FeCl_3/\gamma$-Al_2O_3 were not effective, Cr_2O_3 was also shown to catalyze the conproportionation of HCFC-123 and HFC-125 to make HCFC-124 [72] (eq 19).

$$\underset{123}{CF_3CHCl_2} + \underset{125}{CF_3CHF_2} \longrightarrow \underset{124}{2CF_3CHClF} \qquad (19)$$

Cr_2O_3 will also catalyze the approach to equilibrium from the reverse direction; HCFC-124 will disproportionate over Cr_2O_3 to give a mixture of HFC-125 and HCFC-123, free of the a and b isomers [73].

4.2.5 HX Elimination Reactions

HFC-245eb can be prepared from hexafluoropropene (HFP) by the sequence shown in eq 20 in which the HFP is hydrogenated over a Pd/C catalyst to HFC-236ea. This is dehydrofluorinated over carbon to HFC-1225ye, which is then hyrogenated over a Cu/Pd/C catalyst to HFC-245eb [74].

$$\underset{\text{HFP}}{CF_3CF{=}CF_2} \longrightarrow \underset{\text{236ea}}{CF_3CHFCHF_2} \longrightarrow \underset{\text{1225ye}}{CF_3CF{=}CHF} \longrightarrow \underset{\text{245eb}}{CF_3CHFCH_2F} \tag{20}$$

Haloalkenes can be prepared by dehydrohalogenating saturated hydrogen-containing polyhalocarbons using liquid alkali metal acid fluoride and/or alkali metal fluoride compositions [75]. HCFC-133a can be converted to $CF_2{=}CHCl$ using these catalyst systems as shown in eq 21.

$$\underset{\text{133a}}{CF_3CH_2Cl} \longrightarrow \underset{\text{1122}}{F_2C{=}CHCl} \tag{21}$$

4.2.6 Hydrodehalogenations/Dehydrohalogenations

The substitution of chlorine with hydrogen by catalytic hydrodechlorination (HDC) is a well known reaction. Earlier work by Bitner et al. [76] and Gervasutti et al. [77] over Pd/C using CFC-114a showed that the reaction proceeded primarily to HFC-134a although a small amount of HCFC-124 was also obtained (eq 22).

$$\underset{\text{114a}}{CF_3CFCl_2} + H_2 \longrightarrow \underset{\text{134a}}{CF_3CFH_2} + \underset{\text{124}}{CF_3CFHCl} + HCl \tag{22}$$

Conversion of the HCFC-124 to HFC-134a also occurred, but at much higher temperature. This has been recently studied and a proposed mechanism for the direct formation of HFC-134a involves a surface carbene intermediate, CF_3CF: [78].

A significant problem with the HDC reaction, using Pd/C is catalyst deactivation, resulting from sintering due to the HCl produced in the reaction [79]. Kellner et al. [80] have found that the sintered Pd can be regenerated by treating the deactivated catalyst with a CFC or HCFC at relatively mild conditions. Morokawa et al. [81], have disclosed a series of multimetallic catalysts based on a Group VIII metal and other metal additives. They have interpreted the extended life of these catalysts to chemical adsorption energy and geometric factors.

The effect of catalyst support has also been reported to be have a pronounced effect. Kellner and Rao [82] reported 100% selectivity to HFC-134a from HCFC-124, using the unconventional support AlF_3. Acid washing of the carbon to remove trace quantities of metal impurities was reported by Rao [83] to increase conversion of the CFC-114a and to significantly increase the HFC-134a : HCFC-124 ratio.

Catalytic HDC has been reported for the synthesis of several other CFC alternatives or their intermediates. Particularly interesting is the report by Ichikawa and co-workers [84] that Pd catalysts promoted with additives such as Tl, Bi, Sn, Cu, or

In result in dehydrochlorination (DHC) to give fluoroalkenes (eq 23).

$$\underset{113}{CF_2ClCFCl_2} + H_2 \longrightarrow \underset{1113}{CF_2{=}CFCl} + \underset{1123}{CF_2{=}CFH} + HCl \tag{23}$$

It is reported that a Pd/Bi catalyst gave excellent selectivity to the desired 1123, which is important since addition of HF readily provides HFC-134a [85] using Cr-based catalysts. Pd modified with Au, Te, Sb, and As [86], and Re catalysts modified with Group VIII metals [87] have also been reported.

The hydrodechlorination of CFC-115 to HFC-125 (eq 24) has been carried out using a variety of Group VIII metal catalysts [88].

$$\underset{115}{CF_3CF_2Cl} + H_2 \longrightarrow \underset{125}{CF_3CF_2H} + 2\,HCl \tag{24}$$

Bitner et al. reported that the HDC of CFC-12 (CF_2Cl_2) over Pd/C yields primarily methane. More recently, it has been reported that the desired HFC-32 (CH_2F_2) is formed in about 5% yield using similar catalysts based on Pd/C [89]. These same workers report that higher yields can be obtained starting with HCFC-22 ($CHClF_2$), although methane is still the major product.

4.2.7 Oxygenates as Reactive Intermediates

With world regulatory pressures on the use of chlorinated compounds increasing, the synthesis of fluorine-containing hydrocarbons from oxygen-containing substrates has received much attention in recent years. Reacting simple molecules such as formaldehyde and carbonyl fluoride over an activated carbon catalyst [90] gives an 86% yield of HFC-32 (eq 25).

$$H_2CO + COF_2 \longrightarrow \underset{32}{CH_2F_2} + CO_2 \tag{25}$$

Separating the steps and reacting bisfluoromethyl ether, made from formaldehyde and HF over a Cr_2O_3 catalyst, gives 57% HFC-32 and some methyl fluoride. The approach of decomposing ethers over heterogeneous catalysts has been extended to make other fluorocarbons of interest such as HFC-134a, HFC-125, etc. [91].

γ-Al_2O_3 was used to catalyze the conversion of fluorinated alcohlols (R_fCH_2OH) to fluorinated alkanes (R_fCH_2F). For example, the reaction of CF_3CH_2OH with $COCl_2$ and HF (eq 26) gives HFC-134a in a single step over γ-Al_2O_3 [92].

$$CF_3CH_2OH + COCl_2 + 2HF \longrightarrow \underset{134a}{CF_3CH_2F} + CO_2 + 2HCl \tag{26}$$

HFC-245ca ($CH_2FCF_2CHF_2$), HFC-449pccc ($CHF_2CF_2CF_2CF_2CH_2F$), and HFC-338q ($CF_3CF_2CF_2CH_2F$) were also made this way via the corresponding alcohol. This catalyst will also convert corresponding carbonate esters to fluorinated alcohols. $(CF_3CF_2CH_2O)_2CO/COCl_2/HF$ is converted to HFC-236cb ($CF_3CF_2CH_2F$) in this manner. It is also possible to carry this out in two steps. The alcohol is first reacted with either SO_2Cl_2 or $COCl_2$ to make $R_fCH_2OSO_2Cl$ or

R_fCH_2OCOCl, which is then reacted with HF, catalyzed by either γ-Al_2O_3 or AlF_3 to give the fluoroalkane [93].

4.2.8 Coupling

Polyfluoroalkenes can be easily prepared by the addition of polyfluoroallylic fluorides to fluoroethylenes in the presence of catalysts of the structure AlX_3 where X is one or more of Br, Cl or F, provided that X cannot be entirely F. Of particular interest are reaction products of hexafluoropropene (HFP), where R_1, R_2, R_3, and R_4 of (**1**) in eq 27 are all F, and tetrafluoroethene (TFE) where R_5 of (**2**) is F, which are reacted in the presence of an AlF_xCl_y catalyst.

$$\underset{(\mathbf{1})}{R_1R_2C{=}C(R_3)CF_2R_4} + \underset{(\mathbf{2})}{R_5FC{=}CF_2} \longrightarrow R_1R_2C{=}C(R_3)CF(R_4)CF(R_5)CF_3 \qquad (27)$$

$F(CF_2)_2CF{=}CFCF_3$ is obtained by reacting HFP:TFE in a 1:1 molar ratio. The major products obtained in 32% and 39% yields from the reaction of HFP:TFE in a 1:2 molar ratio are $F(CF_2)_2CF{=}CFCF_3$ and $F(CF_2)_3CF{=}CF(CF_2)_2F$, respectively [94]. These alkenes can then be reduced with hydrogen over a palladium catalyst to afford HFCs which can replace CFCs as cleaning liquids [95] or as solvents for fluoromonomer polymerizations [96].

HCFC-225aa, HCFC-225ca and HCFC-225cb are solvents which are useful for the cleaning of electronic circuit boards and can provide a temporary replacement for CFC-113 which has been extensively used for this application. The 225 isomers are readily prepared by the addition of HCFC-21 to TFE using an $AlCl_xF_y$ catalyst [97] as shown in eq 28.

$$\underset{21}{CHCl_2F} + \underset{TFE}{CF_2{=}CF_2} \longrightarrow \underset{225ca}{CHCl_2CF_2CF_3} + \underset{225cb}{CHClFCF_2CClF_2} + \underset{225aa}{CHF_2CCl_2CF_3} \qquad (28)$$

4.2.9 Chlorinations/Chlorofluorinations

Chlorination of C–H bonds in readily available precursors, provides yet another route to CFC alternatives. Conversion of either HFC-143a or HCFC-133a to the desired HCFC-123 (eq 29) has been reported using activated carbon or chromium oxide [98] and photochemical techniques [99]. The first chlorination is the most difficult, thus progressive chlorinations become faster, so that selectivity is best achieved at low conversion to minimize the formation of CFC-113a.

$$\underset{143a}{CF_3CH_3} + Cl_2 \longrightarrow \underset{133a}{CF_3CH_3Cl} \longrightarrow \underset{123}{CF_3CHCl_2} \longrightarrow \underset{113a}{CF_3CCl_3} \qquad (29)$$

A series of patents from DuPont describe the chlorofluorination of propane and propylene [100] directly to perhalogenated products such as perfluorocarbon PFC-

218 and CFC-216aa (eq 30). PFC-218 is a potential etching agent for silicon and CFC-216aa is an intermediate, through hydrodechlorination (eq 31), to HFC-236fa, a potential foam-blowing and fire extinguistant agents. It is also a precursor to fluoromonomers such as hexafluoro-propylene. The catalysts used for chlorofluorinations are AlF_3, Cr_2O_3 and $CrCl_3/C$.

$$\underset{}{CH_3CH{=}CH_2 + Cl_2 + HF} \longrightarrow \underset{216aa}{CF_3CCl_2CF_3} + \underset{218}{CF_3CF_2CF_3} + HCl \tag{30}$$

$$\underset{216aa}{CF_3CCl_2CF_3} + H_2 \longrightarrow \underset{236fa}{CF_3CH2\,CF_3} + HCl \tag{31}$$

The chlorofluorination reaction typically gives perhalogenated products although it is now possible through new catalysts to prepare HFCs and HCFCs such as the HCFC-225s [101], and HFC-125 [102] (eqs 32 and 33).

$$CH_3CH_2{=}CH_2 + Cl_2 + HF \longrightarrow \underset{225}{C_3HF_5Cl_2} \tag{32}$$

$$CF_3CH_3 + Cl_2 + HF \longrightarrow \underset{125}{CF_3CF_2H} \tag{33}$$

4.3 CFC Destruction

A variety of methods have been proposed for removing CFCs from the environment. These include combustion in air, oxygen, ammonia, or water atmospheres. CFCs can also be reacted with HCl to afford starting materials for the preparation of some of the HFCs and HCFCs discussed in this review.

The combustion of CCl_2F_2 in oxygen at 500 °C was studied in the presence of more than 20 different catalysts. The combustion products are CO, CO_2, and F_2. Because of fluorine's tremendous reactivity, metallic and silicon-based catalysts degraded rapidly. The most durable catalyst was found to be BPO_4; however, it also did not have a satisfactory life because of fluorine reaction with boron [103].

CFC-12 was reacted with NH_3 (eq 34) over metal catalysts on LaF_3 and activated carbon to form HCN, HF, HCl, and N_2 [104].

$$\underset{12}{CF_2Cl_2} + NH_3 \longrightarrow HCN + HF + HCl + N_2 \tag{34}$$

CFCs were decomposed to HCl, HF, and CO_2 at 150 °C to 350 °C by the reaction of H_2O over amorphous alloy catalysts consisting of at least one element selected from the group of Ni and Co, at least one element selected from the group Nb, Ta, Ti, and Zr, and at least one element selected from the group Ru, Rh, Pd, Ir, and Pt. The alloys were activated by immersion in HF [105]. CFCs are decomposed by the reaction of water vapor at temperatures above 300 °C in the presence of iron oxide supported on activated carbon [106]. They are also decomposed by steam in the presence of catalysts comprising alumina or alumina/silica at 350–1000 °C [107].

The fluorine content of CFCs can be reduced by reacting the CFC with HCl in the vapor phase in the presence of a vapor phase fluorination catalyst (e.g. aluminum fluoride and trivalent chromium compounds) [108]. A CFC compound such as $CClF_2CClF_2$ can be converted to CCl_3CF_3 (CFC-113a) and $CCl_2{=}CCl_2$ (PCE). CFC-113a and PCE can be used as starting materials for the preparation of HFC-134a (eqs 7 and 8).

4.4 Purification Techniques

Many fluoroalkenes have been shown to be toxic. Fluoroalkenes, as well as other types of alkene, can be removed from hydrofluorocarbons by hydrogenolysis in the presence of a hydrogenation catalyst [109]. Alkenes can be removed from hydrochlorofluorocarbons in the presence of oxygen-containing phase-transfer catalysts with complex hydrides and/or strong bases [110]. A more convenient method involves hydrogenolysis of hydrochlorofluorocarbons containing alkenes over a hydrogenation catalyst [111].

4.5 Opportunities for the Future

Fluorocarbon catalysis is still largely dominated by halogen exchange reactions involving initial formation of a C–Cl bond and halogen exchange with HF. These reactions lead to very large quantities of HCl, as a byproduct, which is under severe environmental scrutiny. New methods are needed to prepare C–F bonds selectively without the use of chlorine. Although electrochemical fluorinations have been practiced commercially for many years, the main products of the reactions are typically perfluorocarbons since C–H bonds rarely survive. There is a need to develop new techniques which are chlorine free.

An intriguing reaction was published many years ago [112] involving the oxidative fluorination of ethylene to vinyl fluoride (eq 35) using Pd/Cu catalysts. The reaction probably involves Wacker chemistry, proceeding through acetaldeyde, so high degrees of fluorination are unlikely. However, it remains one of the few reported examples with which to make C–F bonds, from a hydrocarbon, without going through a chlorinated intermediate.

$$C_2H_4 + O_2 + HF \longrightarrow CH_2{=}CHF + H_2O \tag{35}$$

Another example that has been reported in the patent literature [113] is the reaction of fluorspar with CO and SO_3 (eqs 36 and 37). The mechanism and viability of this process remains to be determined.

$$CaF_2 + SO_3 \longrightarrow Ca(SO_3F)_2 \tag{36}$$

$$Ca(SO_3F)_2 + CH_3CHO \longrightarrow CH_3CHF_2 + CaSO_4 \tag{37}$$

4.6 Conclusions

Heterogeneous catalysis has played a key role in the synthesis of CFC alternatives. However, for the process to be environmentally safer, the HCl must be recycled or sold. Although it is outside the scope of this chapter, it is important to recognize that new catalytic processes based on Deacon chemistry (eq 38) have been commercialized by Mitsui-Toatsa [114] using Cr-based catalysts.

$$HCl + O_2 \longrightarrow Cl_2 + H_2O \tag{38}$$

For practitioners of heterogeneous catalysis, the challenges are to bring some mechanistic understanding and science to the many new reactions that have been reported over the past several years for the development of the CFC alternatives – HFCs and HCFCs. The challenge is not over. Much work is in progress in industrial and government laboratories to develop environmentally safer "third generation" CFC alternatives.

References

1. F. Swartz, *Bul. Acad. R. Belg.* **1892**, *24*, 309.
2. L. E. Manzer, *Science* **1990**, 31–35.
3. M. O. McLinden, D. A. Didion, *ASHRAE Journal*, **1987**, Dec. 3.
4. L. E. Manzer, V. N. M. Rao, *Adv. Catal.* **1993**, *39*, 329–350.
5. S. Morikawa, S. Samejima, M. Yoshitake, N. Tatematsu, Jpn Patent 2-178 237 (Asahi Glass Co.) **1990**, (*Chem. Abstr.*, *113*, 151 815).
6. P. Cuzzato, A. Masiero, EU Patent 408 004 (Ausimont) **1991**, (*Chem. Abstr.*, *114*, 228 352).
7. D. R. Corbin, V. N. M. Rao, US Patent 5 300 711 (DuPont) **1994** (*Chem. Abstr.*, *118*, 105 304).
8. L. E. Manzer, V. N. M. Rao, US patent 4 766 260 (DuPont) **1988** (*Chem. Abstr.*, *110*, 97 550).
9. A. Akramkhodzhaev, T. S. Sirlibaev, Kh. V. Usmanov, *Uzb. Khim. Zh.* **1980**, 29–31 (*Chem. Abstr.*, *92*, 215 810).
10. S. H. Swearingen, J. F. Wehner, M. G. Ridley, US Patent 5 105 033 (DuPont) **1992** (*Chem. Abstr.*, *114*, 101 118).
11. N. Schultz, H. J. Vahlensieck, R. Gebele, US Patent 3 904 701 (Dynamit-Nobel) **1975** (*Chem. Abstr.*, *74*, 3310).
12. R. E. Fernandez, US Patent 5 084 190 (DuPont) **1992** (*Chem. Abstr.*, *116*, 155 038).
13. P. Hopp, U. Wirth, EU Patent 562 509 (Hoechst) **1993** (*Chem. Abstr.*, *119*, 270 624).
14 S. P. v. Halasz, Ger Patent 2 712 732 (Hoeschst) **1978** (*Chem. Abstr.*, *89*, 214 885).
15. S. Morikawa, S. Samejima, M. Yoshitake, S. Tatematsu, Jpn Patent 2-172 932 (Asahi Glass), **1990**.
16. K. Yangii, S. Yoshikawa, K. Murata, Jpn Patent 2-157 235 (Central Glass), **1990**.
17. S. Morikawa, S. Samejima, M. Yoshitake, S. Tatematsu, Jpn Patent 4-029 940 (Asahi Glass), **1992**.
18. S. Morikawa, S. Samejima, M. Yoshitake, and S. Tatematsu, Jpn Patent 4-029 942A2 (Asahi Glass), **1992**.
19. S. Morikawa, S. Samejima, M. Yoshitake, S. Tatematsu, Jpn Patent 4-029 943A2 (Asahi Glass), **1992**.
20. W. H. Gumprecht, W. G. Schindel, V. M. Felix, US Patent 4 967 024 (DuPont) **1990**; V. N. M. Rao, US Patent 5 015 791 (DuPont) **1991**.
21. G. Fernschild, C. Brosch, DE (German patent application) 4 005 944Al (Kali-Chemie) **1990**.
22. D. R. Coulson, J. Catal. **1993**, *142*, 289.

23. D. R. Corbin, V. N. M. Rao, WO92/16482 (DuPont) **1992**.
24. H. S. Tung, A. M. Smith, US Patent 5 155 082 (Allied Signal) **1992**.
25. B. Cheminal, E. Lacroix, A. Lantz, AU (Australian patent application) 76053/91 (Atochem) **1991**.
26. M. Hoeveler, M. Schnauber, M. Schott, CA (Canadian patent application) 2 068 832 (Hoechst) **1992**.
27. V. M. Felix, W. H. Gumprecht, B. A. Mahler, US Patent 5 334 787 (to DuPont) **1994**.
28. W. N. Milks, US Patent 2 744 147 (Dow) **1956**.
29. R. P. Ruh, R. A. Davis, US Patent 2 744 148 (Dow) **1956**.
30. R. P. Ruh, R. A. Davis, US Patent 2 745 886 (Dow) **1956**.
31. R. Firth, G. Toll, US Patent 3 755 477 (ICI) **1973**.
32. L. J. Belf, US Patent 2 946 827 (National Smelting, LTD.) **1960**.
33. X. Guo, Z. Cai, CN (Chinese patent application) 85/105080 (Zhejiang Chemical Industry Research Institute) **1985**.
34. W. Wanzke, G. Siegemund, T. Müller, EU Patent 407 961 (Hoechst) **1991** (*Chem. Abstr.*, *114*, 121 449).
35. P. Cuzzato, EU Patent 583 703 (Ausimont) **1994** (*Chem. Abstr.*, *120*, 269 636).
36. H. W. Swidersky, W. Rudolph T. Born, EU Patent 492 386 (Kali-Chemie) **1992** (*Chem. Abstr.*, *117*, 150 539).
37. D. R. Corbin, V. N. M. Rao, PCT International Application WO 92/16480 (DuPont) **1992** (*Chem. Abstr.*, *117*, 253 904).
38. J. D. Scott, M. J. Watson, US Patent 5 281 568 (ICI) **1994** (*Chem. Abstr.*, *117*, 236 287).
39. D. R. Corbin, V. N. M. Rao, US Patent 5 321 170 (DuPont) **1994** (*Chem. Abstr.*, *117*, 253 903).
40. J. Barrault, S. Brunet, B Requieme, M. Blanchard, *J. Chem. Soc., Chem. Commun.* **1993**, 374–5.
41. S. L. Bell, US Patent 4 129 603 (ICI) **1978**.
42. W. H. Gumprecht, US (statutory invention application) H1129, **1993**.
43. L. E. Manzer, US Patent 5 051 537 (DuPont) **1991**.
44. H. Ohno, T. Arai, K. Muramaki, T. Ohi, H. Nakayama, Y. Shoji, US Patent 5 276 223 (Showa Denko) **1994**.
45. P. K. Dattani, J. D. Scott, EU Patent 475 693 Al (ICI) **1990**.
46. J. D. Scott, A. Oldroyd, WO (PCT International application) 93/10898 (ICI) **1993**.
47. S. Morikawa, S. Samejima, M. Yoshitake, S. Tatematsu, Jpn Patent 2-179 223 (Asahi Glass) **1990**.
48. J. D. Scott, M. J. Watson, US Patent 5 281 568 (ICI) **1994**.
49. L. E. Manzer, US Patent 4 922 037 (DuPont) **1990**; D. R. Corbin, V. N. M. Rao, WO 92/16481 (DuPont) **1992**.
50. L. E. Manzer, US Patent 5 185 482 (DuPont) **1993** (*Chem. Abstr.*, *114*, 810 16).
51. W. R. Buckman, US Patent 3 644 545 (Allied) **1972**.
52. T. R. Fiske, D. W. Baugh, Jr., US Patent 4 147 733 (Dow) **1979**.
53. M. Yoshitake, S. Tatematsu, S. Morikawa, Jpn Patent 5-339 179A2 (Asahi Glass) **1993**.
54. X. Guo, Z. Ye, X. Ma, CN (Chinese patent application) 1 076 686 A, (Zhejiang Chemical Research Institute) **1993**.
55. S. Takayama, H. Nakayama, H. Kawasaki, N. Hashimoto, N. Kawamoto, EU Patent 128 510 (Showa Denko) **1984**.
56. W. T. Miller, Jr., E. W. Fager, P. H. Griswold, *J. Amer. Chem. Soc.*, **1950**, *72*, 705.
57. A. L. Henne, M. S. Newman, *J. Amer. Chem. Soc.* **1938**, *60*, 1697.
58. K. Okuhara, *J. Org. Chem.* **1978**, *43*, 2745.
59. O. Paleta, L. Frantisek, A. Posta, V. Dedek, *Collection Czechoslov. Chem. Commun.* **1980**, *45*, 104.
60. W. H. Gumprecht, W. J. Huebner, M. J. Nappa, WO (PCT International application) 94/03417 (DuPont) **1994**.
61. R. C. Zawalski, US Patent 5 017 732 (Dixie Chemical) **1991**.
62. S. Morikawa, M. Yoshitake, S. Tatematsu, Jpn Patent 1-172 347 (Asahi Glass) **1989**.
63. L. E. Manzer, V. N. M. Rao, US Patent 4 902 838 (DuPont) **1990**.

64. S. Morikawa, S. Samejima, M. Yoshitake, S. Tatematsu, T. Tanuma, Jpn Patent 2-115 135 (Asahi Glass) **1990**.
65. G. J. Moore, H. M. Massey, US Patent 5 091 600 (ICI) **1992**.
66. S. Morikawa, M. Yoshitake, S. Tatematsu, Jpn Patent 2-040 332 (Asahi Glass) **1990**.
67. W. H. Manogue, V. N. M. Rao, US Patent 5 030 372 (DuPont) **1991**.
68. S. Okazaki, M. Ogura, Y. Mochizuki, EU Patent 450 467 (Du Pont-Mitsui Fluorochemicals) **1991**.
69. P. Cuzzato, A. Masiero, EU Patent 537 760 A2 (Ausimont) **1993**.
70. T. Muller, G. Siegemund, US Patent 4 547 483 (Hoechst) **1992**.
71. R. F. Sweeney, B. F. Sukornick, US Patent 4 192 822 (Allied) **1980**.
72. L. E. Manzer, F. J. Weigert, WO (PCT International application) 92/02476 A1 (DuPont) **1992**.
73. P. Cuzzato, EU Patent 569 832 A1 (Ausimont) **1993**.
74. H. Aoyama, E. Seki, WO PCT International Application 93/25510 (Daikin) **1993** (*Chem. Abstr.*, *120*, 298 045).
75. R. E. Fernandez, R. B. Kaplan, US Patent 5 180 860 (DuPont) **1993** (*Chem. Abstr.*, *118*, 212 469).
76. J. L. Bitner, et. al., US Dept. Comm. Off. Tech. Serv. Rep. 136 732 **1958**, *25*.
77. C. Gervasutti, L. Marangoni, W. Marra, *J. Fluorine Chem.* **1981**, *19*, 1.
78. Boyes, E. D., Coulson, D. R., Coulston, G. W., Diebold, M. P., Gai, P. L., Jones, G. A., Kellner, C. S., Lerou, J. J., Manzer, L. E. et al, *Proceedings of American Chemical Society, Division of Petroleum Chemistry* **1993** *38*(4), 847–9.
79. J. I. Darragh, UK Patent 1 578 933, **1980**.
80. C. S. Kellner, J. J. Lerou, K. G. Wuttke, V. N. M. Rao, US Patent 4 980 324, **1990**.
81. S. Morikawa, S. Samajima, M., Yoshitake, S. Tatematsu, Eu Patent 317 981 **1989** (*Chem. Abstr.*, *111*, 194 096).
82. C. S. Kellner, V. N. M. Rao, US Patent 4 873 381, **1989**.
83. V. N. M. Rao, DuPont: WO (PCT International application) 92/12113, **1992**.
84. R. Ohnishi, I. Suzuki, M. Ichikawa, *Chem. Lett.* **1991**, *841*.
85. S. P. Von Halasz, Ger. Patent 3 009 760 (*Chem. Abstr.*, *95*, 186 620) **1981**.
86. T. Saiki, M. Suida, S. Nakano, K. Murakami, US Patent 5 282 379, **1994**.
87. C. S. Kellner, V. N. M. Rao, F. J. Weigert, US Patent 5 068 473, **1991**.
88. S. Morikawa, S. Samejima, M. Yoshitake, N. Tatematsu, Jpn Patent 3-099 026 (Asahi Glass) **1991**.
89. J. Moore, J. O'Kell, EU Patent 508 660 (ICI) **1992**.
90. T. A. Ryan, UK Patent 212 6216 A (ICI) **1984**.
91. L. Burgess, J. Butcher, T. Ryan, P. P. Clayton, WO (PCT International application) 93/12057 (ICI) **1993**.
92. M. J. Nappa, A. C. Sievert, US Patent 5 274 189 (DuPont) **1993**.
93. M. J. Nappa, A. C. Sievert, US Patent 5 274 190 (DuPont) **1993**.
94. C. G. Krespan, US Patent 5 162 594 (DuPont) **1992** (*Chem. Abstr.*, *117*, 69 439).
95. C. G. Krespan, V. N. M. Rao, US Patent 5 171 902 (DuPont) **1992** (*Chem, Abstr.*, *117*, 233 426).
96. A. E. Feiring, C. G. Krespan, P. R. Resnick, B. E. Smart, T. A. Treat, R. C. Wheland, US Patent 5 182 342 (DuPont) **1993** (*Chem. Abstr.*, *118*, 213 753).
97. A. C. Sievert, C. G. Krespan, F. J. Weigert, US Patent 5 157 171 (DuPont) **1992** (*Chem. Abstr.*, *115*, 70 904).
98. R. F. Sweeney, B. R. Sukornick, US Patent 4 192 822 **1980** (Allied Chemical).
99. R. F. Sweeney, US Patent 4 060 469 **1977** (Allied Chemical).
100. J. L. Webster, E. L. McCann, D. W. Bruhnke, J. J. Lerou, L. E. Manzer, W. H. Manogue, P. R. Resnick, S. Trofimenko, US Patent 5 043 491 **1991**; US Patent 5 068 472 **1991**; US Patent 5 057 643 **1991**; US Patent 5 220 083 **1993**.
101. D. W. Bruhnke, J. J. Lerou, V. N. M. Rao, W. C. Seidel, F. J. Weigert, US Patent 5 177 273 **1993**.
102. M. J. Nappa, V. N. M. Rao, US Patent 5 258 561 **1993**.
103. S. Imamura, *Catal. Today* **1992**, *11*(4), 547–67.

104. Y. Takita, T. Imamura, Y. Mizuhara, Y. Abe, T. Ishihara, *Appl. Catal. B: Environ.*, **1992**, *1*(2), 79–87.
105. K. Hashimoto, H. Habazaki, US Patent 5 220 108 (K. Hashimoto and Yoshida Kogyo) **1993** (*Chem. Abstr.*, *116*, 151 333).
106. S. Okazaki, A. Kurosaki, US Patent 5 118 492 (DuPontMitsui) **1992** (*Chem. Abstr.*, *114*, 191 662).
107. S. Okazaki, A. Kurosaki, US Patent 5 151 263 (DuPontMitsui) **1992** (*Chem. Abstr.*, *115*, 135 502).
108. V. N. M. Rao, S. H. Swearingen, WO PCT International Application 94/13608 (DuPont) **1994**.
109. R. E. Fernandez, V. N. M. Rao, US Patent 5 001 287 (DuPont) **1991** (*Chem. Abstr.*, *114*, 142 640).
110. G. Engler, U. Gross, D. Prescher, J. Schulze, East German Patent 160 718 (Academy of Sciences) **1984** (*Chem. Abstr.*, *101*, 170 689).
111. T. W. Fu, V. N. M. Rao, WO PCT International Application 93/14052 (DuPont) **1993** (*Chem. Abstr.*, *120*, 220 857).
112. T. Kuroda, Y. Takamitsu, Jpn Patent 52-122 310 (Onoda Cement Co., Ltd.) **1977** (*Chem. Abstr.*, *88*, 38 312).
113. R. K. Jordan, US Patent 4 087 475 (R. K. Jordan) **1978** (*Chem. Abstr.*, *89*, 108 083).
114. M. Ajioka, S. Takenaka, H. Itoh, M. Kataita, Y. Kohno, EU Patent 283 198 (Mitsui Toatsa) **1988** (*Chem. Abstr.*, *109*, 213 174).

5 Heterogeneous Catalysis in the Troposphere

V. N. Parmon and K. I. Zamaraev, Boreskov Institute of Catalysis,
Prospect Akademika Lavrentieva, 5, Novosibirsk 630090/Russia

5.1 Introduction

Heterogeneous processes involving natural liquid or solid aerosols play an important role in the chemistry of the Earth's atmosphere. For example, they can provide the downstream flux of many atmospheric compounds via sedimentation in the form of the absorbed or adsorbed species, facilitate recombination of free radicals, as well as provide topochemical, thermal catalytic and photocatalytic chemical reactions in the atmosphere.

The relatively low temperatures and partial pressures of most of the reagents, low catalyst concentration, and the intensity of the solar light flux do not favor high rates of heterogeneous catalytic and photocatalytic reactions in the atmosphere. However, due to the enourmous total volume of the atmosphere, even very slow reactions can result in chemical transformations of huge amounts of some atmospheric components. Note that most heterogeneous reactions are expected to occur in the lowest layer of the atmosphere, i.e. the troposphere, since the concentration of aerosols and their surface area are much higher here than in the upper layers of the atmosphere.

Due to the low temperatures, the thermal heterogeneous catalytic processes which prevail in the atmosphere are simple hydrolytic reactions (such as the hydrolysis of N_2O_5 in acidic water droplets to form nitric acid). However, photocatalysis provides more complicated redox reactions (such as ammonia synthesis from N_2 and H_2O over aerosols containing TiO_2).

The existence of heterogeneous reactions involving aerosols in the Earth atmosphere was recognized many years ago. The role of such reactions, both thermal and photochemical, has only been appreciated recently. The role of heterogeneous catalytic reactions in the atmosphere was reassessed when trying to understand the phenomena of the ozone holes and now they are considered to make an important contribution to the global chemistry of the atmosphere [1–3].

Direct experimental studies on catalytic reactions using real atmospheric aerosols are just beginning. Therefore, the anticipated role of such reactions in the Earth's atmosphere is based mainly on estimates from experiments made with model catalysts, together with known data that characterizes the atmosphere as a sort of global catalytic reactor. For example, a drastic acceleration of chemical transformations in the atmosphere after volcanos eruptions has been observed [1]. Also, the possible impact on the global chemistry of the Earth's atmosphere due to photocatalytic reactions involving continental aerosols at their background concentration levels in the troposphere has been estimated [2]. The most probable candidates as tropospheric thermal catalysts and photocatalysts, as well as the main types of expected tropospheric catalytic reactions are discussed below.

5.2 The Atmosphere as a Global Catalytic and Photocatalytic Reactor

The principal components of the atmosphere are nitrogen 78.09%, oxygen 20.95%, argon 0.932%, and carbon dioxide 0.03% (vol %, dry atmospheric air). The water content varies from 0.1 to 2.8 vol %. However, there are some other components which in spite of their low concentrations exert strong influence on atmospheric chemistry [4]. Table 1 shows the natural content (i.e. average stationary concentrations) of the principal trace components, their average lifespans and rates of supply and removal from the atmosphere. Two latter values are equal to each other and are calculated as the ratio of the stationary concentration of an atmospheric component to its residence time in the atmosphere.

Note that the period of vertical stirring of the atmosphere is ≈ 80 days [4]. Thus, those trace gases whose residence time is less than 80 days are distributed non-uniformly in the atmosphere. Examples of such gases are CO, NO_x, NH_3, SO_2, etc. In the vicinity of their sources their concentrations and the removal (supply) rates may be higher than the values shown in Table 1.

Atmospheric aerosols may be both in the liquid (droplets of water or water solution) and solid states. Solid aerosols, i.e. ensembles of ultrasmall particles that sometimes are embedded into liquid droplets can be roughly divided into two categories:

(i) primary particles which consist of dispersed materials extracted from the earth surface (land and ocean);
(ii) secondary particles that consist of materials which are formed in the course of chemical reactions in the atmosphere.

The average composition of aerosol particles (primary and secondary) in normal (background) atmospheric conditions at various altitude is considered to be:

(i) below 3 km – 50% $(NH_4)_2SO_4$, 35% soil (53% SiO_2, 17% Al_2O_3, 7% Fe_2O_3, 23% other minerals), 15% sea salt (NaCl);
(ii) above 3 km – 60% $(NH_4)_2SO_4$, 40% soil [1, 3–5].

Local variations in the composition of aerosols can also occur. Of importance to heterogeneous chemistry is the observation of high (≈ 100-fold) enrichment of solid aerosols with transition metals such as Zn, Cd, Pb, etc. over their average (Clark) content in the Earth's crust [6].

Table 1. The average natural concentration c of main trace gases in the Earth's atmosphere, their lifespan, and rate of removal (supply) [4]. The quantum yield $\varphi_{0.1}$ is the rate of photocatalytic removal (supply) of these gases to achive 10% of their total removal (supply) rate [2].

Gas	Concentation		Lifespan	Rate of removal ($mol\,l^{-1}\,s^{-1}$)	Quantun yield $\varphi_{0.1}$		
	ppb	$mol\,l^{-1}$ (1 bar, 273 K)			Fe_2O_3	TiO_2	ZnO
CO_2	3.3×10^5	1.5×10^{-5}	4 years	1.2×10^{-13}	1.7×10^{-2}	1.1	60
CO	100	4.5×10^{-9}	0.1 year	1.4×10^{-15}	2.0×10^{-4}	1.3×10^{-2}	0.7
CH_4	1600	7.1×10^{-8}	3.6 years	6.3×10^{-16}	9.0×10^{-5}	5.7×10^{-3}	0.32
CH_2O	0.1–1	$4.5 \times 10^{-(11-12)}$	5–10 days	$(1.0–20)\times10^{-17}$	$(0.1–2.9)\times10^{-5}$	$(0.9–18)\times10^{-4}$	$(0.5–10)\times10^{-2}$
N_2O	0.3	1.3×10^{-11}	20–30 years	$(1.4–2.0)\times10^{-20}$	$(2.0–2.9)\times10^{-9}$	$(1.3–1.8)\times10^{-7}$	$(0.7–10)\times10^{-5}$
NO	0.1	4.5×10^{-12}	4 days	1.3×10^{-17}	1.9×10^{-6}	1.2×10^{-4}	6.5×10^{-3}
NO_2	0.3	1.3×10^{-11}	4 days	3.9×10^{-17}	5.7×10^{-6}	3.6×10^{-4}	2.0×10^{-2}
NH_3	1	4.5×10^{-11}	2 days	2.6×10^{-16}	3.7×10^{-5}	2.4×10^{-3}	0.13
SO_2	0.01–0.1	$4.5 \times 10^{-(12-13)}$	3–7 days	$(0.7–17)\times10^{-18}$	$(1–24)\times10^{-7}$	$(0.6–16)\times10^{-5}$	$(0.4–8.5)\times10^{-3}$
H_2S	0.05	2.2×10^{-12}	1 day	2.6×10^{-17}	3.7×10^{-6}	2.4×10^{-4}	1.3×10^{-2}
CS_2	0.02	8.9×10^{-13}	40 days	2.6×10^{-19}	3.7×10^{-8}	2.4×10^{-6}	1.3×10^{-4}
COS	0.5	2.2×10^{-11}	1 year	7.0×10^{-19}	1.0×10^{-7}	6.5×10^{-6}	3.5×10^{-4}
$(CH_3)_2S$	0.001	4.5×10^{-14}	1 day	5.2×10^{-19}	7.4×10^{-8}	4.8×10^{-6}	2.6×10^{-4}
H_2	550	2.5×10^{-8}	6–8 years	$(9.9–13)\times10^{-17}$	$(1.4–1.9)\times10^{-5}$	$(9.2–12)\times10^{-4}$	$(4.9–6.5)\times10^{-2}$
H_2O_2	0.1–10	$4.5 \times 10^{-(10-12)}$	1 day	$(0.5–52)\times10^{-16}$	$(0.7–7.4)\times10^{-5}$	$(0.5–48)\times10^{-3}$	$(0.3–26)\times10^{-1}$
CH_3Cl	0.7	3.1×10^{-11}	3 years	3.3×10^{-19}	4.8×10^{-8}	3.0×10^{-6}	1.7×10^{-4}
HCl	0.001	4.5×10^{-14}	4 days	1.3×10^{-19}	1.9×10^{-8}	1.2×10^{-6}	6.5×10^{-5}

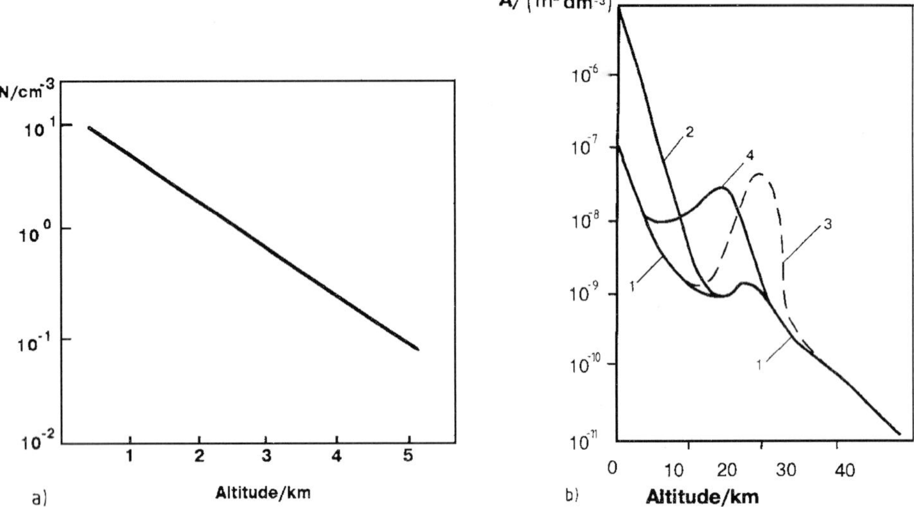

Figure 1. (a) Averaged results of concentration measurements at various altitudes for tropospheric particles of radius $>2 \times 10^{-7}$ m in continental air at mid-latitudes (a summary of data from Ref. 5). (b) Altitude distribution of the specific surface area A of solid atmospheric aerosols (per unit volume of air): 1 background atmosphere; 2 urban areas; 3 stratospheric volcanic clouds; 4 polar stratospheric clouds (adapted from Ref. 3).

The average concentration of aerosol particles in the troposphere varies from 10^{-3} g m^{-3} under arid areas to 10^{-4}–10^{-5} g m^{-3} under ordinary background conditions [7]. Distribution with respect to altitude is illustrated by Fig. 1(A), but at local spots, such as big industrial areas or erupting volcanos, aerosol concentration may be dramatically higher.

Particle size and surface area as well as vertical distribution are also important for heterogeneous reactions. The majority of the mass of atmospheric aerosol matter is represented by particles of size 10^{-7}–10^{-6} m [5], i.e. their surface area must be $\approx 10^{-1}$ m^2 g^{-1}. The specific surface area A of solid aerosols near the Earth surface is considered to be $\approx 10^{-7}$ m^2 dm^{-3} of air under background conditions and can increase by a factor of 100 in urban areas [3]. The vertical distribution of A is shown in Fig. 1(B). Due to sedimentation, almost all of aerosols are located in the lower layer of the troposphere.

For photocatalytic processes, spectral characteristics and the intensity of solar irradiation at various altitudes are important (Fig. 2). It is well known [2], that solar radiation of wavelengths below 300 nm is almost totally absorbed above the troposphere. Thus, to a first approximation, in the troposphere one should consider only reactions that occur over photocatalysts, which absorb light with $\lambda > 300$ nm. Only in the stratosphere and mesosphere will reactions over photocatalysts that absorb light with $\lambda < 300$ nm become important.

Thus, the Earth's atmosphere can be considered as a global heterogeneous catalytic and photocatalytic reactor with known estimates for

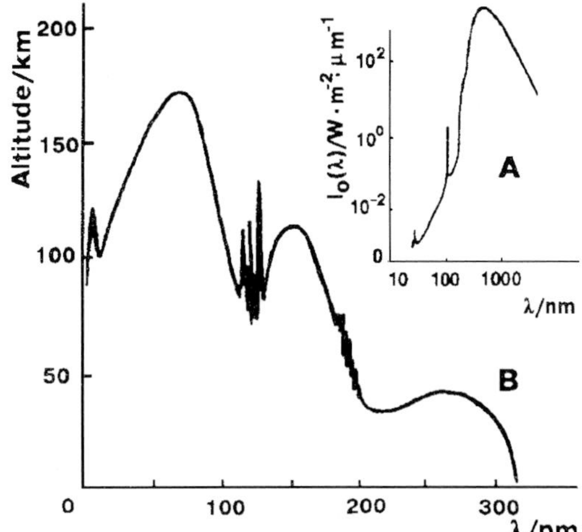

Figure 2. (a) The spectral density of the flux of virgin incident solar radiation $I_o(\lambda)$ at the upper border of the atmosphere. (b) The altitudes at which the intensity of virgin solar radiation at various wavelengths decreases by a factor of 2 [2].

(i) the average composition of the gas phase;
(ii) the average composition, concentration, and surface area of solid aerosol particles that can serve as catalysts or photocatalysts;
(iii) the distribution of intensity and spectral characteristics of the light flux, temperature, and pressure with altitude.

5.3 Thermal Heterogeneous Catalytic Processes over Ice Particles, in Water Droplets or Sulfuric Acid Aerosols

5.3.1 Hydrolytic Reactions

An example of a hydrolytic reaction which proceeds very efficiently in the presence of ice [1] is the transformation of relatively inert atmospheric forms of chlorine (HCl and $ClONO_2$) into Cl_2:

$$HCl + ClONO_2 \rightarrow Cl_2(g) + HNO_3(s).$$

Other examples are the reactions [1, 3]

$$HCl + HOCl \rightarrow Cl_2(g) + H_2O(s)$$

$$HCl + N_2O_5 \rightarrow ClNO_2(g) + HNO_3(s)$$

which are expected to be one of the causes of ozone holes formation in polar areas. These reactions take place in water droplets and/or water layers that cover some dust aerosols. HCl vapor has a strong affinity for liquid water, where it converts to hydrochloric acid and promotes acid-catalyzed hydrolytic reactions. High concentrations of absorbed HCl can lead to the melting of ice particles. Thus, hydrolytic reactions can proceed in the liquid phase even at very low (down to ≈ 200 K) temperatures.

An important atmospheric heterogeneous reaction is the hydrolysis of N_2O_5 which originates from gas-phase recombination of NO_2 and NO_3 radicals [1, 3]:

$$N_2O_5 + H_2O \rightarrow 2\,HNO_3$$

This reaction can proceed in droplets of water or, more probably, of water solutions of sulfuric acid (which are present in the atmosphere in large quantities) even at temperatures as low as ≈ 220 K [1]. Volcanic eruptions producing huge clouds of sulfur-containing aerosols are assumed to make a large contribution to the above hydrolytic processes [1].

5.3.2 Redox Reactions

Possible low-temperature heterogeneous redox catalytic reactions are the following transformations of NO [1]:

$$NO + N_2O_5 \rightleftharpoons 3NO_2$$

$$NO + 2HNO_3 \rightleftharpoons 3NO_2 + H_2O.$$

They are assumed to proceed in water or sulfuric acid aerosols via acid catalysis.

Prehydrolysis of atmospheric SO_2 is also considered as an important step for the acceleration of SO_2 oxidation to SO_3 with atmospheric ozone or hydrogen peroxide, due to much higher reactivity of ionized forms of SO_2 in watercontaining aerosols [1]. Traces of such transition metals as Fe or Mn that are typical for the atmospheric aerosols are expected to promote these oxidation reactions [1, 3]. High catalytic activity is also expected from water-covered particles of elementary carbon or soot which originate from incineration of organic fuels, wood fires, etc. and are present in the troposphere in large amounts [8]. So far there is no evidence for the direct catalytic oxidation of SO_2 with atmospheric oxygen, probably, due to a very low reactivity of molecular oxygen with respect to SO_2 at low temperatures over conventional atmospheric catalysts.

5.4 Heterogeneous Photocatalytic Reactions in the Troposphere

For the major components and impurities of the earth atmosphere (such as nitrogen, oxygen, water, carbon dioxide, methane, methane halides, etc.), electronically excited states are formed only upon absorption of light quanta with the energy $h\nu \geq 5$ eV (i.e. with wavelengths of ≤ 200 nm). Only a small portion of solar energy is found in this spectral region, meaning that most of the solar flux cannot participate in direct photochemical transformations of these compounds.

Photocatalysts, however, can absorb much smaller light quanta, from the visible, near ultraviolet, and even near-infrared spectral regions that constitute the main part of the solar energy flux. Upon absorption of such quanta, they become chemically active and can participate in numerous chemical reactions, even at low temperatures [2, 8–13]. Dust particles of soil, that contain metal oxides or ions of transition metals captured by water aerosols, may serve as photocatalysts for atmospheric reactions. Reactions involving such photocatalytically active particles may occur in all layers of the atmosphere, in contrast to direct photochemical processes which are restricted to far ultraviolet radiation that occurs only in the upper layer of the atmosphere.

5.4.1 Estimation of the Role of Heterogeneous Photocatalytic Processes in the Global Chemistry of the Atmosphere

The possible impact on the global chemistry of the atmosphere by abiotic photocatalytic reactions involving dust particles has been estimated in Ref. 2.

The most important solid materials with photocatalytic properties, that are present in the atmosphere, are metal oxides and sulfides. Many of them are semiconductors, and some of them absorb visible and near-infrared light.

When a semiconductor particle absorbs a light quantum with an energy that exceeds its band gap, an electron from the valence band is excited into the conduction band. As a result, this electron can act as a chemical reductant with respect to the molecules adsorbed on the surface of the particle, while the hole in the valence band can play the role of a chemical oxidant (Fig. 3).

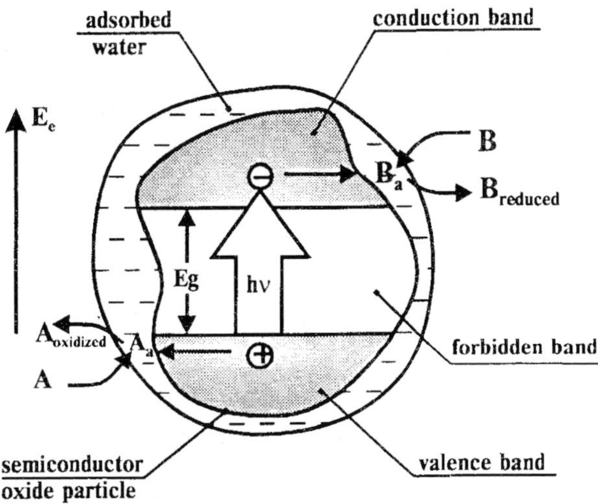

Figure 3. Basic scheme of the possible photocatalytic action of an atmospheric dust particle with semiconductor properties in which the particle is considered to be covered with a layer of adsorbed atmospheric water. A and B are atmospheric components undergoing, respectively, oxidation and reduction by light-generated holes ⊕ and electrons ⊖; A_a and B_a are forms of A and B that have been adsorbed on the particle surface (or absorbed by the water layer). E_e shows the direction of energy change for electrons in the semiconductor particle; E_g is the width of the forbidden band. For simplicity, possible interactions of electrons and holes with water are not shown.

Table 2. Band gaps E_g and red boundaries λ_o of the optical absorption of typical atmospheric semiconductor and dielectic oxides [2].

Oxide	E_g (eV)[a]	λ_o (nm)
Cr_2O_3	1.4	890
CuO	1.7	735
CdO	2.1	595
Fe_2O_3	2.2	570
TiO_2	3.0	420
ZnO	3.2	390
MgO	7.2	178
SiO_2	8.6	145
Al_2O_3	9.0	138

[a] The value refers to the bulk material and does not take into account the particle size.

The width E_g of the forbidden band (i.e the band gap) of a semiconductor serves as the threshold energy that determines the red, i.e. long-wave, boundary λ_o, of the optical absorption spectrum of the material. This parameter is very important to the photocatalytic properties of a semiconductor particle. E_g data for some oxides that are present in the atmosphere, are given in Table 2.

All oxides listed in Table 2, except pure MgO, SiO_2, and Al_2O_3, absorb light with $\lambda > 300$ nm and are able to act as photocatalysts in all layers of the Earth's atmosphere. The values E_g and λ_o listed in Table 2 refer to pure bulk semiconductor materials. With particles smaller than $\approx 10^{-8}$ m, these values can be shifted to wider band gaps due to quantum effects, whereas, impurities are known to shift E_g to smaller values. To a first approximation, in the estimates given in Ref. 2, these phenomena, that balance each other, were not taken into account.

In order to estimate quantitatively the possible influence of heterogeneous photocatalysis on the composition of the atmosphere, one can compare the rates of the expected photochemical reactions of various atmospheric components over aerosol photocatalysts with the rates of the natural removal of these components from (or supply into) the atmosphere (see Table 1).

The rate V of a photocatalytic process on solid semiconductor particles is determined by the spectral density $I_a(\lambda)$ of the absorbed light power and the quantum yield, $\varphi(\lambda)$, of the reaction at a particular wavelength λ:

$$V = \int I_a(\lambda) \times (hc/\lambda)^{-1} \times \varphi(\lambda) \times d\lambda$$

Here, h is Planck's constant and c is the velocity of light in vacuum, so that $I_a(\lambda) \times [hc/\lambda]^{-1}$ is the number of quanta with the wavelength λ that are absorbed per second.

$I_a(\lambda)$ was evaluated in Ref. 2 using data on the spectrum and intensity of solar radiation, concentrations of photocatalytically active particles in the atmosphere, and their absorption spectra. Among various oxides that are listed in Table 2, Fe_2O_3, TiO_2, and ZnO deserve special attention, since they are present in the atmosphere in rather large quantities and are capable to provide various photocatalytic reactions [2, 8–13]. The spectral density of the absorbed light power (i.e. radiation power, absorbed at a wavelength λ in 1 l) for not very turbid air can be evaluated as

$$I_a(\lambda) = I_o(\lambda) \times (1 - e^{-\varepsilon(\lambda)l}) \approx I_o(\lambda) \times \varepsilon(\lambda) \times l$$

where $I_o(\lambda)$ is the spectral density of the flux of the incident solar light (see Fig. 2), l is the total thickness of the oxide layer in cm, and $\varepsilon(\lambda)$ is the extinction coefficient.

$I_a(\lambda)$ was evaluated in Ref. 2 under the assumptions that the average background concentration of dust in the atmosphere is equal to the lower limit of its estimate for the continental area, i.e. $10^{-5}\,g\,m^{-3}$ at altitudes less than 3 km, and that 35% of the overall dust quantity are soil particles. It was also assumed that the average content of chemical elements in soil particles corresponds to that in the Earth's crust (4.65% Fe, 0.57% Ti, 8.3×10^{-3}% Zn [14]). This gives the following concentrations for atmospheric dust: Fe_2O_3 $2.3 \times 10^{-7}\,g\,m^{-3}$, TiO_2 $3.3 \times 10^{-8}\,g\,m^{-3}$, ZnO $3.6 \times 10^{-10}\,g\,m^{-3}$. This means that the quantities of these oxides in 1 l of air are equal to those in a layer with an area of $10\,cm^2$ and thickness 4.4×10^{-13} cm for Fe_2O_3, 8.3×10^{-14} cm for TiO_2, and 6.3×10^{-16} cm for ZnO. The above figures for the thickness of the hypothetical oxide layers are very small. However, these photochemically active layers can provide a sufficient influence due to the enormous size of the atmosphere and the huge total flux of solar light coming to Earth.

The value of ε is approximately constant and equal to $2.6 \times 10^4\,cm^{-1}$ in the absorption band of TiO_2, i.e. at $\lambda < \lambda_o = 420$ nm, and decreases sharply at longer wavelengths. In rough estimations, the value of $\varepsilon(\lambda)$ for Fe_2O_3 and ZnO can also be assumed constant and equal to $\approx 10^5\,cm^{-1}$ in their absorption bands (i.e. at $\lambda < \lambda_o = 570$ nm for Fe_2O_3 and $\lambda < \lambda_o = 390$ nm for ZnO, see Table 2).

Taking into account that the spectral density of the flux of incident solar light can be approximated by black body radiation with a temperature of 5800 K, and that the annual average density of solar light power on the upper surface of the Earth's atmosphere (over the hemisphere exposed to light) is $0.139\,W\,cm^{-2}$, the quantity N of light quanta absorbed by oxides in 1 l of air per second can be calculated. For light with wavelengths between λ_o (the red boundary of the oxide absorption band) and 300 nm (the ultraviolet boundary for the light that penetrates into the troposphere where the major part of the atmospheric dust is located), the following values of N were obtained: 4.2×10^{11} quanta $l^{-1}\,s^{-1}$($7.0 \times 10^{-13}\,mol\,l^{-1}\,s^{-1}$) for Fe_2O_3, 6.4×10^9 quanta $l^{-1}\,s^{-1}$ ($1.1 \times 10^{-14}\,mol\,l^{-1}\,s^{-1}$) for TiO_2, and 1.2×10^8 quanta $l^{-1}\,s^{-1}$ ($2.0 \times 10^{-16}\,mol\,l^{-1}\,s^{-1}$) for ZnO [2]. Remember that these figures for N refer to the lower limit, at the background concentration of aerosols in the troposphere. In clouds of volcanic dust or over large industrial regions the respective figures can be higher by several orders of magnitude. Remember also [6], the concentration of Zn in the air is usually 100 times greater than the value taken for the estimation in Ref. 2; this should make N for ZnO also 100 times larger.

At present it is impossible to forecast *a priori* quantum yields φ of photocatalytic reactions on solid surfaces. For this reason, the estimation of the role of photocatalysis in the global chemistry of the atmosphere was achieved as follows [2]. A value for φ was estimated, sufficient to provide the removal or supply of a noticeable part (e.g. 10%) of an atmospheric component via a photocatalytic route. This evaluation was based on the removal (supply) rates of Table 1 and has shown at which values of φ the photocatalytic formation or decomposition of a particular compound should become important for composition of the atmosphere. These estimates were then compared with the values of φ that are actually known for the formation or decomposition of the same compound. From this comparison one can draw conclusions about whether the known values of φ are high enough for a particular photocatalytic process to be significant to global chemistry of the atmosphere.

The quantum yield $\varphi_{0.1}$ that is needed to provide a 10% contribution of the photocatalytic pathway to the observed removal (or formation) rate was estimated [2] using the equation

$$\varphi_{0.1} = 0.1 \times \frac{\text{total rate of removal (or formation of the substance (mol l}^{-1}\,\text{s}^{-1})}{\text{number of the light quanta absorbed per second (mol l}^{-1}\,\text{s}^{-1})}$$

$$= 0.1\frac{R}{N}.$$

The corresponding values of $\varphi_{0.1}$ are given in Table 1.

5.4.2 Experimental Data on Photocatalytic Reactions of Atmospheric Components

A Photocatalysis on Semiconductor Oxides

According to the expected mechanisms of heterogeneous photocatalysis (Section 5.4.1), this should provide mainly redox chemical reactions.

For subsequent discussion of the possible role of photocatalytic reactions in the chemistry of the atmosphere, it is convenient to classify them into two groups:

(i) reactions of the main atmospheric components that lead to the formation of trace compounds;
(ii) reactions of trace compounds that lead to their removal from the atmosphere [2]

In the latter reactions, other trace compounds may be formed.

B Reactions of the Main Atmospheric Components

The following reactions belong to this group:

(i) Formation of ammonia and hydrazine from nitrogen and water:

$$N_2 + H_2O \xrightarrow[TiO_2]{h\nu} NH_3 + O_2;$$

$$N_2 + H_2O \xrightarrow[TiO_2]{h\nu} N_2H_4 + O_2.$$

These processes were observed in many experiments with TiO_2 powder in a nitrogen atmosphere in the presence of water adsorbed on the TiO_2 surface. Addition of small amounts of α-Fe_2O_3 to TiO_2 essentially increases the reaction rate. Nitrogen is also known to be reduced with water to NH_3 on Fe_2O_3 or vanadium-promoted hydrous ferric oxide.

(ii) Oxidation of nitrogen:

$$N_2 + H_2O \xrightarrow[ZnO-Fe_2O_3]{h\nu} H_2 + HNO_2$$

(iii) Water decomposition:

$$2H_2O \xrightarrow[TiO_2]{h\nu} 2H_2 + O_2$$

Photocatalytic water decomposition on TiO_2 was reported in both liquid and gas phases. In the latter case decomposes water adsorbed on the TiO_2 surface.

(iv) Oxidation of water by oxygen into hydrogen peroxide:

$$H_2O + O_2 \xrightarrow[TiO_2]{h\nu} H_2O_2$$

This reaction most probably proceeds through the formation of OH radicals on the surface of TiO_2.

(v) Formation of organic compounds from CO_2 and H_2O, for example:

$$CO_2 + H_2O \xrightarrow[TiO_2]{h\nu} HCOOH, CH_2O, CH_3OH$$

These processes occur in TiO_2 suspensions in water. ZrO_2 and other oxides were also found to be active in the photocatalytic reduction of carbon dioxide.

According to the available experimental data, φ can take values of 10^{-3}–10^{-4} for photocatalytic formation of H_2 and O_2 from H_2O and of organic compounds from CO_2 and H_2O over TiO_2. These values exceed or are of the same order of magnitude as the values of $\varphi_{0.1}$ for H_2 and CH_2O formation over TiO_2 under normal conditions (Table 1). Thus, photocatalytic reactions on TiO_2 may be important as a source of hydrogen and organic compounds in the atmosphere.

For the more abundant Fe_2O_3 semiconductor, quantum yields of 10^{-3}–10^{-4} would be sufficient to make the photocatalytic pathway important for the regular supply into (removal from) the atmosphere of all the gases that are listed in Table 1, except CO_2.

C Reactions of Trace Components of the Atmosphere

Photocatalytic reactions of this group are intensively studied for the purposes of waste water and air purification and photochemical syntheses. Several reviews [8–13, 15–17] have been published on this subject. Here, only typical examples of such processes are presented.

(i) Complete oxidation of hydrocarbons:

$$C_2H_6 + O_2 \xrightarrow[TiO_2]{h\nu} CO_2 + H_2O$$

$$C_6H_6 + O_2 \xrightarrow[TiO_2]{h\nu} CO_2 + H_2O$$

Such reactions may be important for removal from the atmosphere and soil of various hydrocarbons after oil spills.

(ii) Complete oxidation of halogenated hydrocarbons:

$$C_6H_5Cl + O_2 \xrightarrow[TiO_2]{h\nu} CO_2 + HCl + H_2O$$

$$CHCl_3 + O_2 + H_2O \xrightarrow[TiO_2]{h\nu} CO_2 + HCl + H_2O$$

Photocatalytic decomposition of CCl_4, $CHCl_3$, $CFCl_3$, CF_2Cl_2, $C_2F_3Cl_3$, and $C_2F_4Cl_2$ in the presence of O_2 over the particles of ZnO, Fe_2O_3, TiO_2, volcanic ash, chalk, and desert sands has been reported. Such reactions may remove from the atmosphere a certain amount of ozone-depleting halides of methane and ethane. They can also lead to the formation of acid rains containing HCl.

(iii) Complete oxidation of oxygen-containing organic compounds, for example alcohols, salicylic acid, phenol, benzoic acid, α-naphtol, etc.

(iv) H_2S and NO_x decomposition:

$$H_2S \xrightarrow[CdS,\ ZnS]{h\nu} H_2 + S$$

$$NO_x \xrightarrow[\text{metal halides, aluminosilicates}]{h\nu} N_2 + O_2$$

Such reactions may be important for H_2S and NO_x removal from the atmosphere.

(v) SO_2 oxidation:

$$SO_2 + O_2 + H_2O \xrightarrow[\text{dust}]{h\nu} H_2SO_4$$

This reaction may be one reason for the formation of acid rain, from SO_2 that has been emitted into the atmosphere, since under atmospheric conditions the rate of this photocatalytic reaction can be higher than for the corresponding therimal catalytic reaction (Section 5.3).

(vi) Reactions of partial oxidation that are observed on disperse TiO_2 and ZnO, e.g.:

$$C_6H_6 + O_2 \rightarrow C_6H_5OH.$$

The photocatalytic reactions listed above were studied mostly in water suspensions or colloids of semiconductor particles. However, taking into account that aerosol particles in the troposphere are often covered with a layer of adsorbed water, it can be supposed that the same reactions may occur in the atmosphere too.

Unfortunately, there are only few reliable data on the quantum yields of the above-mentioned photocatalytic reactions on aerosol particles in both water suspensions and gas phase. This does not allow unambiguous judgement on their real role in the removal of various trace compounds from the atmosphere. Some quantitative data are available in this field, however, such as the photocatalytic oxidation of isoprene and monoterpene hydrocarbons on particles of ZnO, TiO_2, chalk, sand, and ash, studied under laboratory conditions. The quantum yield 0.1 was reported for isoprene oxidation in the gas phase on solid ZnO upon irradiation with $\lambda > 300$ nm [18]. For $CHCl_3$ oxidation to CO_2 and HCl on TiO_2, φ was 2×10^{-2} for near UV light, while on α-Fe_2O_3 φ was less than 10^{-3} [19]. Note, that experimental conditions in the cited works differed markedly from those in the real atmosphere; therefore, the values of φ in the atmosphere may also differ from the above data. However, the published values substantially exceed the critical values of $\varphi_{0.1}$ for many of the photocatalytic reactions on TiO_2 (Table 1); this may be regarded as evidence for rather than against the importance of photocatalysis in the global chemistry of the atmosphere.

D Photocatalysis over Compounds with Wide Band Gaps

In addition to semiconductor oxides, some other aerosol species may also contribute to photocatalysis in both the troposphere and the stratosphere.

First are alkali halides particles, which are generated in huge amounts over the surface of the World's oceans. Alkali halides have been known for nearly 60 years to be photocatalysts that assist a large variety of reactions [20]. Among them the photocatalytic oxidation of H_2, CO, and CH_4, and the decomposition of H_2O and CO_2 are of particular importance [21, 22]. The fundamental absorption bands of alkali halides are at wavelengths below 250 nm, i.e. they correspond to the adsorption of stratospheric UV. However, alkali halides often contain color centers (generated either by impurities or by rigid UV-light) which may exhibit photoadsorptive and photocatalytic activity at wavelengths up to 800 nm [21]. Similar behavior can be expected of sulfate aerosols which are the main component of stratospheric aerosols (the so called Aitken particles).

Those oxides with wide band gaps, such as γ-Al_2O_3, MgO, etc., are also known to be efficient photocatalysts for the oxidation of ammonia, nitrogen oxides, and hydrocarbons [23–26]. When evaluating the possible role of these most abundant components of continental aerosols in the tropospheric photochemistry, their importance may be due to their photoactivity under irradiation not only in their main absorption bands, but also in the bands of surface chemisorbed species. For example, for dispersed SnO_2, it was found that surface carbonates were formed upon CO_2 adsorption, which in turn sensitized some surface photoreactions to light quanta with energies less than 2 eV (below 600 nm) [27].

An important contribution to cleaning the atmosphere from halogenated hydrocarbons, including freons, may be expected from MgO- and CaO-containing aerosols. These are present in large amounts in volcanic clouds as well as in fire or industrial aerosols [3, 6]. Indeed, these oxides were found recently to be efficient photoabsorbents of halogenated compounds, most probably performing stoichiometric substitution of the oxides oxygen anions for halogen anions with quantum yields near unity, even under visible or mild UV light.

Recently, it was found that carbon and other species such as the fullerenes are also active in UV photocatalysis. In addition, these can experience photocatalytic destruction [28, 29]. This suggests a possible contribution to atmospheric photocatalysis by soot particles.

A contribution to UV-induced photocatalysis is also expected from the cations of transitions metals such as iron, copper, and manganese dissolved in water droplets or the water layer that may cover a solid aerosol. Probable photoreactions in this case would be water splitting, or redox processes involving atmospheric contaminants that are easily absorbed by the water layer [30]. The primary step in these reactions is expected to be redox transformations of the electronically excited metal compounds [8].

Another feature of photocatalysis involving water-coated solid oxides in the atmospheres is expected to be the generation of hydroxyl and/or peroxide radicals. These species can react rapidly with many water-soluble trace compounds that are present in the atmosphere, thus providing a pathway for their photocatalytic removal.

E Photogenerated Catalysis

UV is able to generate, on solid surfaces, new states of transition metal ions that can serve as active sites for thermal catalytic reactions. A typical example here is generation of long-lived low valence states of vanadium and copper on the surface of SiO_2 [31, 32]. These states catalyze the thermal oxidation of alkenes even at room temperatures. Other examples of generation with UV light of active catalytic sites on the surface of dispersed oxides are reviewed in Ref. 33.

Unfortunately, to date no data are available for the quantitative estimation of the role of photogenerated catalysis and photocatalytic reactions over halides and oxides with wide band gaps in the global chemistry of the atmosphere. Therefore, in the estimates made above, only photocatalytic reactions over the oxide semiconductors Fe_2O_3, TiO_2, and ZnO with relatively narrow band gaps were taken into account. However, new compounds may be added in the future to the list of photocatalysts that are important to the chemistry of the atmosphere.

5.5 Role of Absorption and Adsorption in Tropospheric Catalysis and Photocatalysis

The important role of *absorption* into water droplets, for thermal hydrolytic atmospheric reactions, was mentioned in Section 5.3. There is experimental evidence that absorption of many substances in this case proceeds with conventional values for the Henry's coefficients, i.e. the impurities are strongly concentrated in the droplets.

Particles of solid atmospheric aerosols may be also covered by a layer of atmospheric water in the liquid or frozen state (see Fig. 3). Many of the compounds listed in Table 1 (e.g. H_2S, SO_2, CH_2O) are readily soluble in water, and may be strongly concentrated in this layer. Their concentration in this way must serve to enhance photocatalytic processes over aerosols covered with water.

To be converted catalytically or photocatalytically involving a solid aerosol that is not covered by a water layer, atmospheric components have to be *adsorbed* onto their surface. Experimental data on adsorption for real atmospheric aerosols, or their models under natural conditions, are still very scarce. However, as suggested by the estimates made in Ref. 2 with CO_2 and H_2S as particular examples, even at a very low pressures of a trace component (e.g. 10^{-10} bar for H_2S) they still may be adsorbed on a solid aerosol with a high enough coverage θ.

5.6 Conclusions

The above overview suggests that both thermal catalytic and photocatalytic reactions on solid and liquid aerosol particles play an essential role in the global chemistry of the atmosphere. Atmospheric photocatalytic reactions are, perhaps, more numerous than the thermal catalytic reactions since the latter are more sensitive to the low temperatures of the atmosphere. In contrast to noncatalytic photochemistry, which takes place mostly in the stratosphere and mesosphere under the action of far UV light, most photocatalytic chemistry is expected take place in the troposphere under near-UV, visible, and even near-IR light. However, its possible contribution to the chemistry of the stratosphere and mezosphere, should not be neglected.

This leads to the suggestion that the dust in the atmosphere may play an important role, catalytically and photocatalytically cleaning the atmosphere from harmful compounds. Large desert areas, which are the main generators of continental dust, possibly play the role of the kidneys of the planet.

At present it is not possible to suggest more precise estimates than those made in this article, of the actual role of particular photocatalytic reactions in the atmosphere. To improve present knowledge, it is necessary to study in laboratories the quantitative characteristics of heterogeneous photocatalysis and thermal catalysis over natural aerosols, and under conditions that would be more close to those in the atmosphere. The most important characteristic to be measured is the quantum yield of photocatalytic reactions of atmospheric components on atmospheric aerosols containing Fe_2O_3, TiO_2, and ZnO, since these are the most plausible candidates for the role of photocatalysts due to their appropriate photochemical properties and rather high concentration in the troposphere.

Heterogeneous photocatalytic and thermal catalytic processes may seriously influence the atmospheric chemistry in several ways. First, tropospheric photocatalytic processes may accelerate complete oxidation of sulfur and nitrogen oxides (into H_2SO_4 and HNO_3) as well as mineralization (also through complete oxidation into H_2O, CO_2 and HCl) of halogenated organic molecules, thus affecting the strength of acid rains. Catalytic hydrolysis of chlorine-containing compounds facilitates the evolution of Cl_2 that may stimulate the subsequent reactions depleting the ozone layer. Photocatalytic mineralization of halogenated organics and photocatalytic production of dihydrogen and ammonia (or even hydrazine) may also influence the fate of the ozone layer. Photocatalytic oxidation of hydrocarbons may reduce hazardous consequences of oil spills. Photocatalytic oxidation of methane and mineralization of the halogenated organics may decrease the concentration of these greenhouse gases in the atmosphere. Photocatalysis also influences the chemical composition of aerosol particles by, for example, accelerating the formation of sulfates and ammonium salts from oxides and salts of alkali metals.

An important problem is identification of the products that are actually formed from various chemicals (in particular from complex organic compounds such as pesticides or refrigerants) on solid photocatalysts under conditions similar to those in nature. The available data here are also quite scarce. However, the possibility is not excluded that under certain unfavorable conditions in the presence of oxygen, such compounds are oxidized not only to CO_2 and H_2O, but also to more toxic or environmentally unfriendly compounds. This may be particularly true for compounds that contain such elements as Cl, P, or F.

The answer to the question as to whether heterogeneous photocatalysis and thermal catalysis are important for global and local chemistry of the atmosphere, therefore seems to be "yes" rather than "no". However, to make this conclusion unambiguous, more experimental studies are still needed.

References

1. J. G. Calvert (Ed.), *The Chemistry of the Atmosphere: Its Impact on Global Change*, Blackwell Scientific, Oxford, **1994**.
2. K. I. Zamaraev, M. I. Khramov, V. N. Parmon, *Catal. Rev.-Sci. Eng.* **1994**, *36*, 617–648.
3. Yu. M. Gershenson, A. P. Purmal', *Uspekhi Khimii*, **1990**, *59*, 1729–1756 (in Russian).
4. P. Brimblecomb, *Air Composition & Chemistry*, Cambridge University Press, Cambridge, **1986**.
5. V. E. Zuev, G. M. Krekov, *Optical Models of the Atmosphere*, Gidrometeoizdat, Leningrad, **1986** (in Russian).
6. S. G. Malakhov, E. P. Makhon'ko, *Uspekhi Khimii* **1990**, *59*, 1777–1798 (in Russian).
7. J. Heicklen. *Atmospheric Chemistry*, Academic Press, New York, **1976**.
8. V. A. Isidorov, *Organicheskaya Khimiya Atmosphery (Organic Chemistry of the Atmosphere)*, Khimiya, St-Petersburg, **1992** (in Russian).
9. D. R. Schryer (Ed.), *Heterogeneous Atmospheric Chemistry*, American Geophysical Union, Washington, **1982**.
10. K. I. Zamaraev, V. N. Parmon, *Catal. Rev.-Sci. Eng.* **1980**, *22*, 261.
11. M. Gratzel (Ed.), *Energy Resourses Through Photochemistry and Catalysis*, Academic Press, New York, **1983**.
12. E. Pelizzetti, N. Serpone (Eds), *Photocatalysis. Fundamentals and Application*, Wiley, New York, **1990**.
13. E. Pelizzetti, M. Sciavello (Eds), *Photochemical Conversion of Solar Energy*, Kluwer, Dordrecht, **1991**.
14. *Chemical Encyclopedic Dictionary*, Sovetskaya Entsiklopediya, Moscow, **1983**, (in Russian).
15. M. Sciavello (Ed.), *Photocatalysis and Environment. Trends and Applications*, Kluwer, Dordrecht, **1988**.
16. D. F. Ollis, H. Al-Ekabi (Eds), *Photocatalytic Purification and Treatment of Water and Air*, Elsevier, Amsterdam, **1993**.
17. Yu. A. Gruzdkov, E. N. Savinov, V. N. Parmon in *Fotokataliticheskoe Preobrazovanie Solnechnoi Energii*, (Eds: V. N. Parmon, K. I. Zamaraev), Novosibirsk, Nauka, **1991**, p. 138 (in Russian).

18. V. A. Isidorov, E. M. Klokova, P. V. Zgonnik, *Vestnik Lenigradskogo Universiteta, ser. 4*, (**1990**), *1*, 71; (**1990**), *3*, 61 (in Russian).
19. C. Kormann, D. W. Bahnemann, M. R. Hoffmann, *J. Photochem. Photobiol., A* **1989**, *48*, 161.
20. A. N. Terenin, *Zh. Fiz. Khim.* **1935**, *6*, 189.
21. V. K. Ryabchuk, L. L. Basov, Yu. P. Solonitzyn, *React. Kinet. Catal. Lett.* **1988**, *36*, 119.
22. L. L. Basov, Yu. P. Efimov, Yu. P. Solonitzyn *Advances in Photonics*, Leningrad State University, Leningrad, **1974**, Vol 4. p. 12 (in Russian).
23. A. V. Alekseev, D. V. Pozdnyakov, A. A. Tsyganenko, V. N. Filimonov, *React. Kinet. Catal. Lett.* **1976**, *5*, 9.
24. C. Yun, M. Anpo, Y. Mizokoshi, *Chem. Lett.*, **1980**, *7*, 788.
25. A. Mansons, *J. Chim. Phys. et Phys. Chim. Biol.* **1987**, *84*, 569.
26. H. Balard, A. Mansons, *J. Chim. Phys. et Phys. Chim. Biol.* **1987**, *84*, 907.
27. V. S. Zakharenko, A. E. Cherkashin, *React. Kinet. Catal. Lett.* **1983**, *23*, 131
28. M. Gevaert, P. V. Kamat, *J. Chem. Soc., Chem. Commun.* **1992**, 1470.
29. L. Stradella, *React. Kinet. Catal. Lett.* **1993**, *51*, 299.
30. V. N. Parmon in *Photocatalytic Conversion of Solar Energy* Part 2 (Eds: K. I. Zamaraev, V. N. Parmon), Nauka, Novosibirsk, **1985**, p. 6–106 (in Russian).
31. E. V. Kashuba, L. V. Lyashenko, V. M. Belousov, *React. Kinet. Catal. Lett.* **1986**, *30*, 137.
32. V. M. Belousov, E. V. Kashuba, L. V. Lyashenko, *React. Kinet. Catal. Lett.* **1986**, *32*, 33.
33. B. N. Shelimov, V. B. Kazanski in *Photocatalytic Conversion of Solar Energy*, (Eds: K. I. Zamaraev, V. N. Parmon), Nauka, Novosibirsk, **1991**, p. 109–137 (in Russian).

Index

T